FAO IRRIGATION AND DRAINAGE PAPER 37

arid zone hydrology for agricultural development

based on the work of
k.r. jones

with contributions from
o. berney
d.p. carr
e.c. barrett

and
fao staff

FOOD AND AGRICULTURE ORGANIZATION OF THE UNITED NATIONS
Rome 1981

The designations employed and the presentation of material in this publication do not imply the expression of any opinion whatsoever on the part of the Food and Agriculture Organization of the United Nations concerning the legal status of any country, territory, city or area or of its authorities, or concerning the delimitation of its frontiers or boundaries.

M-56

ISBN 92-5-101079-X

PREFACE

The problems of hydrology and water development planning in arid zones are often very different from those in more humid areas. To take but one example, the total annual flow of a stream in an arid area may take place in a few hours, while a stream in a humid area is likely to be perennial. In the past there was no great interest in the hydrology of arid areas and most hydrological research and literature were concerned with humid zones. Hydrologists trained under these conditions found, on moving to arid areas, that many assumptions and techniques which they had accepted as standard were no longer useful. So either time would be lost in re-thinking the application of hydrological principles and adjusting them to arid conditions or humid zone concepts would be used where they were not truly applicable with consequently poor results.

The purpose of this paper is to present some of the experience gained in recent years by a number of arid zone hydrologists. This paper is not intended as a definitive text book on the subject. The World Meteorological Organisation is the agency in the United Nations system which is responsible for preparing standards and guidance on routine hydrological data collection, networks and services. However, such guides take a long time to prepare and, in agreement with the WMO, the FAO is offering this interim contribution to the subject. At this point it is a pleasure to record the assistance which the British Overseas Development Administration has given to this work.

Concerning the technical content of the paper, the aim has been to describe the characteristics of the main hydrological elements - rainfall, evaporation, runoff, sedimentation and groundwater of particular significance in arid zones, and then, for each element, to give useful information on field data collection and analysis. As far as possible, information on those matters which are already well covered in existing literature has been kept to a minimum to avoid duplication, while at the same time giving examples in sufficient detail for the reader to follow through a problem without having to refer to a library.

In conclusion, this paper has been written by field hydrologists for their colleagues in the field, and is based on personal experience. For this reason it is unavoidably somewhat limited in its treatment of problems and in the examples which are necessarily taken from those areas of the world of which the writers had most experience. However, this experience is offered to practising engineers and hydrologists in the hope that it will be of some help in their work and will take our knowledge and understanding of hydrological problems in arid areas one step forward. The Water Resources, Development and Management Service of FAO, which is responsible for the paper, would welcome any constructive comments, ideas or suggestions from its readers.

TABLE OF CONTENTS

LIST OF FIGURES

Page

ANNEX FIGURES

LIST OF TABLES

WATER AND AGRICULTURAL DEVELOPMENT

1. INTRODUCTION

This booklet is concerned with hydrology for agriculture in arid zones. However, there are other uses for water in arid zones and it may be useful first to look briefly at the place of agriculture in arid zone development.

One third of the world's land surface has been classified as arid or semi-arid. The world's total population has increased four times in the past 150 years and may double again in the next 30 years. Easily-developed land is already exploited. Attention therefore is increasingly turning to arid areas to relieve population pressures and to provide more food.

The natural resources of arid and semi-arid regions - particularly soil and water - are limited, and are often in a delicate environmental balance. Desert encroachment due to a lack of conservation planning, and the dangers of destroying or depleting beyond recovery these productive resources, are evident at the present time and may be disastrous if development is based on short term expediency rather than long term stability. In some areas these problems may already have arrived; land developed for agriculture in the last ten years is already being abandoned due to over-pumping of the very limited groundwater resource or due to increasing salinity.

At this point the planner is obliged to question the basic aim of development. Is it to provide homes for a maximum number of people? or a maximum amount of food? or an increase in living standards for the existing population? Should arid areas be developed along lines that have been successful in humid areas, or should the special potentials of arid areas be developed to complement activities in humid zones?

"What then has the arid zone to offer to the world's humid lands? In agriculture it can provide a certain well-known range of crops adapted to the range of semi-arid and arid climates. Furthermore, it can offer those crops which, due to the difference in phase in climatic seasonality, can be produced during low latitude humid winters. In industry, plants and industrial complexes can benefit from the relatively easy availability of extensive areas at comparatively low prices, at least initially, and of a warm climate with clear skies." (Amiran, 1977) Ample sunlight and largely unused solar energy are also positive features.

The most obvious negative features are poor distribution or lack of water (abundant groundwater may be available in some areas but with small natural recharge this raises the problems of mining a non-renewable resource) and high evaporation.

Agriculture is a high user of water, and clearly therefore has to be adapted to the priorities of the situation in relation to other needs, for example domestic and industrial. Indeed in many areas a combination of specialized agriculture, light industry and domestic use may be the optimum balance.

This is not the place to describe agricultural techniques which, by minimising water use, are most suitable for arid zones. It is sufficient to mention that research is proceeding along many such lines - use of artificial environments under plastic or other types of cover; spraying crops with transpiration reducing agents, and breeding varieties having low water requirements. Runoff farming may also be mentioned here - that is, the diversion of runoff to provide an adequate supply to small areas, either on a field plot or small farm basis. This has long been practised in such areas as Qatar, where nearly a thousand small internal drainage basins are found of a few square kilometres catchment area each which often have at their low points an area of a few hectares containing good soil and watered by the runoff from the surrounding area. Runoff farming has also been re-introduced on a pilot scale in the Negev Desert with considerable success (Evenari, 1977).

Other problems in the development of arid zones may arise from the sudden availability of modern technology to societies unprepared for it. Traditional farming methods by irrigation from groundwater in parts of the Near East were based on animal power to raise the water. The rate of withdrawal of groundwater was limited by the source of power. Over the centuries, the balance between rate of water withdrawal and natural recharge was achieved without substantial use of groundwater storage and field efficiency was relatively high. Electrical or engine-driven pumps have increased the available power and modern tube well construction methods have greatly increased the capacity of wells. The absence of effective controls on drilling and pumping has resulted in groundwater mining with the consequences of decreasing yields, deteriorating water quality, saline intrusion in coastal areas and wastage of water.

It usually takes time for social awareness of a new problem to be aroused and effective controls introduced. By this time an aquifer may be exhausted and large investments in irrigated agriculture lost. Cases of this sort can be cited for a number of countries. It may be concluded that a higher level of competence - not only technical but social and political - is needed in arid than in humid zones to achieve sustained progress in economic development.

This point links up with the need for the maintenance of the environmental balance and the long term conservation of resources already discussed. Traditional forms of water law needs revision to cope with new forms of land and water development.

Moreover, land and water need to be considered together; for example the apparent under-use of grazing land in times of good rainfall in semi-arid areas may leed to an expansion of herds and the drilling, in times of drought, of new wells to supply them. This increased density of grazing means that the safety margin in the land system will be used up so that the next drought will start with no reserves of unused grazing; the subsequent loss of stock will be caused by lack of food rather than water. In this case it can be seen that the unconsidered development of water resources alone has led to an abuse of land resources. Many other examples could be quoted to illustrate the effects of lack of understanding of the consequences of one developmental action on other resources. As Maddock (1977) says "Obviously, with so many interlocking systems - social, political, and economic as well as physical - there are times when adverse effects will be accepted. But whenever the acceptance of an adverse effect requires a new project to correct it, one begins to wonder about the validity of the original decision. Thus one questions the deliberate mining of groundwater in view of the number of high-priced projects to rescue areas of falling water tables or declining availability of water of usable quality. The poor return on the investment from small irrigation projects on streams with a highly variable water supply also leads to questions about such decisions".

One could add the increasingly common construction of surface storage reservoirs whose wastage of water by evaporation, and rapid sedimentation renders their short lives effective only in encouraging a departure from traditional agricultural practices and increased downstream development. Such development cannot be sustained in the long-term without further periodic injections of finance to protect the original investment by maintaining the water supply, and the lives and property of those who have come to depend upon the development, from the inevitable progressive reversion to a more violent natural discharge regime.

This leads to one of the questions of risk in arid or semi arid zone agriculture. In rain-fed farming one of the main causes of risk is the variability of the rainfall. Although population, social and economic analyses are necessary, the first point to establish is that, though at present it may not be possible to control the weather or even to predict it very far ahead, there are many ways in which it is possible to adapt to it by research into new crop varieties, moisture conserving farming methods and, of course, irrigation. But for any of these to be effective a good understanding of the variability of the rainfall in relation to crop growth is essential. Studies of the effects of variability in annual, seasonal and monthly rainfall have yielded useful results in many areas, but with improved crop/water models the need now is for more specific studies of shorter duration rainfall variabilities; while further research is also needed on longer time-scale changes, measured in tens of years.

The difference between climatic variability and climatic change is a delicate point that arouses much debate. However, in the real world what matters is the fact that the climate is indeed never static, and while a study of the past may still be the best basis on which to plan for the future, this is because the prediction of climate is too uncertain. However, even quite small "changes" of climate, such as an increase in rainfall variability about the mean even if the mean remains unchanged, can be sufficient to tip the balance against - or for - the cultivation of a certain crop in an area where rainfall is marginal. Although irrigation provides a form of risk insurance, the effects of even a small climatic "change" may be significant here also. Surface runoff in rivers and recharge to groundwater represent a small residual between two much larger terms in the hydrologic equation, rainfall and evaporation. It follows that a small change in one of these factors will have a significantly greater effect on the water availability.

In planning for water development the quantity of water available is not the only important variable; the quality may be equally or more important, particularly in arid zones where the natural input to the system from rainfall is small. The problem in such areas is one of inputs from surface water flow and groundwater recharge that always contain some salts, dissolved from rocks and soil, and increased salt loads from fertilizers and sewage effluents - and yet the main outflow - evapotranspiration - is of pure water only. The salts inevitably accumulate in the soil and groundwater unless there is some through-flow; and this is, in a sense, a "loss" of good water. The deterioration of groundwater, particularly in irrigated areas, may be slow but must be considered at the project planning stage and its effects controlled from the start by suitable agricultural practices, such as high efficiency irrigation methods combined with adequate drainage. The recycling of water from sewage effluents or irrigation return flow is one way of obtaining the maximum benefit in regard to limited water quantity; but quantity without regard for quality can lead to disaster.

2. THE ARID ZONE ENVIRONMENT AND HYDROLOGICAL MEASUREMENTS

2.1 THE ENVIRONMENT

The distinctive hydro-climatic features of arid zones are:

- high levels of incident radiation
- generally high diurnal and seasonal temperature variations
- low humidity a short distance inland from the sea
- strong winds with frequent dust and sand storms
- sporadic rainfall of high temporal and spatial variability
- even greater variability of short-duration runoff events in ephemeral drainage systems
- generally high infiltration rates in channel alluvium
- high sediment transport rates and
- relatively large groundwater and soil moisture storage changes.

Although many definitions of aridity and semi-aridity appear in the literature none can be considered entirely satisfactory and the terms "arid" and "semi-arid" remain somewhat imprecise. The term is applied herein to those parts where rainfall will not support regular rain-fed farming. This definition encompasses all the seasonally hot arid and semi-arid zones classified by means of rainfall, temperature and evaporation indices (Walton, 1969).

If crop failure two years in ten due to inadequate rainfall represents the marginally acceptable risk for dry-farming; in lands where the rainfall season, and thus that for rain-fed farming, occurs in winter - such as Jordan, Iraq and Northern Saudi Arabia - this puts the dividing line of the arid zone at about the level of the 250mm isohyet.(de Brichambaut and Wallen, 1963) In more southerly lands, where the rainfall season occurs at the time of highest temperature - as in the Sahel - the limits to dry-farming define the boundary of the arid zone between 300-400mm.

Much of the arid land of the world falls within the sub-tropics where high-pressure systems, associated with subsiding air masses undergoing dry-adiabatic warming, are the dominant meteorological conditions. Their climates are influenced to a greater or lesser extent by variations in the latitudinal position of the inter-tropical convergence zone (ITCZ) the low-pressure "thermal equator" at which northern and southern hemisphere winds converge. Davy et al (1976) have demonstrated that its movements have an important influence on rainfall in the Sahel. The seasonal position of the ITCZ corresponds with monthly rainfall maxima in many arid areas on the fringes of the tropics, and the pole-ward extent of its movement in any year is a significant determinant of the rainfall which they experience.

The geomorphology of arid zones is characterized by marked breaks in topographic slope where mountains, hills or major outcrops with little or no true soil-cover give way to typically flat and deltaic alluvial land forms. Up-gradient of this break the dominant geomorphological process is erosion by runoff from denuded slopes which have well-developed dendritic drainage patterns although rectangular patterns exist in crystalline rock zones. Basins are frequently endoreic (without surface outlet) and there is active channel capture by younger water-courses. Drainage lines, which may start as simple eroded, boulder-filled gullies high up on immature slopes, become successively gravelly and then sandy as bed slopes flatten; valley bottoms tend to have more or less deeply incised drainage and Quaternary alluvial terraces are common. Reduction of flow due to infiltration and lateral dispersion in wider channels causes abrupt deposition of bed-load and much suspended load with resultant formation of gravel/boulder bars. Fluvial channel processes are especially apparent in the transition zone at the emergence of well-defined drainage channels from hard-rock mountainous regions.

Considerable thickness of coarse alluvium is commonly found flanking the active channel in this zone.

The high areal variability of storm rainfall causes differential runoff response within the complex drainage networks of large basins. Systematic patterns of runoff generated locally within the basin coalesce to produce apparently random runoff sequences further downstream. These infrequent spate flows are often highly charged with the sediments readily available within the basin.

As these spates traverse the more gently-sloping alluvial land forms they are depleted by infiltration through channel beds, velocities reduce and ill-sorted gravels, sands and silts are carried down onto the flood plain. The more poorly-sorted material is in general deposited by the larger floods. Re-working of these deposits takes place and there is a more or less apparent tendency for the proportion of finer material to increase progressively downstream. Recession flows often lay down thin silt deposits which, on drying, suffer appreciable cracking and flaking; they are readily removed by the advancing waters of the next flood.

If the primary drainage channels of arid zones do not terminate at inland surface-water depressions or the sea, they eventually braid into a complex of micro-channels and are lost. It is unusual for the active reach of an ephemeral water-course to extend more than 100kms downstream of its point of emergence from the mountain zone onto a outwash fan or alluvial plain.

For the purposes of this booklet, only endogenous drainage systems (those originating within the arid or semi-arid zone as defined) are considered. Exogenous rivers (those originating in humid zones) traversing arid zones do not present comparable problems of measurement and hydrologic analysis.

2.2 HYDROLOGICAL NETWORK DESIGN CONSIDERATIONS

In the arid zone environment, problems of hydrometry, hydrologic analysis and interpretation, and water resource assessment and development are severe. Difficulties encountered in measuring hydrologic cycle components and methods recommended for dealing with them are described in Section II to VI. Possibly the most difficult problem in relation to network design is that of deciding what to measure and where to measure it in order most effectively and economically to meet the objectives of the study.

WMO (1974) definitions of optimum and minimum networks are:

- Optimum network; "... by interpolation between values at different stations, it should be possible to determine with sufficient accuracy for practical purposes the characteristics of the basic hydrological and meteorological elements anywhere in the country. By characteristics is meant all quantitative data, averages, and extremes that define the statistical distribution of the element studied."

  Optimum networks should comprise long-term principal stations whose purpose is to furnish the basis for statistical studies, medium-term secondary stations whose purpose is to establish correlations with principal stations and/or with terrain characteristics, and short-term special stations such as project-oriented stations intended to provide data not available from Principal and Secondary stations or from correlations between them.

- Minimum network; "...is one which will avoid serious deficiencies in developing and managing water resources on a scale commensurate with the overall level of economic development of the country." and "...is intended as a first step to fill only the most serious existing gaps, from the standpoint of the requirements for water-resource developments".

  Minimum networks should be of high quality and are considered to be part of the long-term elements of the optimum network towards which they should develop.

Most arid zone national networks fall into the minimum network category or below and may consist of a small number of long-term daily-rainfall stations, some tens or hundreds of relatively recently installed daily, totalising and recording raingauges, some tens of runoff stations; often a somewhat surprisingly high number of climate or agrometeorological stations and a groundwater observation network whose existence and numbers usually result from earlier, ad hoc, general water resource surveys or investigations into specific aquifers.

If such networks consist of well-sited, well-designed and well-operated elements they are capable of producing, in some 20 to 30 years, reasonably reliable assessments of point rainfall characteristics for the geographic regions sampled, runoff and groundwater level statistics for the particular runoff and groundwater station locations and climatic elements whose values can be extrapolated within the region.

Such networks cannot normally provide areal rainfall estimates for periods shorter than seasons or years. Runoff values or coefficients may not readily be interpolated or extrapolated, unless alluvial aquifer storage changes are small or can be quantified. Such values are uniquely related to their particular location unless drainage areas are small enough for rainfall probabilities to be assessed. Groundwater observation data for areas of actively recharged aquifers remote from pumping effects are most effective in revealing the hydrologic - hydrogeologic response to virtually unknown inputs, but are difficult to quantify. Climatic data, other than those of humidity, are relatively easy to obtain and are readily transposable.

Hydrologic processes operating within major basins, whose size and apparent resources indicate development potential, generally cannot be monitored sufficiently well to improve upon resource estimates based on direct measurement of the runoff or groundwater resources themselves. Thus, the quickest and most accurate assessment of runoff resources at a point is produced by accurate runoff measurement coupled with rainfall data for index stations whose main purpose is to assess the typicality of the runoff period. Groundwater resource estimates are best produced by mathematical modelling techniques allied to aquifer testing and hydrochemistry.

Rainfall measurement is required for estimates of point rainfall probability and, by means of dense recording rain-gauge clusters, for estimates of areal rainfall on small representative basins whose runoff and storage change response is capable of being measured to the required accuracy. Climate data may be used to estimate the potential evapotranspiration rate of vegetation in alluvial aquifer studies, and are required for the design of irrigation schemes.

To summarise, one might say that when financial resources are limited, which will often be the case, then these should be used to establish:-

1. Minimum networks incorporating representative basins as well as existing storage reservoirs.

2. Specific project stations to determine the water resources available for development.

3. TRADITIONAL FORMS OF WATER USE IN ARID ZONES

3.1 RUNOFF CONTROL

Runoff control methods may be generally classified as follows:

- terracing on side-long slopes to intercept dispersed runoff, with one or more contoured walls of stone behind which downwashed sediments gradually accumulate to form narrow strip fields;

- cross-terracing in small hillside drainage channels, where a series of fields are formed from sediment retained behind the terrace walls, primitive provision for spill being provided at the lower end of each field;

- groups of terraced fields located in small valley bottoms, and watered by long catch-water channels which intercept the runoff on the surrounding hills and conduct it to the cultivated areas;

- water-spreading on terminal deltas by either "wild flooding" or using ad hoc measures for guiding the waters, such as semi-permanent checks, bunds or spurs to give some initial direction to the floods;

- primary spate irrigation systems, whereby terraced field complexes of tens or hundreds of hectares which flank the main channels or valley bottoms, alluvial plains or deltas are watered by diverting flood flows into them.

- storage dams, sometimes with canals constructed in rock now mainly defunct.

In addition to using runoff for agriculture the traditional dwellers in these lands also divert it to cisterns to meet both domestic and stock watering needs. An excellent description of ancient practices for runoff management typical of most of the Middle East and North Africa - as well as a penetrating account of much else that is important in understanding the traditional regime of arid zones - has been given by Evenari et al (1971).

Of the six traditional types of runoff-based agricultural systems described above, in general only those dependent on the diversion of runoff from the larger ephemeral rivers have much scope for improvement. Spate systems of this type usually represent a significant part of the capital base of an arid region and they often constitute a viable focus for further development.

Nevertheless the upland terrace systems of traditional agriculturists do need to be taken into account in water resource investigations. Where they are well-maintained they effectively control runoff; once allowed to fall into disuse they take on a particular hydrological significance. Their individual elements are so closely interdependent that the failure of one part, if not repaired, will eventually lead, by gully erosion and progressive collapse, to the destruction of the whole. Therefore, whatever the economic virtues of such terraces, it should be recognised that their abandonment on a large scale can upset the hydrological conditions within a basin for a considerable period of time. The consequence of terrace abandonment, which is for all practical purposes irreversible, is the rapid release of vast quantities of fine alluvium patiently accumulated over the centuries.

3.2 USE OF FLASH FLOODS

As noted previously, spate irrigation is one of the important traditional ways of using flash floods for agriculture in arid lands. However, it can only be practiced where floods can be expected to occur reliably every year. In Section VII an approach to the use of ephemeral flows for agriculture will be discussed which relies on the conjunctive use of groundwater to remove some of the risk inherent in runoff-based irrigation in arid zones. Traditional spate systems, where they exist, will be shown to constitute key elements in the implementation of this approach.

3.2.1 Traditional Spate Irrigation

Spate irrigation systems can be sub-divided into those located in upper catchments and foothils which serve at most a few hectares located on the inner bends of river beds or at the junction of minor channels which the main stem, and the much larger systems, often located downstream of the point of emergence of the channel onto an alluvial fan or plain. Major spate systems of the latter type annually irrigate many hundreds of hectares in some arid regions and systems irrigating upwards of a few thousand hectares are not uncommon.

The former systems consist of modest dykes of dry masonry or boulders with gravel, which divert water from the water course to groups of small, more or less carefully levelled basins. Floods often cause severe damage to these off-take works and, sometimes result in the erosion of soil from the basins. Moreover, unstable channel beds often necessitate re-location of the off-take to ensure command of small floods. It is not uncommon for the availability of soil or suitable sites for new basins to be the main constraint to the further development of these minor spate systems; and in general little can economically be done to improve matters. On occasion there is some scope for technical assistance by way of improving diversion facilities with, for example, gabions or by making better use of natural spillways to lessen flood damage. However the limited nature of such work can rarely justify hydrological studies and design is usually best tackled on an ad hoc basis by engineers with practical experience of arid zone conditions. Therefore in the remainder of this section emphasis will be directed for the most part to major spate systems.

The off-take and canal systems of the large schemes are anything but modest, although often rudimentary. Typically, one to four metre high bunds divert large quantities of water to a complex of basins through conveyance channels which are sometimes serving several hundreds of hectares, feed groups of individual basins which themselves may range up to one hectare in size.

Modern practice would favour lower weirs incorporating carefully designed sand sluices. Almost independent of height, sedimentation will generally ensure that the weir is rapidly filled to the extent that it becomes in effect a drop structure.

Settlement of the finer sediment carried by the flood waters occurs in the basins, and average sedimentation rates of the order of one centimetre a year have been observed on regularly watered fields. The temporary constructions employed for diversion weirs and off-takes have the advantage that they can be re-built progressively farther upstream to keep pace with basin aggradation. If this were not the case irrigation command would ultimately be lost.

Although now being built in concrete, the traditional diversion works on these large spate systems were usually constructed by primitive means from boulders and brushwood with gravel, and sand excavation from the river bed. The primary function of these is to provide head at off-take and they effect little or no storage regulation on flood peaks. Although rudimentary efforts are made to protect them from erosion and scouring they are in practice only effective for peak discharges up to a few hundred cubic metres per second; damage by over-bank spills occurs increasingly at discharges much greater than this. Such bunds are washed out by large floods with the resultant loss of crops and often considerable damage to the river bed. Their occasional reconstruction and the continual necessity for maintenance and repair consumes much labour and they represent a significant drain on the resources of the local agricultural community.

When spates arrive at these off-takes, water flows into the uppermost basin. This is allowed to fill to a depth of about half a metre before a breach is made in its lower bund so that water can pass to the next downstream basin. Individual groups of basins are thus irrigated by water running through the upper basins to those lower down. Little flow regulation is possible and the basins receive very different water applications.

Additional waterings after planting are, as a rule, not possible because of the danger of washing out the crop. Hence the arrival of a shortduration flood early in the season, which results in only the upper basins in any group being unirrigated basins in that season, even if further floods materialise. This difficulty was overcome even in very early times and in certain areas by the use of rather small primitive canals often running close beside the flood channel. Temporary closures are used to divert water serially to groups of fields. Second waterings are assured by using such closures only for downstream fields or only for very brief periods for fields already watered. Alternatively this limitation can be overcome to some extent by using groundwater drawn from underlying alluvium as a supplementary source of irrigation water. This is an important development and will be discussed more fully later.

On spate systems crop yields are closely related to the moisture retention capacity of the soil. In practice this is usually high because the soil is predominantly made up of silt which has been deposited over the years. Siltation has been observed to have raised land surface elevations by as much as ten metres in some old spate systems.

Because of the high level of variability inherent in the ephemeral runoff of arid zones, upstream lands have a higher probability of receiving spate irrigation in any particular year than those lying further downstream; the total area of land watered being dependent on both the distribution and amount of runoff. Complex arrangements of Water Rights are usually strictly enforced on traditional spate systems (FAO, 1973) and indeed these are often a major issue in planning for the further development of these systems. In theory, a probability of receiving irrigation water in any year can be assigned to each group of basins within a spate system. This will depend on the stochastic properties of the runoff in the main channel, the nature of the diversion and distribution works and the system operating procedure as embodied in the Water Rights attached to each land holding. As a very general guide, it would appear that about fifty to sixty percent of the cultivatable land commanded by a reasonably large traditional spate system may be cropped every year; with the remainder receiving irrigation water at less frequent intervals. It would also appear that on the fringes of a thriving spate system the local cultivators traditionally limit their essentially speculative labours for field preparation to land that can be irrigated on average at least once every four to five years.

3.2.2 Ephemeral Runoff and Agricultural Development

In general, only irrigation development based on ephemeral flows in major water courses is likely to justify a detailed hydrological investigation of surface runoff for agriculture in arid zones. For the most part however it will be found that where such water resources do exist, traditional spate irrigation systems have been developed by the local inhabitants - indeed the very existence of such systems is a reliable indicator that appreciable volumes of runoff frequently occur.

From the foregoing discussion of large traditional spate systems, it will be obvious that they are not particularly efficient uses of either the available land or water resources. In particular there is a tendency to waste water (from the irrigation view-point) in precisely those years then most runoff occurs. The damage caused by the high peak discharges can result in the loss not only of the means to control effectively the flood waters for irrigation but also of life and property.

Clearly, improvements in the operation of spate irrigation schemes would be brought about by introducing additional storage into their hydrological systems. This would be required to serve two functions:

- detention storage for flood control to damp out short duration discharge peaks;

- seasonal carry-over storage to hold water that is in excess of immediate requirements until it is needed to make up for deficits in future inflows.

In temperate climates both these storage functions are simultaneously fulfilled by conventional reservoirs. The hydrological environment in arid zones is such however, that it is usually best to separate these functions when storage is required for the efficient utilization of ephemeral runoff.

In practice, traditional methods of flood diversion on spate systems are only effective for peak discharges not exceeding the structural and hydraulic capacity of the water provided; damage to canals and by over-bank spill occurs increasingly at higher discharges. Many of the flood events experienced are often within these limits. It is a distinguishing feature of arid zone floods that their peaks are very sharp and it is rare for discharges in excess of 500 m^3/sec to persist for more than one to two hours. In these circumstances relatively small volumes of detention storage will afford significant regulation of flood peaks, provided that there is provision for rapid evacuation of the storage (through outlets or leakage) in order to accommodate any subsequent floods.

Catchment geomorphology in arid zones is often such that strategically located, relatively cheap, detention structures can be used successfully as "spate breakers" to bring about drastic reductions in flood peaks. Such spate breakers enable maximum discharges to be reduced to manageable levels, so that the total volume of flood runoff can be regulated by downstream diversion weirs without the need for these to be of massive construction.

In addition to the obvious reductions in the amount of flood damage and the provision of flows that are manageable for spate irrigation, the regulation of peak discharges by upstream spate breakers also allows the downstream diversion weirs to be used for the effective control of recharge by bed infiltration into the underlying alluvium.

On considering the need for storage to smooth out the variability in the ephemeral flow of arid zones, it becomes obvious that the high degree of variability that characterises this runoff requires relatively large reservoir volumes if it is planned to utilize anything approaching the long term mean annual flow. In addition, water would have to be stored long enough to allow distribution to large areas without excessive canal discharges.

Potential sites for conventional reservoirs on ephemeral water courses are often small and steep and, if developed, they usually have high construction costs per unit volume of storage provided. As flood waters are highly charged with suspended sediment, the useful life of conventional surface water reservoirs would be short. Moreover, evaporation losses would be high and the inherent variability of seasonal and annual discharges would result in these reservoirs being virtually empty for a considerable proportion of time. The construction of such reservoirs can also upset strongly regarded water rights, particularly when these are associated with traditional spate irrigation systems. A detailed economic analysis reported by Bottomley et al (1974) clearly shows that proposals for dams in ephemeral water courses should be examined very carefully if they are to be more than diversion structures or spate breakers, since the level of water control that they afford may well not justify the extra cost. As a general guide, it would appear that the conventional surface storage approach to flow regulations will only be economic in those rare situations in arid zones where, without it, significant quantities of water would be irretrievably wasted.

A better form of storage to make efficient use of arid zone runoff may be the unconsolidated alluvium typically associated with the ephemeral water courses of these zones, which are often extensive and usually have water tables which lie at shallow depths. The total volume of good quality water stored in these aquifers is in some cases one or two orders of magnitude greater than the long term mean annual runoff of their associated ephemeral water courses. By contrast, it would rarely be feasible in arid zones to provide a conventional surface reservoir with a storage volume equal to this amount. Moreover, such groundwater reservoirs - which are neither subject to siltation problems nor appreciable evaporation losses - are readily recharged by infiltration throuth the bed of the channel system which carries the ephemeral flow. Under natural conditions the bulk of this recharge often does not occur in the most desirable location. Nevertheless, if peak flows are controlled by upstream spate- breakers, a series of diversion weirs can be used for the spatial regulation of recharge to the alluvium.

Hence, in general, a rational approach to irrigation development based on major ephemeral channel systems in arid zones is the adoption of a conjunctive use strategy which utilises both the sporadic flood runoff and the shallow groundwater of the underlying alluvium. The basic elements of such an approach being:

- spate breakers to control the level of flood peaks;
- weir structures to regulate flow in the channel both for direct diversion to cultivated aras and for managing recharge;
- wells and ancillary equipment to draw on groundwater storage during periods when there is no flow in the ephemeral channels.

This approach, which is inherently in tune with the hydrological processes at work in such systems, can be particularly attractive from the economic view point. As it allows traditional spate irrigation systems to be integrated into new developoments - thus enabling potentially contentious water rights issues to be circumvented - the appreciable costs already sunk in existing distribution canals and land terracing can be mobilised. Moreover, the partial reliance of such developments on groundwater storage means that they can be scheduled so that appreciable benefits are realised more or less from the beginning of a project as these are not necessarily dependent, in the first instance, on the completion of all project works. Thus, with imaginative planning, attractive cash flow patterns can be ensured.

4. CHARACTERISTICS OF ARID ZONE RAINFALL

4.1 RAINFALL CHARACTERISED BY CONVECTIVE STORMS

Precipitation in arid zones results largely from squall lines and convective cloud mechanisms producing storms typically of short duration, relatively high intensity and limited areal extent. However, low intensity frontal-type rains are sometimes experienced - usually in the winter season. Where most precipitation occurs at this time of year, as in Jordan and in the Negev, relatively low-intensity rainfall may represent the greater part of annual rainfall.

Arid zone rainstorms may develop in the presence of an irregular cloud cover associated with large-scale air masses, or they may develop within band-type cloud patterns. Storm clouds are made up of a number of individual "thunder" clouds; each of which contains several convective cells which actually produce the rain. The life of individual cells consists of a developing stage, a mature stage and a dissipating stage and is limited to some tens of minutes. However, the downdraft associated with dissipation tends to form new cells, and this process of cell regeneration often produces convective storm rainfall durations of the order of one or two hours. Squall lines may travel over very large distances, as for example in the Sahel.

Geomorphology plays a significant role in controlling the occurrence of convective rainfall in arid zones invaded by moist air. Differential heating effects induced by slopes or rock outcrops and convergence of winds induced by local topography provide initial lift to buoyant air. These lifting mechanisms result in the formation of convective clouds which rapidly build to heights that are sufficiently cool for the formation of raindrops. Low topographic relief and variations in surface reflectance can be significant in determining the distribution of rainfall in arid lands.

The spatial distribution of rainfall produced by convective storms is determined primarily by the characteristics of cell clusters and their movement relative to the local terrain. These typically result in a central core area of 1 to 5 km in diameter receiving relatively uniform amounts of rainfall surrounded by more or less elliptical isohyets grading off to zero rainfall over a distance of less than 10 km (Sharon, 1972). The amount of rain falling in the central core is partly dependent on the movement of the cell cluster during the storm. Recorded core depths for individual storms are usually of the order of a few tens of millimetres, while maximum values appear to be around 70-80 mm to, very exceptionally, 100 mm for arid zones.

In arid lands with low ground level elevations and little topographic relief, it is not uncommon to see rain descending from the base of clouds failing to reach the ground. This phenomenon - known as virgis rainfall - is due to the depletion by evaporation of the falling rain.

Because of direct evaporation from rain prior to its reaching the ground, caution is required when using stable isotope ratios in studies of the evaporation "history" of terrestrial waters in arid lands. Rainfall often has lighter isotopic compositions with increasing elevation (International Atomic Energy Agency, 1968), whereas there is a tendency for scanty rains at lower altitudes to be, from the outset relatively rich in heavy isotopes due to evaporation from falling rain. In areas subject to isolated rainfalls having appreciable differences in stable isotope composition at ground level, any resulting groundwater recharge will reflect this characteristic.

4.2 INTENSITY CHARACTERISTICS OF STORM RAINFALL

Work carried out in the United States, Australia and East Africa suggest an asymmetric average storm intensity profile for storms of short duration, with the highest intensities falling in the first part of the storm.

Analysis of rainfall records for Saudi Arabia reveals considerable variability in the timing of maximum intensities for various durations during storm rainfall. In general, where the occurrence of individual storm cells can be discerned from the record, the maximum short-duration intensity occurs during the first half of the corresponding period. Where a storm sequence comprises a number of such cells, those with the highest intensity generally occur first. The longer the duration considered the more likely is the maximum intensity to be randomly distributed within the storm.

The statistical characteristics of high-intensity, short-duration, convective rainfall are essentially independent of location within a region and are similar in many parts of the world. Analysis of short-term rainfall data suggests that there is a reasonably stable relationship governing the intensity characteristics of this type of rainfall. Studies carried out in Saudi Arabia (Raikes & Partners, 1971) suggest that, on average, around 50% of all rain occurs at intensities in excess of 20mm/hour, and 20-30% occurs at intensities in excess of 40mm/hour. This relationship appears to be independent of the long-term rainfall at a particular location.

4.3 VARIABILITY OF ANNUAL RAINFALL

Water development planning and management in arid zones presents difficulties which are due less to the limited amount of rainfall than to the inherent degree of variability associated with it.

In temperate climates, the standard deviation of annual rainfall is about 10-20% and in 19 years out of 20, annual amounts are between 75% and 125% of the mean. In more arid climates the ratio of maximum to minimum annual amounts is much greater and the annual rainfall distribution becomes increasingly skewed with increasing aridity. With mean annual rainfalls of 200-300mm the rainfall in 19 years out of 20 typically ranges from 40% to 200% of the mean and for 100mm/year, 30% to 350% of the mean. At more arid locations it is not uncommon to experience years with no rainfall.

Halcrow (1969) have reported that at one site in the Desert of Thar in North West India with a mean annual rainfall of about 130mm, some 860mm of rain fell in one two-day period. Similarly, it was noted that a site in Northern Chile which received no rainfall in four consecutive years, experienced 15mm of rain in the fifth year, giving a notional mean of 3mm for the five years, whilst on a subsequent occasion 64mm were recorded in a single shower.

The coefficient of variation of annual or seasonal point rainfall amount is, in general, negatively correlated with the mean annual rainfall. Figure 4.1 illustrates this feature on a regional scale for the Near East and Northern Africa. It was derived from an analysis of annual rainfall data from localities with mean annual rainfalls of less than 600mm and embraces the semi-arid to arid belt that runs from Iran in the East, to Mauritania in the West.

4.4 LONG-TERM TRENDS IN ANNUAL RAINFALL

There has been considerable speculation in recent years on the question of climatic change in general, and the subject of shifts in the rainfall patterns of semi-arid zones has been a focus for much of the associated debate. The evidence available does not appear to be conclusive on the issue of permanent changes in climate on a historical time-scale. However, it is now widely recognised that significant variations do occur in the "average" climatic conditions locally experienced in successive periods of thirty to fifty years duration.

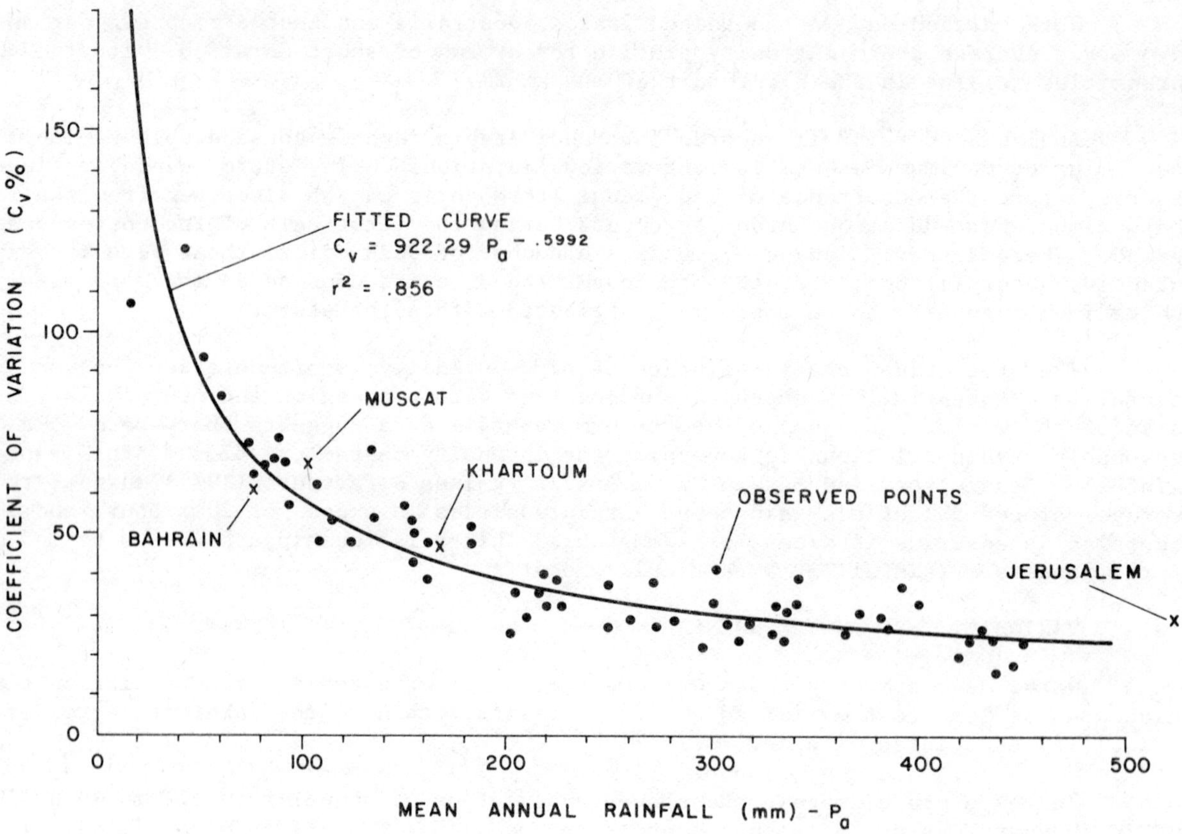

Fig. 4.1 Coefficient of variation of annual rainfall as a function of mean annual amount

The data points shown in the figure each relate to a sample of gauging station records in the Near East and North Africa. The percentage coefficient of variation (C_v %) of annual rainfall at a station is the standard deviation ($\hat{\sigma}$) of annual totals divided by their mean and expressed as a percentage. The figure shows the greater variability of annual rainfall with decreasing mean rainfall amount at a point. C_v values for Bahrain, Muscat, Khartoum and Jerusalem are approximated reasonably well by the fitted curve.

On the time scale of single decades, rainfall and temperature records are characterised by sequences of consecutive years lying above or below the medium term average value. This feature, termed "persistence", has also been observed in the time series of many other geophysical phenomena. A reasonably long (50-100 years) record of annual rainfall or temperature may be characterised by the presence of a number of oscillations having short to medium term periods of 5 to 25 years duration on which is superimposed a greater degree of random inter-annual fluctuation. Figure 4.2a shows the long term annual rainfall series observed in Jerusalem to demonstrate this feature.

The existence of a long-term trend component in the Jerusalem record is revealed by a plot of cumulative departures from the mean for the record period, indicating a change at about the turn of the century from an increasing to a decreasing rainfall average (Fig 4.2b).

4.5 STATISTICAL ANALYSIS OF ANNUAL POINT RAINFALL

In statistical terms, arid zone annual rainfall data have a positive skew - in contrast to temperate zone data which tend to follow the Normal distribution with little or no skew. This means that the probability distributions of arid zone station annual rainfall typically exhibit the following characteristics:

- there are more annual values below the mean than above it,
- The value with the greatest probability of occurring (the mode) is less than the value exceeded in 50% of all years (the median)
- the mode and median are both less than the mean.

The difference between these three statistics increases with skewness, as may be seen in Figure 4.3 which illustrates the probability distributions of three arid and semi-arid zone rainfall stations together with that of a high rainfall station.

Analysis of the data for Khartoum, Jerusamel and Muscat produces the results shown in Table 4.1. These show the number of years required to produce estimates of annual mean rainfall within ± 10% and ± 20% of the station mean for the record period. Thus, a 5-year mean annual rainfall of 130mm could in theory come from a population mean in the range 85mm to 180mm; for example from Muscat (where the 67-year mean is 101mm) or Khartoum (where the 65-year mean is 165mm), at a 95% confidence level.

At this juncture a word of caution is appropriate in relation to the use of classical, univariate statistics to describe annual rainfall amounts in arid lands. Many such localities, particularly near the fringes of these areas, derive their rainfall from more than one major feature of the general circulation of the atmosphere in such a manner that one mechanism tends to dominate the rainfall pattern in any particluar year. In consequence, the annual rainfall series may appear to be drawn from a mixed distribution and not have the single-peak characteristic. This is apparent in the bi-modal frequency distribution of annual rainfall amounts recorded at Bahrein in the Arabian Gulf. This is thought to be attributable to the relative importance in different years of two large-scale meteorological phenomena (Pike et al, 1975). Bi-modality in the parent population may be assessed by study of seasonal rainfall patterns and by statistical analysis of rainfall on a seasonal basis, defined in accordance with the likely periods of occurrence of the different meteorological conditions responsible for producing rain.

In temperate climates the arithmetic mean annual rainfall is usually used to express the prevailing, or normal, characteristics of local rainfall. In arid lands, however, the mean annual value is strongly influenced by the more extreme events, as indicated by the skewness of the distribution. In consequence the arithmetic mean is of limited practical value in characterising the rainfall regime in arid lands.

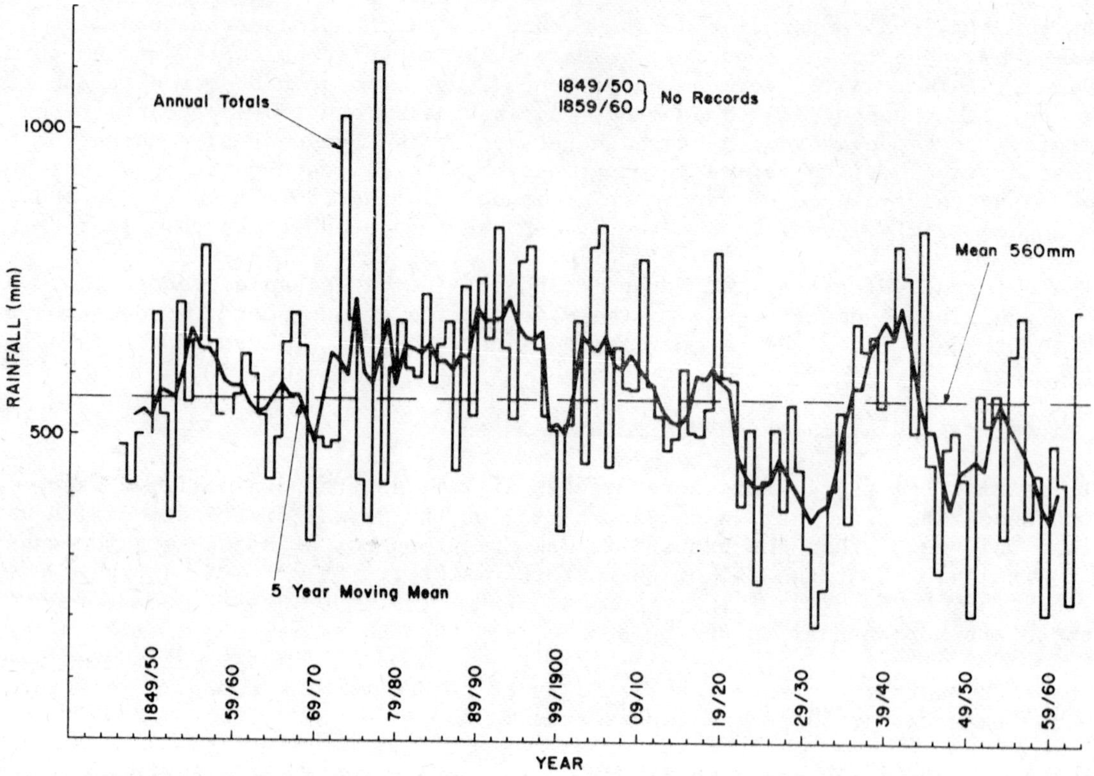

(a) Annual rainfall at St. Anne's Jerusalem 1846 to 1964

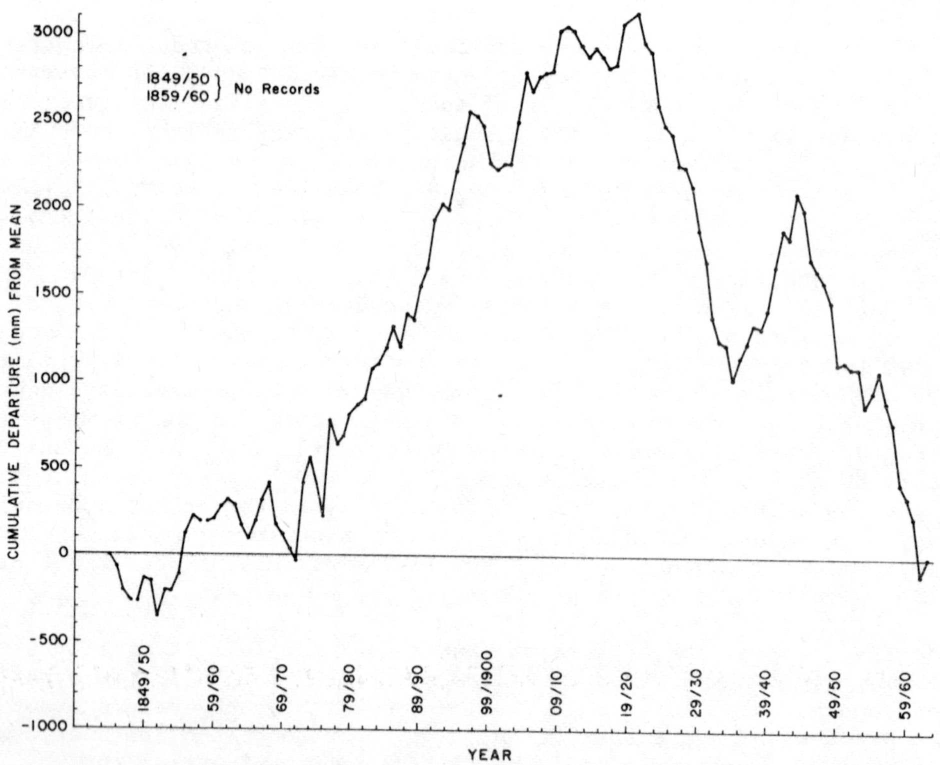

(b) Cumulative departures from 1846 to 1964 mean (560 mm) for St. Anne's Jerusalem

FIGURE 4.2

a. Annual Rainfalls at St. Annes, Jerusalem

The annual rainfall totals illustrated on the Figure suggest persistence, notably between 1920/21 and 1944/45 when periods of below and above average annual rainfall amounts are clearly recognisable.

By calculating moving means, the existence of any underlying oscillatory, non-random element of a time series may be revealed. In the case illustrated the 5-year moving mean accentuates the oscillatory pattern suggested by the annual time series.

Persistence is absent from the Jerusalem annual rainfall time series but it is present in the series of 5-year moving means. The most significant oscillation is that whose period and amplitude are greatest for a range of n-year moving means. The practical implications revealed by the analysis are:

- that estimates of long-term mean rainfall based on only 5 years of data range from 350 to 730mm for Jerusalem and

- that agricultural projects must be designed in the knowledge that such oscillations may affect the reliability of crop production and storage requirements.

b. Cumulative Departures from 1846-1964 Average for Jerusalem

Periods of upward and downward movement of the graph correspond respectively with those of above and below average rainfall on Figure 4.2(a). The dominant characteristics of the graph are:

- the trend of above average rainfall up to about 1911/12 and

- the consecutive 14-year period of below average rainfall from 1922/23 to 1935/36 when the average of 409mm was some 73% of the 116-year average of 560mm.

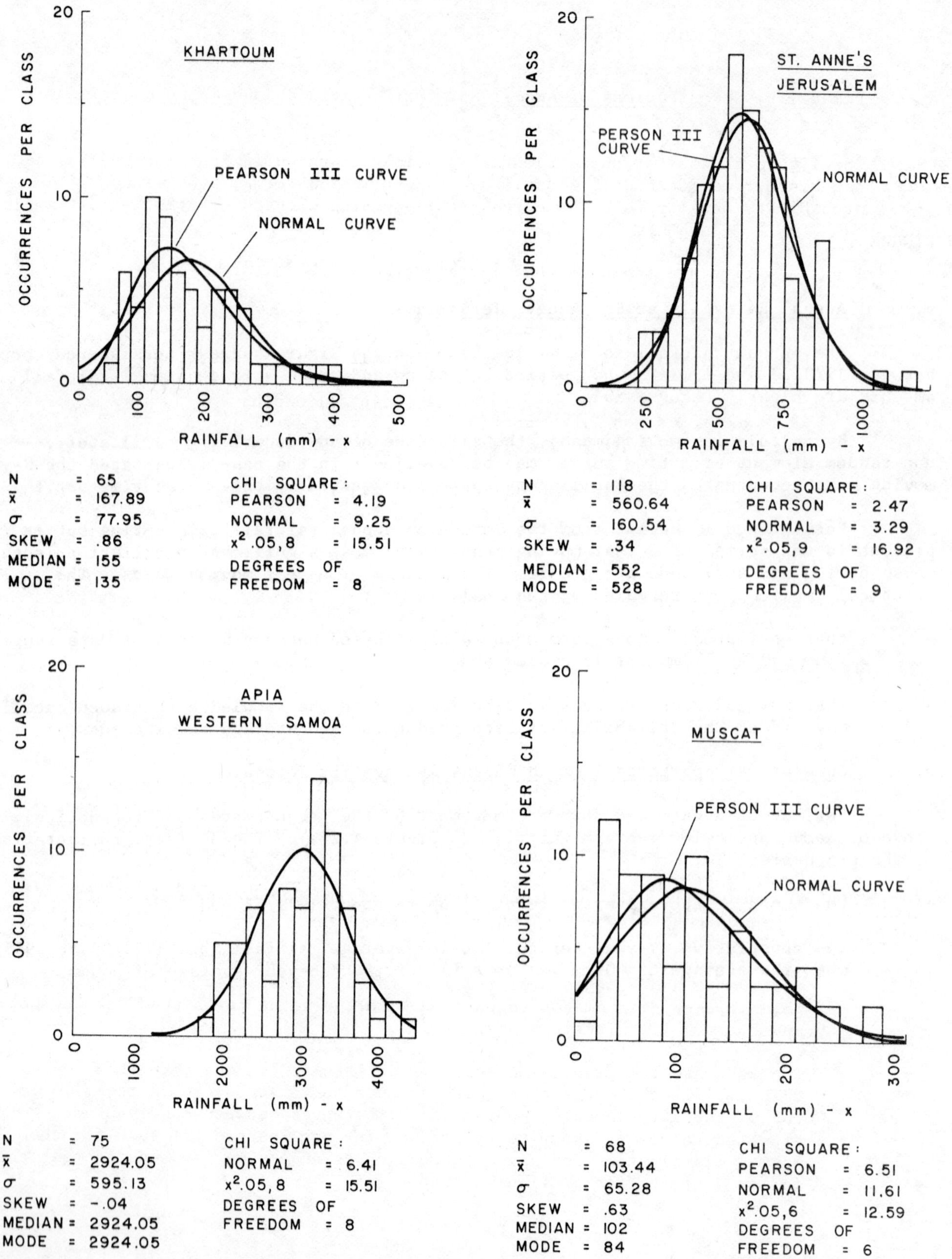

Fig. 4.3 Probability distributions of annual rainfall at arid and humid zone stations

FIGURE 4.3

Probability Distributions of Annual Rainfall at Arid and Humid Zone Stations

The figure shows, for three rainfall stations in the arid zone and one in the humid zone, histograms of annual rainfall with superimposed Normal and Pearson III (Gamma) probability density function curves corresponding with the statistics of sample data.

The statistics are shown on the figure as follows:

$\bar{x}$ - mean annual rainfall depth

Median - the annual rainfall depth exceeded in half of all years.

Mode - the most frequently occurring annual rainfall depth.

N - the number of years of record.

$\hat{\sigma}$ - The standard deviation of the N annual rainfall depths

Skew - the skewness coefficient, estimated from the data sample, indicating the degree and direction of asymmetry of the distribution with respect to the Normal. Positive skew coefficients indicate a predominance of annual rainfall depth less than the mean and are typical of arid zone rainfall data.

The skewness coefficient, commonly denoted by g, is calculated as:

$$g = \left[\frac{N}{(N-1)\ (N-2)} \right] \Sigma(x-\bar{x})^3$$

The data for the four arid zone stations exhibit marked positive skew; those for Apia a slight negative skew. The skewness coefficient is an unstable quantity when calculated for small samples since, as may be seen from the equation, its magnitude is strongly influenced by extremes of the sample.

Chi square the statistic which provides a measure of the "goodness-of-fit" of the sample data to the expected distribution.

The chi-square value χ^2 corresponding with the α probability level of exceedance and ν degrees of freedom (taken from statistical tables) is shown beneath each diagram for comparison with the value calculated for the Normal and Pearson III distributions.

In each case the calculated value of Chi-square is less than the $\chi^2_{\alpha,\nu}$ value. This indicates that neither distribution type can be rejected at the .05 probability level. However the differences between observed and expected values is shown to be smaller for the Pearson III than for the Normal distribution for all arid zone cases and it may be assumed that the former distribution provides a better fit.

TABLE 4.1

ANNUAL RAINFALL VARIABILITY

STATION	RECORD PERIOD	YEARS n	MEAN mm	MEDIAN mm	MODE mm	SD σ mm	SE mm	95% CI N = n, N = 5 (2)	Nr 95 (3) 10%	Nr 95 (3) 20%
JERUSALEM	1846 - 1964	116	560	540	470	162	16	(1) 528-592 (2) 395-725	33	8
KHARTOUM	1908 - 1972	65	165	140	110	78	10	(1) 146-184 (2) 72-262	90	22
MUSCAT	1894 - 1976	67	101	98	110* 25*	64	8	(1) 85-117 (2) 22-180	164	41

<u>Notes:</u>

* Bi-modal distribution caused by winter or summer rainfall predominance

(1) 95% CI N = n is the 95% confidence interval of the mean for the record period of n years

(2) 95% CI N = 5 is the 95% confidence interval of any 5-year mean

(3) Nr 95 is the number of years of record required for estimating mean annual rainfall within ± 10% and 20% of the long-term mean at 95% confidence level.

The Table shows the increasing relative variability with decreasing mean annual rainfall amount and the possible error associated with estimation of mean, mode or median from a short record period.

Where an adequate period of record is available the median or other quantiles may be preferred as an index of the effectiveness, in terms of agriculture or hydrology, of arid zone rainfall. The median is not greatly influenced by extremes but, when data are markedly skewed, even the median is not necessarily a good measure of central tendency. Some have, therefore, preferred to use the sample mode, the amount observed to occur most often in the past, as the most significant annual rainfall statistic for arid zones. Insofar as the mode is representative of the most frequently experienced conditions it has attractions for resource planning. Davey, et al (op.cit) also consider the rainfall corresponding with 80% exceedence probability to be appropriate for many planning purposes in the Sudano-Sahelian zone of West Africa.

4.6 SPATIAL VARIABILITY OF ANNUAL RAINFALL

Besides its marked variability through time, annual rainfall in arid zones is also highly variable in space as a result of the inherently local nature of convective storms, as discussed in 4.1. After a storm it is not uncommon for appreciable amounts of rain to be observed at a single gauge site while all neighbouring gauges record little or even no rain. Although over the years these effects tend to balance out, it is nevertheless true that annual rainfall recorded at gauge sites a short distance apart can vary significantly.

Davey et al (op.cit) give illustrations of the spatial variability of annual point rainfalls in the arid and semi-arid regions of West Africa; Figure 4.4 shows an example of the results of their analysis. However, the results of inter-station regression analysis of annual rainfall in Saudi Arabia indicate that statistically significant cross-correlations between stations up to about 150km apart may exist.

4.7 SEASONAL RAINFALL

Although it is possible to make generalisations on the characteristics of annual or seasonal rainfall experienced in arid lands, variations in the distribution of rainfall during the season are less easy to describe. However, it is this variability that is of supreme importance in determining the exploitable water resource for development.

In general, seasonal or monthly rainfall exhibits more temporal and spatial variability than that of annual rainfall. In view of the short rainfall record normally available, reliable statistics for assessing the likelihood of receiving specified rainfall amounts for periods of one month or less, can seldom be derived from the records at a single gauge site.

The most profitable approach to the analysis of rainfall in arid zones is undoubtedly to tackle the problem on an event-by-event basis. This of course implies that the analysis will be more complex than that usually adopted for temperate regions. However, if the time series of precipitation in arid zones is examined in terms of the statistical properties of alternating periods of wetness and dryness, and a probabilistic structure is employed to describe the combination of rainfall events in time and space, it is possible to extract useful information for planning purposes from the available records (Duckstein et al, 1973).

A major advantage of adopting an event-based probabilistic approach to the analysis of arid zone rainfall is that the regionalisation process allows information collected at one location to be utilized elsewhere within a meteorologically homogeneous area. It thus goes some way towards reducing the uncertainties inherent in transposing hydrological relationships in arid zones.

Work carried out at the Southwest Watershed Research Centre in Tucson, Arizona (Smith and Schrieber, 1973; Osborn and Lane, 1972) has demonstrated the value of detailed analysis of temporal and spatial rainfall patterns.

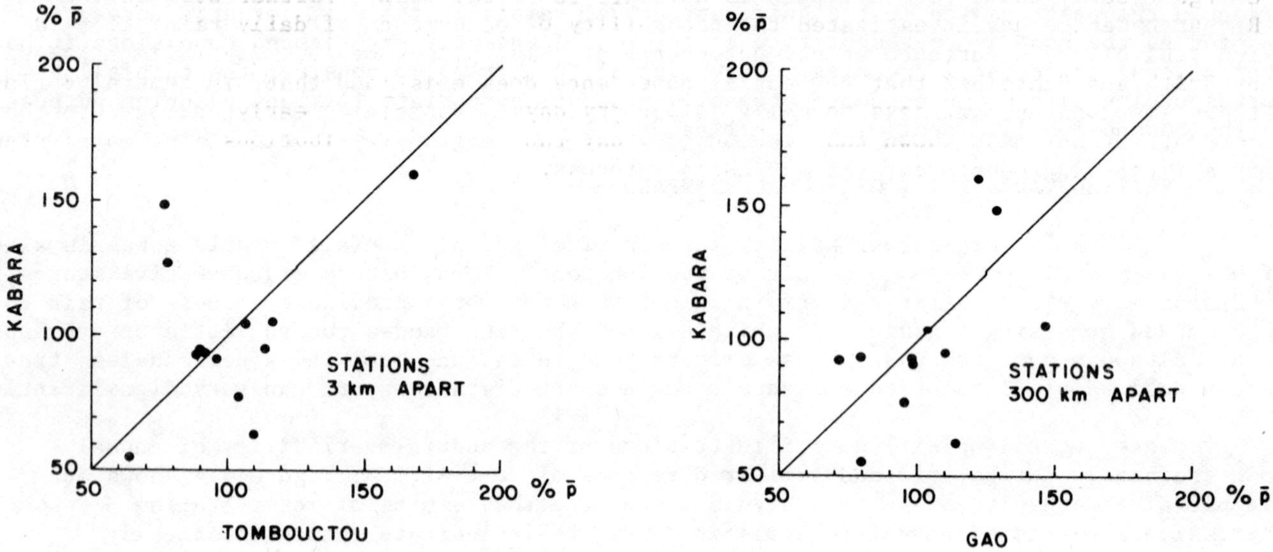

Fig. 4.4 Spatial variability of annual point rainfalls (Examples from Sahelian zone of West Africa)

Each point indicates the annual total observed at the two stations as a percentage of the period mean.

Mean annual period rainfall for the period 1951-1965 was:

Tombouctou	:	226 mm
Kabara	:	227 mm
Gao	:	292 mm

The two graphs illustrate the lack of correlation between annual rainfall total for stations 300 km and 3 km apart. Analysis of the records for the two stations shows that:

- The mean annual rainfalls at Tombouctou and Kabara for the period illustrated (1951-1965) were 226 and 227 mm respectively whereas the average difference between annual totals at the two stations was 45 mm, about 20%.

- The differences between rainfall at each station result from the nature of the rainfall distribution. Individual storms affect one station more than the other, the effect being proportionately greater for short durations than over a period of many years during which the probability of each station receiving the same number of storms and distribution of storm depths is greater.

A daily point rainfall frequency analysis of a 73 year record for the four-month (June-September) summer season during which two-thirds of all rainfall and most runoff occurs, was carried out. The results (Figure 4.5) show that, over a long time period, there is a well-defined daily rainfall probability pattern within the season. Such patterns undoubtedly exist within other arid and semi-arid zones but are unlikely to emerge clearly until record length is adequate to reveal them. Further work at the Research Centre has investigated the probability of occurrence of daily rainfall as a function of the occurrence or non-occurrence of rain on the previous day. It was shown by Smith and Schreiber that sequential dependence does exist and that, in general during the summer season, wet days normally follow dry days - especially early and late in the season. It was also shown that wet and dry day run length distributions are best fitted by a first-order dependent (Markov chain) process.

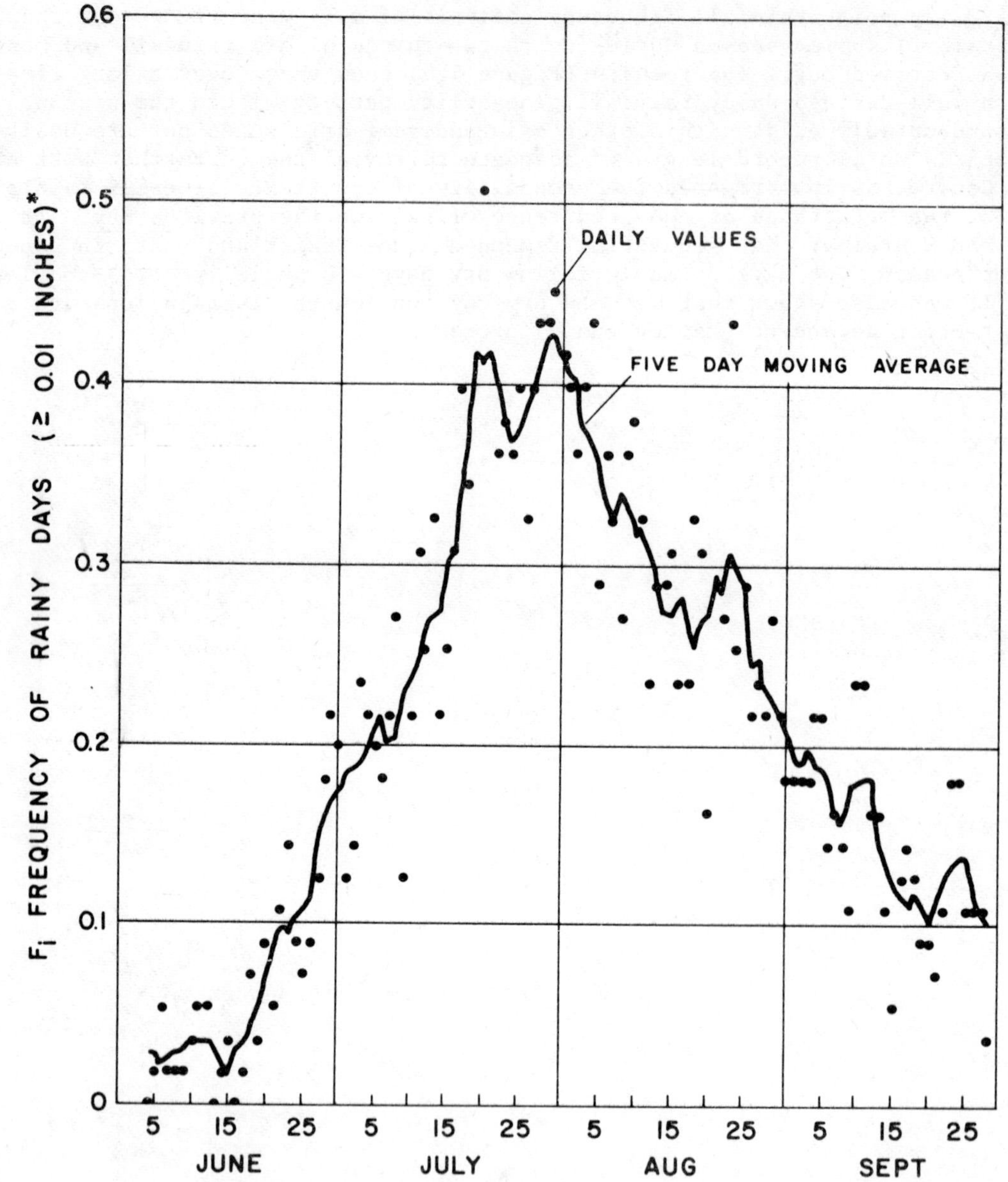

Fig. 4.5 Pattern of variation of historical rainy day frequency within a 4-month season at Tombstone (Smith and Schreiber, 1973)

The plotted points on the graph show the frequency of occurrence of daily rainfall of 0.01 inches (2.5 mm) or more for each calendar day in a 73-year record period. The solid line shows the result of smoothing by calculating 5-day moving means.

From the graph, the probability of significant daily rainfall occurrence throughout the summer rainfall season may be inferred. Thus, we would expect about 2 rain days in June, 10 in each of July and August, and some 4 or 5 in September. Estimates could similarly be made for individual weeks or pentads during the rainy season.

The distribution of rainfays within individual years (and the amounts between years) is likely to be strikingly different from the mean distribution.

5. FIELD MEASUREMENT OF RAINFALL

5.1 RAINGAUGE NETWORK

5.1.1 General Considerations

In Section 2.2 the World Meteorological Organisation specification for optimum and minimum networks was outlined. The practical purposes for which an arid zone raingauge network might be required are:-

(a) Rainfall-runoff models,

(b) Rainfall-runoff-recharge models,

(c) Rainfall-recharge models,

(d) Agriculture (crop requirements),

(e) To study point-rainfall characteristics,

(f) To develop relationships between long-term annual rainfall and long-term runoff.

Deterministic rainfall-runoff models cannot be developed for arid regions in which convective rainfall processes are dominant, except in respect of very small representative areas. This is due to:

- the lack of precision in basin rainfall measurement due to extreme spatial variability of intensity and amount,
- the short duration of runoff-producing storm rainfall elements relative to basin response time at the point of interest,
- the infinite complexity of infiltration, recharge and runoff processes over the surface area of typical arid zone basins,
- the large retentive capacity of fissured and fragmented rock surfaces and of wind-blown and alluvial sand and gravel deposits resulting from arid zone geomorphological processes,
- the storage capacities of alluvial aquifers relative to storm runoff,
- the low proportion of rainfall which runs off,
- the errors due to severe problems of runoff measurement in conditions of highly unsteady flow rates and mobile alluvial channel regimes.

The number of gauges required to measure the spatial and temporal distribution of runoff-producing storm rainfall over areas much larger than typical storm size, with the required precision for deterministic modelling, would be prohibitively costly in terms of capital outlay and operation. The results achieved would be of academic interest only unless model development were successful in predicting runoff-recharge-storage responses which were themselves amenable to accurate measurement for calibration and testing purposes and unless both rainfall processes and responses were transposable to other areas.

Models which attempt to simulate the direct recharge response and agriculture interests require only the measurement of point rainfall processes and the means with which to monitor the vertical movement of infiltrated water.

For basins larger than 100 km^2, the development of relationships between long-term mean areal rainfall and runoff is possible but is likely to produce results that are virtually meaningless. Network density requirements for estimating annual rainfall with specified precision are discussed in Section 5.1.2. Even if the required density is attainable and annual rainfall is measured to within say ± 10%, the complexity of surface and sub-surface storage and runoff processes is such that simple relationships would be unlikely to emerge.

In arid zones, national networks should consist of a relatively sparse distribution of recording and totalising gauges, to be regarded as long-term elements for statistical analysis of point rainfall. This network should be designed to sample geographically and climatically distinctive domains and no attempt should be made to infer areal rainfall from it. Whenever feasible, daily gauges may be included.

Within the minimum network, a number of dense networks of recording raingauges should be installed as part of representative basin networks. These should cover small areas within which detailed study of hydrological processes would be undertaken. These gauges will yield useful information on the areal extent of storms.

Where major aquifer outcrops, or basins with apparent development potential exist there might be a need to install a recording and totalising raingauge network capable of providing accurate areal rainfall estimates on an annual or seasonal basis. Such networks and those of representative basins should be designed in accordance with one or other of the principles and techniques described below.

5.1.2 Network Density Requirements

5.1.2.1 The Need for Network Analysis

The adequacy of the raingauge network to meet criteria imposed by the purposes for which the network has been designed (whether "minimum", "optimum" or "project") should be tested as soon as possible after installation. The hydrologist is normally concerned more with areal rainfall during particular time-intervals on gauged basins or aquifer outcrops, than with point rainfall values as such. He may, however, be interested in rainfall in the probabilistic sense in order to be able to infer, by interpolation between gauged points, the characteristics of rainfall at an intermediate, ungauged location. In this case, the efficiency of the network may need to be tested only as to the quality of measurement of point rainfall processes over a range of meteorological conditions and the adequacy of the network density and distribution.

In arid zones the areal variability of predominantly convective rainfall is such as to make routine areal, basin-wide rainfall estimates impossible other than for periods of a year or more. This is because of the uneconomically large number of gauges required to estimate monthly, daily or storm rainfall on a routine national network basis.

In either case it is essential to test objectively the efficiency of an existing raingauge network and to increase the density (or improve the distribution) of gauges where shown to be necessary to attain the required average precision. In the case of ungauged or inadequately gauged areas some idea of the number of gauges required must be estimated.

In much of the arid zone, especially in mountainous regions, it may be possible to improve the network only with great difficulty and it will usually not be economic to provide the optimum density. However, even in the most adverse conditions it is necessary to assess the limitations of existing networks. It has to be decided whether data requirements for future water resources developments may be met by short or medium-term raingauge installations or whether existing stations and derived correlations, and further developments of them, will suffice. In the context of prospective multi-million dollar projects the installation and maintenance of raingauges in remote locations might well be justified, provided the results possible to achieve are capable of influencing connclusions as to project viability.

The questions to be answered are:-

- What data is the raingauge network required to produce?
- If a network exists, can it produce the data required?
- If not, can it be upgraded to do so?
- What additional gauges, if any, are required?
- Where should they be installed.
- Could the money be spent to better advantage on other hydrometric network stations, e.g. runoff stations.
- Will it be possible to operate the network effectively for an adequate period to produce the required results.

The answers to these questions might, in some cases, result in there being no raingauge network additions.

5.2 THE DISTRIBUTION OF RECORDING, TOTALISING AND DAILY RAINGAUGES

The required proportions of recorders, totalisers and daily raingauges in a national raingauge network, or parts of it, should be determined on the basis of the following considerations:

5.2.1 Daily Raingauges

In most countries of the world the basic national raingauge network consists of daily gauges, the distribution of which is designed to provide areal coverage such that the temporal and spatial distribution of rainfall (and/or snowfall) may be assessed. These networks usually consist of raingauges operated for their own purposes by Government agencies, public and private organisations and private individuals. Where the network is deficient the public authority responsible for coordination of hydrological data collection may, as part of the optimisation process, install and operate additional stations.

The design of the "standard" gauge is normally determined by the Government meteorological organisation on the basis of experience with alternative designs in the meteorological conditions of the particular country and in accordance with international standards.

Standard gauges with various collector rim areas and stipulated exposure conditions differ as between countries. It is imperative that projects adopt the same standard when installing supplementary gauges.

Unfortunately, the normal site for installation of the standard gauge in some arid zone countries is a rooftop where exposure conditions prevent accurate measurement due to distortion of rain drop paths by wind eddies. In addition to this, the general standard of observation is generally low.

Only those stations equipped with national standard gauges and having first class exposure conditions and observational standards should be regarded as standard. Most of these are at synoptic and climate stations where both these conditions may be approached.

5.2.2 Totalising (Storage) Raingauges

These are normally used to supplement the standard raingauge network in remote, uninhabited areas such as mountain or desert zones. They may be read at weekly, monthly, or seasonal intervals as circumstances permit. They are sometimes used to produce estimates of monthly rainfall but should strictly be regarded as providing a measure of annual and seasonal rainfall amount only.

Comparisons between totalising and standard gauge catches should be made at climate stations in order evaluate correction factors to be applied to totaliser records.

5.2.3 Recording Raingauges

These are normally required only to determine the temporal characteristics of rainfall and in highly developed countries may represent only about 10% of the total network. They are normally distributed in such a way that intensity variations are sampled in different parts of the country where widely varying meteorological factors and storm-producing mechanisms, may exist. When an adequate period of record is available the regional characteristics of point rainfall may be determined by interpolation between stations.

In arid countries, where temporal and spatial rainfall variability is extreme, dense networks or recorders are required to form gauge "clusters" sampling areal rainfall patterns in different meteorological zones.

The optimum balance for raingauge networks in arid zones is biased towards a predominance of totalising and recording gauges. Standard daily gauges are required to provide the basis for correlation of rainfall amounts measured by the network to a uniform standard. It is normal international practice to install standard daily gauges at recorder stations. However, where recorders (mainly monthly operating) and totalising raingauge predominate (as recommended above), it is only necessary to complement a small proportion with standard daily gauges where reliable observers are available.

5.3 RAINFALL STATIONS

Recommendations regarding the installation of raingauges of all types as well as observational procedures will be found in Chapter 2 of the WMO Guide to Hydrological practices. These recommendations should be adhered to as closely as possible, keeping in mind that data from stations installed for specific purposes must be recorded according to the same rules adopted for the national network if these data are to be made available for general use.

Although the subject of site effects is fully discussed in the above WMO publication attention is drawn to the considerable effect of wind speed on the amount of precipitation collected in raingauge when the size of the raindrop is small.

6. PROCESSING, ANALYSIS AND PUBLICATION OF RAINFALL DATA

6.1 QUALITY CONTROL

6.1.1 Field Record Checks

Subjective and objective quality control methods should be applied to the raw data in field and office. It is important that doubts concerning rainfall data are raised early during data processing and referred to the observer while the circumstances of particular record periods are fresh in his mind.

6.1.2 Error Sources

The principal causes of inaccurate rainfall records are:

(a) Observer Errors

(i) Failure to attend the gauge at specified times and frequencies.

In the case of daily raingauges this may lead to accumulation of the rainfall catch over several days.

Crediting rainfall amounts accumulated over several days to a single day can result in frequency analyses which over-estimate the magnitude of daily falls for any return period and under-estimate the frequency of daily rainfall occurrences.

In some areas it may be impossible to find someone willing or able to attend the gauge regularly or to perform the duties of observer fully. In such cases it is preferable to install a monthly recording or totalising gauge, to reduce the level of attendance to that of guarding the instrument only.

The errors resulting from irregular, infrequent attendance lead to systematic under-estimation of rainfall.

(ii) Reading and Recording Errors

Failure of the observer to read the daily raingauge or the recording raingauge measuring glass properly and/or to write correctly what he has read can only be dealt with by intensive training during field trips.

Detection of errors of this nature may be difficult but scrutiny of daily precipitation data and the monthly array of daily rainfall data for the area often reveals anomalies which should attract the attention of the hydrologist. For example:

- Habitual or intermittent lack of a decimal point in the record entries.
- The excessive occurrence of zeros and fives, or any other digit.
- Values given to two decimal places.
- Unexpectedly high or low values (to be amended with caution).

These and similar anomalies should be investigated as soon as possible to prevent the persistence of observer-induced errors.

(iii) Date errors

The observer may have been instructed to enter daily gauge readings on the date of reading or on the previous day. It is preferable that observers should enter the daily value against the date of observation and that crediting of measured values to the previous day should be carried out at the data processing stage. Installation and history sheets in office files should in any case indicate which practice is followed at each station. These errors are systematic at individual stations but their random occurrence over the network tends to obscure occasionally systematic daily rainfall patterns when they occur.

(b) Transcription Errors

These may occur when transferring the record from the observer's form to the daily rainfall summary sheets at the regional office. Each summary sheet should be checked and initialled by the checker who should also ensure that daily rainfall values have been ascribed to the correct day. These errors are normally random but may be systematic if the dating method is incorrect.

(c) Index and Calibration Errors

These may arise from:

- Broken or worn daily raingauge dipsticks;
- Bent collector rims
- Leaking totaliser bottles
- Accumulation of sand in recorder tipping buckets or float chambers
- Measuring cylinder does not accord with gauge rim diameter.

Except in the last case, the magnitude of the errors associated with these instrument defects is not easily assessed and can only be minimised by careful checks and attention to maintenance.

Rainfall recorders should be checked frequently by slowly pouring a measured quantity of water into the gauge collector and noting the chart pen position. Pen position or range adjustments should only be made after the instrument has been thoroughly cleaned.

6.1.3 Missing and Doubtful Data

The estimation of missing rainfall values is an extremely important activity. Errors in an estimated daily value may affect the precision of a monthly total by less than 10% and of an annual total by an insignificant amount. Even an estimate for the whole of a particular month may produce an annual total which is within 10% of actual rainfall. Failure to estimate values for a day, or even several days, may make it impossible to estimate such important parameters as the mean and standard deviation of annual rainfall. Estimation of missing values must be carried out at the quality control stage of data processing. If gaps in the time-series of point rainfall data remain unfilled until the analysis and interpretation stages, synthesis of missing values becomes more difficult.

Missing values may be estimated subjectively through study of the daily rainfall array, by consideration of run-off records and by scrutiny of inspection reports.

It is normally difficult to make reliable estimates of daily values unless the evidence from surrounding gauges is strong - e.g. all gauges showing similar daily amounts which might be the case for widespread rainfall of frontal origin. However, it is often desirable to apportion a known accumulated recorder catch between the days of recorder failure in order to provide a continuous time-series for analysis of the distribution of daily depths. The undoubted errors in such apportionment are likely to be less, ultimately, than would result from omitting the period from the analysis. It should be noted that in some semi-arid areas more than half the annual rainfall may fall in a few hours.

It is important for the user of rainfall data to be made aware of the degree of estimation contained in published figures. For this reason a simple system involving the use of symbols should be adopted.

6.1.4 Annual Summaries of Daily Rainfall

These are completed for each daily and recording raingauge by transferring the quality-checked data from the station daily summary and are usually retained in the station file.

Before the rainfall data are transferred to the summary they should be checked when collected from the observer and a further check and inter-station comparison and reconciliation made when transferring the data from the field record to the monthly array of daily rainfall. The transfer of data to the summary should therefore involve no interpretation but only systematic recording procedure.

6.1.5 Totaliser Records

Totalisers are read at approximately monthly intervals, but with the date of reading rarely coinciding with a month's end. Monthly rainfall totals, even in arid zones, are hydrologically significant and it is therefore desirable to attempt to apportion totaliser catches by months or seasons.

In arid regions it is impossible to be certain of the temporal distribution of rainfall collected over a period of a month or more, even if there is a reliable daily raingauge or recorder in the vicinity, because of highly irregular areal rainfall distribution. Apportionment of totaliser catch between calendar months is therefore aimed only at producing reasonably valid long-term monthly and seasoned averages and not necessarily to represent the actual distribution during a particular period or calendar year.

However, since the random variability of areal rainfall processes has not been retained, monthly areal rainfall estimates using monthly totaliser values should not normally be attempted.

6.2 RAINFALL INTENSITY ANALYSIS

Rainfall intensity analysis should be carried out as routine using either or both of two basic methods.

In the first method, depths of rain falling within selected time intervals (usually 10-minutes) are abstracted from the recorder trace and the corresponding mean intensities calculated. Maximum depths and mean intensities for durations of up to three hours are tabulated for each storm and analysed separately. The instantaneous intensity within discrete time intervals may vary considerably and the derived mean intensity becomes progressively less significant (hydrologically) with increasing duration. The method conforms generally with standard international practice and facilitates valid comparisons with data from other countries.

In the second method the rainfall intensity for any duration, rather than the depth for a fixed duration, is the criterion for interpretation of the recorder trace. In this method the objective is to classify rainfall depth by intensity segments in order to determine the hydrological response of natural surfaces under various conditions of initial loss and infiltration rate.

Both methods are described in detail in Annex 1 to this Paper.

Modern recording systems, on paper or magnetic tape or solid-state "memory" may produce an output in the form of rainfall depth per unit time or time per unit of rainfall depth. Whichever processing method is adopted should aim to meet the users needs. The hydrologist in arid lands is normally more concerned with instantaneous rainfall intensity than with mean intensities over discrete time intervals.

6.3 ESTIMATION OF MEAN AREAL RAINFALL

6.3.1 Rainfall - Area Relations

The rainfall depth measured by raingauge represents a very small sample of the area gauged. The raingauge network density requirement is dependent upon the shape of the "rainfall surface" (i.e. the shape of the notional water surface which would result from accumulated rainfall over the area) for the duration being considered. The degree of irregularity of the rainfall surface is dependent upon this duration. The configuration of the long-term mean surface is much smoother than that of short-duration storm rainfall and fewer gauges are required to measure it.

In arid countries, the optimum national network can only aim to produce areal rainfall input estimates with acceptable precision on a seasonal or annual basis. For this reason it is usual to calculate annual inputs to derive run-off or recharge volume factors for application to unmeasured water-course/aquifer systems and for regional analysis. Inputs from important storm events, especially those covering large areas, may sometimes be quantified with acceptable accuracy for flood studies. The relationship between size of area, gauge density and rainfall surface regularity as estimated by the available network determines the precision of the input estimate.

6.3.2 Methods of Estimating Mean Areal Rainfall

Mean areal rainfall for the required period may be estimated in a number of different ways. The method to be used depends on the network distribution, the relationship of rainfall depth to altitude, aspect and locality, the availability of topographic maps, the skill of the hydrologist and the purpose for which the areal input estimate is required. Whichever method is applied the first requirement is for assessment of the reliability of the data and the weight which should be given to individual values.

The point rainfall data for the period concerned are classified according to their quality as determined during control and checking procedures:-

Class 1

- The station rainfall total includes estimated partial depths representing less than 10% of the whole. The precision of the total is considered to be within 10% of the true rainfall in the area represented by the gauge.

Class 2

- The station rainfall total includes estimated partial depths representing more than 10% of the whole (amount usually bracketted in annual array of monthly values) and the precision of the total is considered to be outside a range of ± 10% about the true value.

  In the case of recorders the bucket catch is taken to be a true measure of rainfall depth where the recording mechanism has failed - provided that evaporation from the bucket has been suppressed by a suitable (eg. diesoline) additive. Blockage of recorder inlets during periods when significant rainfall amounts have been measured in nearby gauges would make estimation necessary.

In the case of totalisers, blockages are likely occur more frequently and the record may contain a higher proportion of estimated rainfall than at recorders. Vandalism and the introduction of foreign material is more prevalent in relation to totalisers.

The annual total contains estimated proportions of rainfall accumulated over periods spanning two or more months and apportioned in relation to adjacent recorder or daily gauge catches. Totaliser records are thus more likely than recorders to be in this class.

In the case of daily gauges classification poses greater problems due to the human element. The precision of annual amounts can only be assessed on the basis of inspection reports, observer reliability etc. Measurement may have been delayed several days or missed altogether or rainfall may have been incorrectly recorded. The recorded amount may be an under-estimation of up to 20%.

Class 3

\- The annual total is doubtful and cannot be used for areal rainfall estimation.

In the case of recording raingauges, malfunctioning of the mechanism at the end of the calendar year. With no reliable means of apportioning the catch between each year, would render the annual total for either year doubtful though the two-year total or the average for several years would be quite acceptable. This situation is rare. In the case of totalisers, repeated vandalism would render the record invalid.

All work of classification requires extensive reference to inspection reports and field documents at the quality control stage when estimates of missing values are also made. The classification should be checked by an experienced hydrologist or climatologist.

Method 1 - Isohyets

Ideally a contour map should be used and the relationships between rainfall depth and altitude, aspect and distance inland from moisture sources determined. In arid countries these relationships emerge only after a relatively long period of record and may only be applied with confidence to long-period isohyetal maps. The sparsity of data in large areas of necessitates interpolation aided by whatever means are available and the most likely position of an isohyet is estimated by interpolation which takes account of geographical controls.

The mean areal rainfall is calculated by planimetry of inter-isohyetal zones to determine their areas, multiplying these areas by the weighted mean rainfall depth corresponding with the isohyetal values and summing the products.

Advantages of the method are:

(i) It permits the analyst to exercise his own judgment and knowledge of average and specific rainfall distribution within the area.

(ii) Suspect or missing data need not be estimated.

(iii) No artificial weighting is applied to raingauge catches.

(iv) Topographic controls may be taken into account.

Disadvantages are:

(i) It is more time-consuming than most other methods.

(ii) It is more subjective - the accuracy depending upon the skill of the analyst.

Method 2 - Thiessen Polygons

In this method the area represented by individual raingauges is related geometrically to their distance apart. With non-uniform gauge spacing, more weight is given to catches of gauges in the lower than in the higher network density parts of the area for which areal input estimates are required. The method is objective in that different analysts will obtain the same result. Once the polygonal network has been established and the station weights computed, the method is much more rapid than the isohyetal method.

Doubtful or missing data must be estimated to avoid reconstruction of the polygonal network.

The method is not recommended for normal use in arid regions but has application for study catchments with dense raingauge networks from which daily inputs are calculated using objective computer techniques.

Method 3 - Triangles

This method is similar to Method 2 and has the same advantages and disadvantages. Less work is involved initially but routine computation is somewhat more laborious.

Method 4 - Arithmetic Mean

Where gauges are uniformly distributed the simple unweighted arithmetic mean of rainfall values in and adjacent to the area under study may be used to estimate mean basin input. The method may give results similar to those of other methods and may be used for rapid estimates. It may be more accurate than the Thiessen or triangle method if the latter give undue weight to remote stations. It is generally less accurate than the isohyetal method applied by a skilled analyst.

Method 5 - Percent Normal

"Normal" rainfall is defined as the long-term annual mean input and distribution to an area. The percent - normal method takes account of the Normal rainfall during the preparation of isohyets. It is thus similar to the isohyetal method with semi-empirical weighting according to orography but is less subjective in application. Rainfall for the required period, which theoretically may be one day or the duration of a major storm, is expressed as a percentage of the Normal.

Normal and iso-percental maps are used in the construction of the rainfall map for the required period. The method is laborious and justifiable only when it is known that short-duration (or even annual) "rainfall surfaces" correspond reasonably closely with the "Normal Surface". The areal variability of storm rainfall from the majority of storms make it unlikely that any improvement in the precision of input estimates would result from use of the method. However, for consistency in estimating the input from heavy widespread rainfall of frontal origin or annual rainfall input, it has some merit.

6.4 RAINFALL DEPTH-DURATION-FREQUENCY RELATIONSHIPS

Rainfall intensity as a function of duration and return period is important when design floods must be estimated. Considerable attention has been paid to the subject throughout the world and many empirical rainfall intensity formulae have been derived from recording raingauge data.

Bell (1969) showed that, world-wide, the ratios of the T-year to 10-year rainfall depths for a range of durations up to two hours are sensibly constant for all return periods T up to 100 years and that ratios of D-duration to 1 hour rainfall depths are similarly constant over a range of T. Table 6.1 shows derived ratios for specified durations based on long recording rainfall records in the United States, USSR and Australia.

TABLE 6.1

COMPARISON OF P_D/P_{60MIN} DEPTH-DURATION RATIOS

Country	5-min	15-min	30-min	120-min	1-day
United States					
Mean	0.29	0.57	0.79	1.25	
Std. deviation	0.03	0.04	0.04	0.08	
Australia					
Mean	0.30	0.57	0.78	1.24	
Std. deviation	0.04	0.04	0.03	0.06	
USSR					
Mean (Maximum)	0.37	0.61	0.87	1.40	
Mean (Minimum)	0.26	0.49	0.71	1.20	
Saudi Arabia					
Station (A) Mean	0.37	0.65	0.85	1.20	1.67
Std. deviation	0.05	0.07	0.09	0.11	0.15
Station (B) Mean	0.31	0.57	0.78	1.19	1.70
Std. deviation	0.03	0.05	0.07	0.30	0.81
Station (C) Mean	0.28	0.53	0.75	1.21	2.28
Std. deviation	0.01	0.04	0.08	0.11	0.65
Station (D) Mean	0.29	0.57	0.81	1.21	2.08
Std. deviation	0.04	0.04	0.06	0.11	0.03

In Saudi Arabia it was found (Wan, 1976) that application of the "station-year" concept to a relatively short period of recording raingauge data for 18 stations, produced similar ratios, as shown in Table 6.2. It was concluded that, for Saudi Arabia, the derived mean ratios could be used to estimate the rainfall depth corresponding with any duration up to 2 hours for return periods up to 100 years - given an estimate of the 10-year/60-minute value. The latter value may be estimated from an approximate relationship with mean annual rainfall. Figures 6.1 and 6.2 illustrate the relationships derived from 18 rainfall stations with up to 6 years of record in Saudi Arabia. The estimate obtained by use of these curves is, at best, only as good as the estimate of mean annual or 10-year/60-minute rainfall. However, in a field of much uncertainty, the results obtained by application of this method are probably at least as good as can be obtained by any other.

TABLE 6.2

COMPARISON OF DEPTH-FREQUENCY RATIOS

Return Period Years	Saudi Arabia*		USA		Australia	
	Mean	Standard Deviation	Mean	Standard Deviation	Mean	Standard Deviation
2	0.62	0.04	0.63	0.05	0.65	0.05
5	0.85	0.02	0.85	0.05	0.85	0.05
25	1.19	0.03	1.17	0.03	1.18	0.05
50	1.33	0.03	1.31	0.05	1.33	0.03

* Average Values for stations B, C and D.

6.5 PROBABILITY OF DAILY RAINFALL OCCURRENCES

6.5.1 Introduction

Daily rainfall frequency is a potentially useful parameter for studies of rainfall-runoff relationships, for direct recharge estimates and for agriculture in marginal rain-fed farming areas. The method used to estimate the probability distribution of the annual daily rainfall depth exceedances, using the "station-year" method, is set out below.

In arid zones events of interest to hydrologists and agricultural development planners include:

- Rainfall amount for single rain-days or rain-periods (consecutive rain-day sequences),
- Number of days or periods per year or season with rainfall amount greater than specified threshold levels.

In the first case it is required to know how rainfall amount is distributed among rainfall periods of one day or longer. In the second case it is the distribution of the number of exceedances per year or season that is required.

6.5.2 Distribution of Daily Rainfall Depths

In studies at the Southwest Watershed Research Centre in Arizona carried out by Smith and Schreiber (op.cit) based on three long record periods (55-73 years) it was shown that the daily rainfall depth exceedance frequency distribution could be fitted by a compound exponential function.

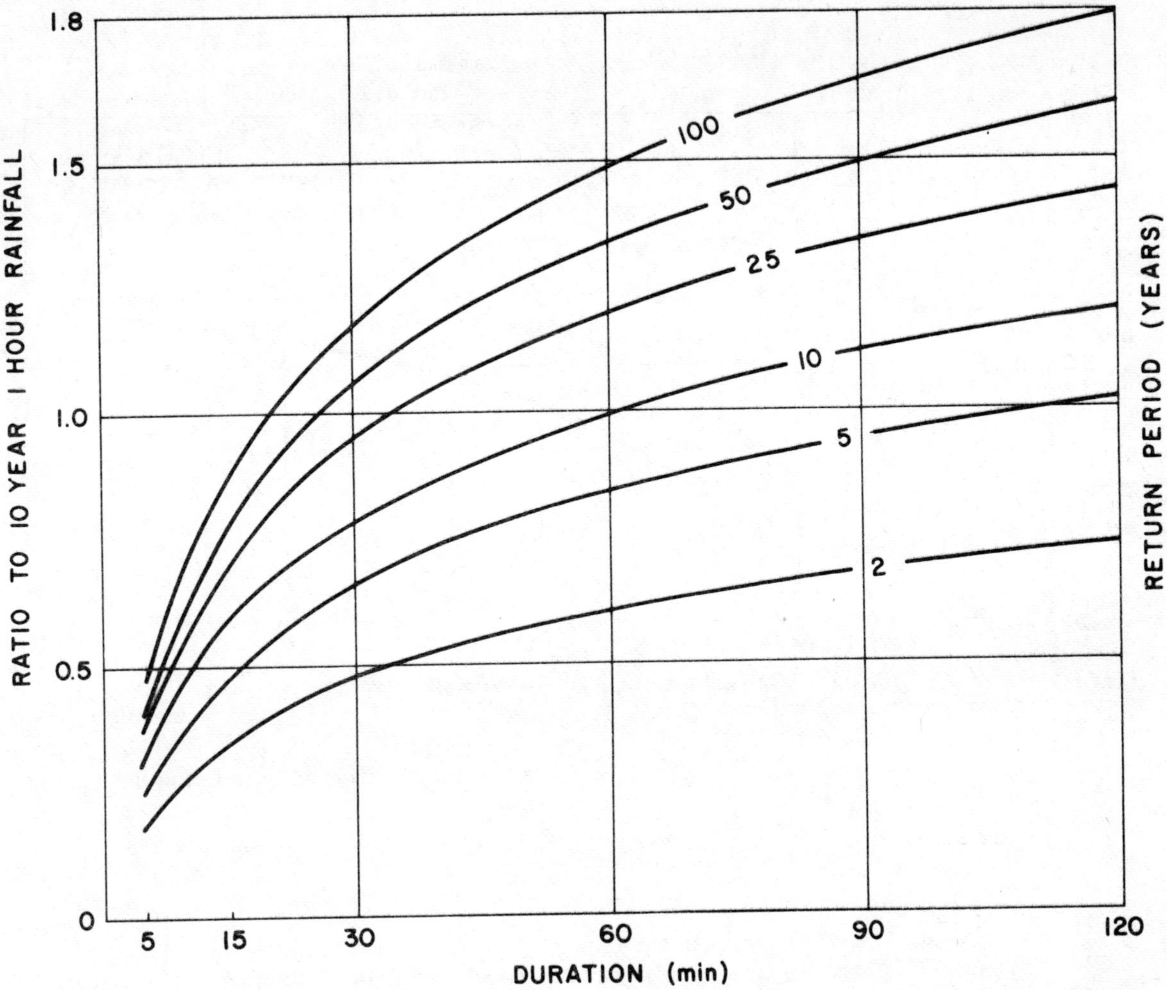

Fig. 6.1 Depth-duration-frequency curves

The curves express the relationships derived from application of the station-year approach to analysis of Saudi Arabia rainfall data covering a maximum of 6 years. It has been found that these curves are closely similar to those developed from long rainfall records in many other parts of the world. If the 10-year 1-hour rainfall value can be estimated, depths for return periods up to 100 years and durations up to 2 hours may be derived by application of the ratio found by use of the figure.

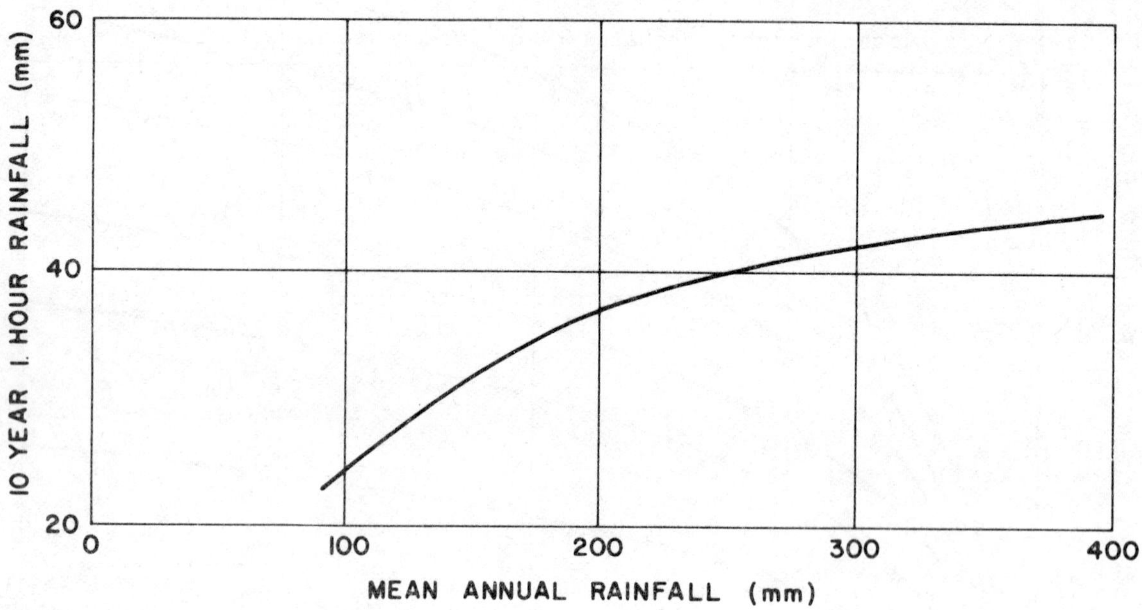

Fig. 6.2 Relationship between 10-year 1-hour rainfall and annual rainfall

The relationship shown by the figure is based on the station-year analysis and applies strictly to Saudi Arabia. Such relationships are relatively easy to derive but much less easy to verify. An increase in the 10-year 1-hour value with increase in mean annual rainfall should be expected. The depth-intensity relationships shown by Figure A.1.4 are virtually independent of mean annual rainfall. However, since the frequency of storms is greater in high rainfall areas, it follows that the average maximum, D-duration rainfall depth in any year would also be higher.

The hypothesis that the daily depth distribution is independent of when a rain-day occurs was tested. It was found that:

- There is no significant difference between the depth distribution for the first days of a rain-period and that for all other days,

- There is a slightly higher probability of larger storms occurring in periods of two or more days than in single days,

- There is little difference between the distributions representing the first, middle and last part of the rainfall season,

- The mean rain-day depths for years with less than 20 storms is slightly higher than those with more than 40 storms.

Where only short-term records are available the distribution of daily rainfall depths may be estimated from groups of stations assumed to be representative of the long-term annual rainfall conditions of their sample means:-

- Tabulate daily rainfall by depth interval classes for each station in each group.

- Calculate the total occurrences in each class for each group,

- Cumulate total occurrences from low to high frequencies,

- Calculate the exceedance probability by dividing each cumulated total by the total number of rain-days for the group.

The procedure is illustrated by Example 6.1 and the results shown on Figure 6.3.

The same procedure may be applied to consecutive rain-day periods of any duration and to seasonal and annual rainfall amounts.

6.5.3 A Statistical Quality Test of Daily Rainfall Records

In arid zones the spatial variability of rainfall is such that it is rarely possible to judge the quality of a daily rainfall record subjectively. Experienced hydrologists may doubt the data, or know them to be wrong, from personal knowledge of the timing and extent of storm rainfall based on their observation of ground conditions and on recording raingauge data. But evaluation of past records is more difficult. The hydrologist usually looks for consistency in the timing of rainfall occurrences by examining the monthly array of daily rainfall for an area, and for a logical distribution of rainfall amounts at a single station. The former test cannot be applied objectively in arid zones since it is quite possible for only one gauge in a large network to record rainfall on a particular day. The latter test can be applied objectively by analysis of the frequency distribution of daily rainfall amounts.

The method used is to tabulate daily rainfall according to depth classes and to check whether the distribution of these depths for the station being tested (the "test station") is statistically different from what we might expect. The expected distribution would normally be based on that produced by analysis of "standard station" recording raingauge records but might be based on those of a well-observed daily gauge. The statistical method appropriate to this form of quality check is the Chi-square test.

The test cannot be applied to short-term data as it is quite possible for two adjacent gauges to have been subject to different temporal distribution patterns during a single year. A five-year period has been found suitable for parts of Saudi Arabia.

EXAMPLE 6.1

PROBABILITY DISTRIBUTION OF DAILY RAINFALL DEPTH

Rainfall Zone 100-200mm

Period 1967-1972 (6-years)

Station	Rainfall Depth Class								
	0-	1-	5-	10-	15-	25-	40-	60-	100-
	Total Occurrences in Record Period								
Ta-215	175	103	48	21	12	4	2	-	-
J-220	106	60	25	11	8	4	2	-	-
J-214	165	108	46	28	18	11	2	1	-
Ta-203	241	128	60	37	22	11	4	2	-
Ta-220	238	125	65	37	21	7	4	-	-
B-207	159	98	47	20	9	3	-	-	-
B-209	172	102	38	17	8	2	1	-	-
TOTAL	1256	724	329	171	98	42	15	3	
MEAN	179	103	47	24	14	6	2.1	0.43	
PROBABILITY	1.00	.573	.253	.134	.078	.034	.012	.002	

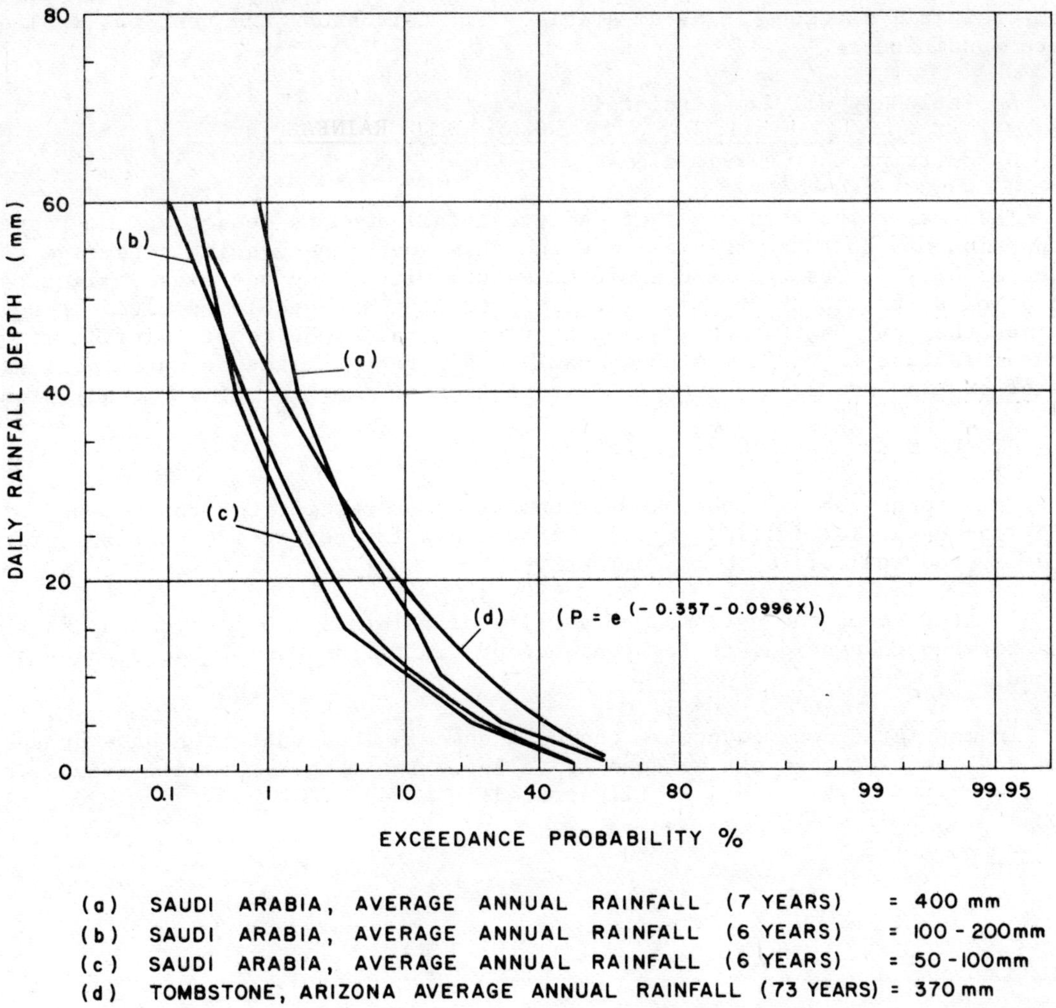

Fig. 6.3 Distribution of daily rainfall depths

The four curves shown on the figure show the distribution of rain-day depth exceedance plotted against a Normal probability scale. Curves (a), (b) and (c) represent mean annual rainfall zones in Saudi Arabia and were developed using the station-year approach. Curve (d) is the relationship derived from a 73-year record for Tombstone, Arizona, and is not greatly different from curve (a).

The curves show that, given a rain-day occurrence, there is a 5% chance that the depth will exceed 15 mm in a 50-100 mm rainfall zone and 25 mm in a 400 mm rainfall zone. 50% of all rain-day depths for all zones are less than 5 mm.

The test and standard stations must be in the same general location and be likely to receive the same mean rainfall. Example 6.2 shows the calculation made to test a number of daily rainfall stations in Saudi Arabia. For this study the test was applied for two separate conditions:

- A including all daily rainfall classes

- B omitting the 0-0.9mm class.

The reason for this was that daily rainfall amounts less than 1mm represent a small portion of the total rainfall and the Chi-square test, being sensitive to large differences between observed and expected values within any one class, might cause rejection of a station which was basically producing acceptable results. However, it was recognised that an observer who failed to record small amounts of rainfall might do so because of failure to visit the gauge daily. His record might be unreliable as to the timing of events even if the overall distribution of them is not seriously wrong.

6.6 DOUBLE MASS ANALYSIS

This technique has been used extensively for testing the consistency of hydrological data, for filling gaps in records and for correcting portions of records revealed by the analysis as being defective.

In arid zones the extreme variability of rainfall and the general lack of long records for enough stations to form base group limits the use of the double-mass technique.

In many arid zone countries the data base is inadequate for base groups to be derived and the best that can be done is to apply the method to small groups or even pairs of stations. The method is illustrated in Annex 2.

EXAMPLE 6.2

Station Details

Station	Type	Lat. °	Lat. '	Long. °	Long. '	Alt. m	Pa mm	Period
1. A-101 Abha	Test, Daily	18	13	43	30	2190	(406)	1966-72
2. A-209 Abha	Standard, Rec.	18	13	42	29	2200	451	1966-72

Occurrences of Daily Rainfall

Station	Period	Daily Rainfall Depth Classes 0-0.9	-4.9	-9.9	-14.9	-24.9	-39.9	-59.9	-99.9	100
1	7.66-6.67	-	13	5	2	4	3	-	-	-
	7.67-6.68	-	13	19	14	12	-	3	2	1
	7.68-6.69	-	8	9	6	5	1	2	-	-
	7.69-6.70	-	13	11	2	4	-	2	-	-
	7.70-6.71	-	25	11	6	2	2	-	-	-
	Total "O"(1)		72	55	30	27	6	7	2	1
2	7.66-6.67	17	25	6	3	2	-	2	-	-
	7.67-6.68	23	39	19	10	5	5	1	1	1
	7.68-6.69	10	37	10	7	1	2	2	-	-
	7.69-6.70	13	20	10	-	4	1	2	-	-
	7.70-6.71	36	25	10	4	2	3	-	1	-
	Total "E"(2)	99	146	55	24	14	11	7	2	1

Chi-square Test

A	$\frac{(O-E)^2}{E}$	99	37.3	0	1.5	12.2	2.3	0	0	0	152.3
B	"	-	37.3	0	1.5	12.2	2.3	0	0	0	53.3

(1) "O" denotes observed occurrences (at the "test" station)
(2) "E" denotes expected occurrences (at the "standard" station)

The degrees of freedom (ν) associated with test are the number of classes minus one. $\nu_A = 8$ and $\nu_B = 7$. We assume a significance level (α) of 5% or 0.05.

From standard tables, $\chi^2_{A,.05,8} = 15.5$ and $\chi^2_{B,.05,7} = 14.1$. As the test statistic is larger in each case the result shows a significant difference between the two distributions and it could be concluded that the two sets of data are not taken from the same population, or that one or other set is wrong.

In this case we may safely reject Station A-101 data as being incorrect.

7. CHARACTERISTICS OF EVAPOTRANSPIRATION IN ARID ZONES

7.1 GENERAL

The following symbols are used herein:

E - Evaporation from any surface.

Eo - The theoretical evaporation which would occur from an infinite expanse of shallow open water without energy storage properties when exposed to specified climatic conditions which remain unaltered by the evaporation itself.

ETp - Potential evapotranspiration. The theoretical evapotranspiration which would occur from an infinite expanse of natural or cultivated terrain, adequately supplied with moisture when exposed to specified climatic conditions which remain unaltered by the evapotranspiration itself.

ETa - The actual evapotranspiration (evaporation and transpiration) occurring from any natural or cultivated terrain.

Epan- The evaporation from an evaporation pan (USWB Class A unless otherwise stated).

ETo - Evapotranspiration from an extensive, 8-15 cm tall green grass cover of uniform height. Used as a reference crop evapotranspiration value for estimating crop water requirements.

Actual open water evaporation depends upon the energy storage properties of the water body. It may be calculated using the water budget method in which inflow, outflow and storage changes are measured and evaporation is the residual

$$E = P + I - O + \Delta S$$

in which P is the precipitation on the water surface, I and O are the surface and subsurface inflow and outflow and Δ S is the change of storage in the water body.

The method is applicable to well-controlled inflow, outflow and storage change measurement at reservoirs or lakes from which seepage losses are known to be small. Open water evaporation may also be estimated from evaporation pan data, either by direct application of a seasonal factor to Epan or by application of formulas or techniques using measurements of pan water temperature and vapour pressure deficit at pan level (Kohler et al, 1955). It may be estimated (as Eo) using any one of a variety of energy budget, mass transfer and combination methods.

For the hydrologist engaged in water resource assessment in arid zones Eo and Eto may be used as evaporation or evapotranspiration reference levels to estimate:

- Reservoir evaporation;
- Evaporation from shallow depressions following heavy rainfall,
- The evapotranspiration component of the water balances of shallow alluvial aquifers under conditions for which these losses are not restricted by the moisture supply.

ETp may only be determined with confidence experimentally by using irrigated plots or lysimeters in which water application, percolation and storage change are measured under controlled conditions. It may be estimated (as ETo) by using energy-balance or eddy-correlation techniques. In semi-arid regions ETp may be used to define the geographic limits of economic rain-fed farming or the range of uncertainty in the economic returns to be derived from good agricultural practice in such areas.

ETa can only be estimated with any degree of confidence as the residual in the water balance of natural drainage areas for which the precipitation, runoff and groundwater outflow and storage changes can be measured. It may also be estimated, using energy balance or eddy correlation techniques and by modified versions of the combination energy budget and mass transfer approach or using soil moisture balance models.

The literature abounds with formulas and variations of them derived from experimental (and some field) data. No attempt is made in this paper to review or summarise the mass of material available on this subject. Instead, the reader is referred to a small sample of such reviews and summaries for further discussion.

(a) Sellers, W.D; Physical Climatology, University of Chicago Press, 1965.

(b) WMO TN.83, Measurement and Estimation of Evaporation and Evapotranspiration WMO-No.201.TP.105, Geneva.

(c) Doorenboos, J. et.al. Crop Water Requirements Irrigation and Drainage Paper No.24 FAO Rome 1977.

(d) UNESCO, Arid Zone Research - XV. Plant-Water Relationships in Arid and Semi-Arid Conditions. 1960.

Reference (c) will be found to be particularly useful to those engaged in the determination of crop water requirements.

7.2 CLIMATIC ELEMENTS AFFECTING EVAPOTRANSPIRATION

Combination aerodynamic and energy balance formulas commonly utilise mean daily, 10-daily or monthly data for the following climatic elements:

- Air temperature Ta (^{o}C) at thermometer-screen height
- Relative humidity RH (%) or wet bulb temperature T_w (^{o}C) at thermometer screen height.
- Incoming radiation Rc (Cal. cm^{-2} . day^{-1}).
- Wind-speed U_2 (m/s, km/day or miles/day) at 2m above ground level.
- Duration of bright sunshine.

For some purposes the soil temperature profile is measured and, in the method of estimating Eta described by Tanner and Fuchs (1968), the temperature of the ground surface is required. Pike (1970) has described short-term studies requiring application of the energy-budget method. Measurement of instantaneous Ta and Tw gradients above ground level and of net radiation fluxes above and a short distance below ground level, using special equipment, are required for this method.

7.3 TEMPORAL AND SPATIAL VARIABILITY OF POTENTIAL EVAPOTRANSPIRATION

Typical diurnal fluctuations in heat fluxes and evapotranspiration rates in an arid zone location is illustrated on Figure 7.1. For most hydrological and agricultural purposes, 10-day mean Eo values are adequate to reveal the systematic seasonal variations underlying the random effects due to the vaguaries of weather. The range of monthly Eo

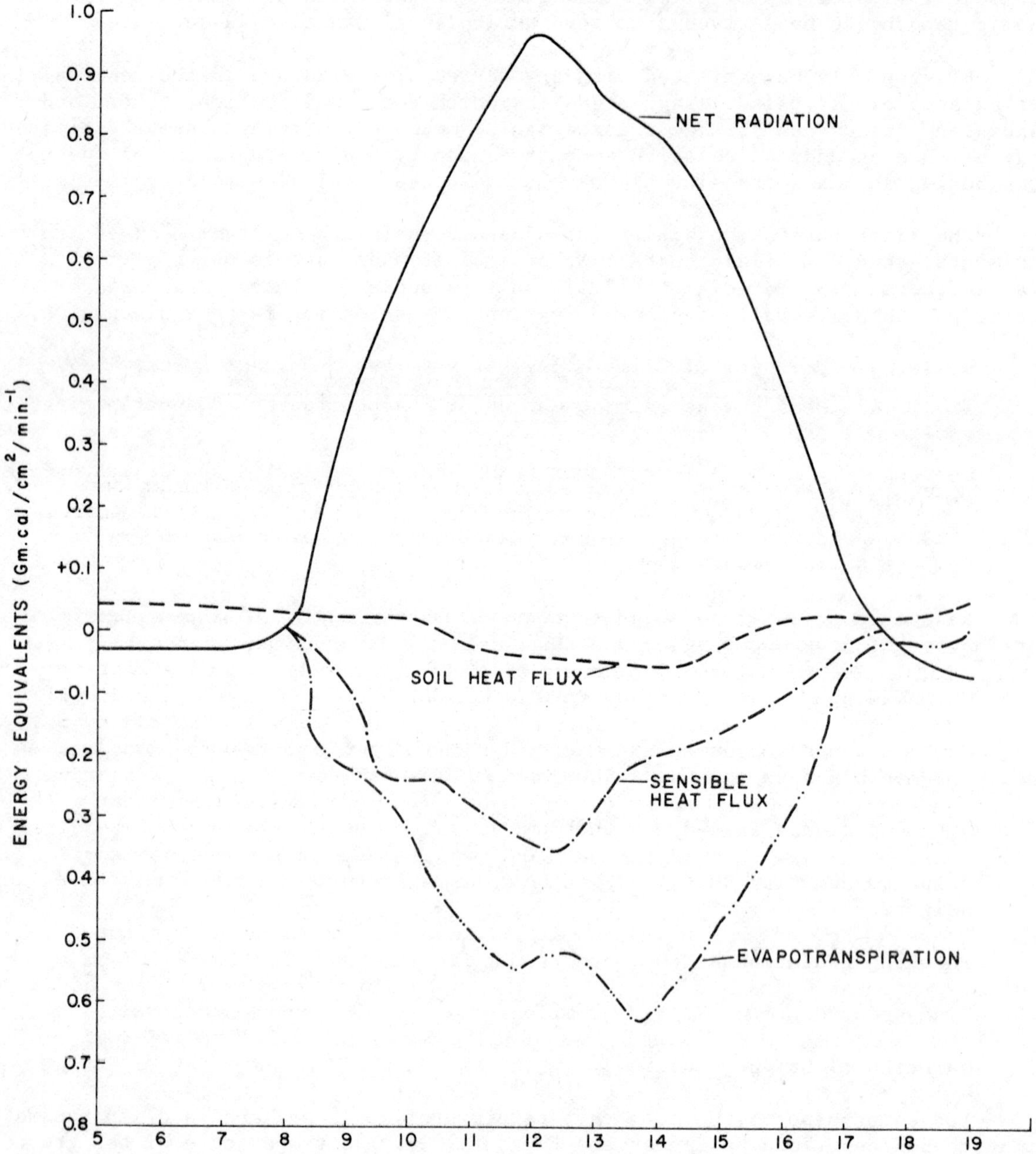

Fig. 7.1 Typical diurnal fluctuations in heat fluxes and evapotranspiration rates at an arid zone location

FIGURE 7.1

Energy Balance over a Stand of Phreatophytes
Khulays 19.2.67

The figure shows the diurnal variation in energy fluxes during one day at a point in an area of phreatophyte infestation in Saudi Arabia. The purpose of the study was to determine the evaporative energy component of the net radiation.

Net radiation is the algebraic sum of the soil heat flux, the sensible heat flux to the air and the heat utilized by evapotransporation. Net radiation was measured by means of a net radiometer, which effectively deducts reflected short wave radiation and surface long wave radiation from the total incoming radiation. Soil heat flux was measured by means of a thermal transducer placed under the soil surface. The residual energy flux obtained by deducting soil heat fluxes from net radiation was apportioned between sensible heat and evapotranspiration heat fluxes using the method described in Section 9.2. This method requires the measurement of temperature and vapour pressure differences above the ground; in this case at height of 0.5 m and 2 m.

The result of this apportionment is shown on the figure. The total evapotranspiration (ETa) during the day was estimated to be 3.7 mm corresponding with about 70% of the energy input at the soil surface.

On the same day ETp was estimated as 4.3 mm, Eo as 5.4 mm and Epan was 8 mm.

over a 6-year period for a station in Saudi Arabia shown in Table 7.1, illustrates the typically marked seasonal pattern and relatively low inter-annual variability of Penman Eo and ETp estimates for monthly periods. Applying Van Bavel's modification of Penman's method to daily data reveals the far greater variability about the seasonal cycle. Variations in the seasonal pattern of climatic elements with latitude in America and Africa are illustrated on Figure 7.2(a) and (b).

It must be stressed here that the random fluctuations, seasonal patterns and latitudinal variability referred to above, relate to the Penman estimate of potential evapotranspiration or evaporation and not to what actually occurs in nature over a region. In practice ETa will equal ETp or Eo only when the conditions stated in 7.1 occur, for instance in irrigated areas and from surfaces wetted by recent rainfall and open water bodies.

Eta over the greater part of arid zones amounts to a very high proportion of rainfall but a very low proportion of ETp.

If groundwater storage changes are relatively small and may be neglected, an estimate of long-duration ETa is provided by deducting runoff from rainfall. However, the ETa estimate thus obtained has little meaning other than for the particular basins and periods concerned. It cannot be generalised, since to do so would be to imply that runoff resources may be estimated from rainfall less a percentage of Penman Eo (as is possible in humid zones), which is not the case.

In view of the inherent errors involved in the measurement of areal rainfall and estimation of ETa, basin runoff cannot be determined by application of the water balance equation. In practice in arid zones, errors in areal rainfall estimation are greater than basin runoff. Errors in ETa estimates are even greater.

Every effort should be made to collect evaporation data from lakes and reservoirs in the region under consideration or from regions having a similar climate. Such data will always be useful for checking estimates made using either pan factors or Eo values which are difficult to transfer to large water bodies. Observed lake or reservoir data even at sites which may be far distant from the projected development will always be more reliable than theoretical open water evaporation estimates.

Lake or reservoir evaporation should normally be estimated only from the recession limb of the water level hydrograph when inflow has virtually ceased. In the case of reservoirs the release rate and water level - storage relationship must be accurately known.

TABLE 7.1

SUMMARY OF PENMAN Eo ESTIMATES - NAJRAN, SAUDI ARABIA

YEAR	J	F	M	A	M	J	J	A	S	O	N	D	ANNUAL
MEAN mm/day	4.8	6.0	7.2	7.9	9.0	9.6	9.6	9.1	8.6	6.8	5.5	4.6	7.4
$\hat{\sigma}$ mm/day	0.74	0.79	0.83	1.00	0.72	0.40	0.48	0.73	0.73	0.75	0.58	0.51	0.56
SE mm/day	0.30	0.32	0.34	0.41	0.29	0.16	0.20	0.30	0.30	0.31	0.24	0.21	0.23
MEAN mm/m	150	170	225	240	280	290	295	280	255	210	165	140	2700
% OF ANNUAL	5.6	6.3	8.3	8.9	10.4	10.7	10.9	10.4	9.4	7.8	6.1	5.2	100

ALGER

Lat. 36° 46' Long. 3° 03'
Alt. 60 m
Np = 30
Ne = 20
Pa = 691 mm
ETp = 1102 mm

TESSALIT (MALI)

Lat. 20° 12' Long. 0° 59'
Alt. 494 m
Np = 10
Ne = 10
Pa = 118 mm
ETp = 2817 mm

DORI (UPPER VOLTA)

Lat. 14° 02' Long. 0° 03'
Alt. 277 m
Np = 30
Ne = 10
Pa = 542 mm
ETp = 2203 mm

KUMASI (GHANA)

Lat. 6° 43' Long. -1° 36'
Alt. 287 m
Np = 30
Ne = 10
Pa = 1459 mm
ETp = 1361 mm

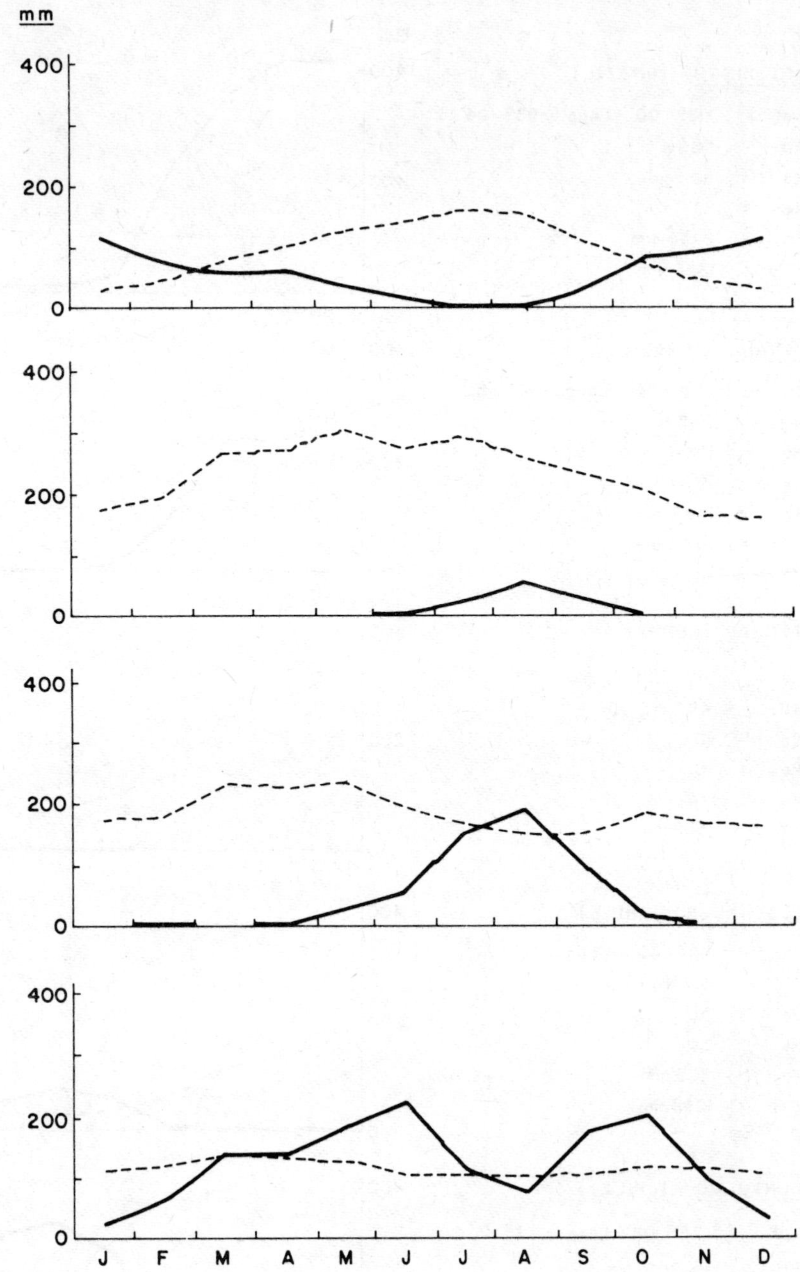

LEGEND

N - No. OF YEARS OF RECORD { p = RAINFALL; e = EVAPOTRANSPIRATION }

Pa - MEAN ANNUAL RAINFALL

ETp - MEAN ANNUAL POTENTIAL EVAPOTRANSPIRATION

—— MEAN MONTHLY RAINFALL

- - - MEAN MONTHLY POTENTIAL EVAPOTRANSPIRATION

Fig. 7.2(a) Seasonal distribution of rainfall and potential evapotranspiration for a range of latitude in Africa

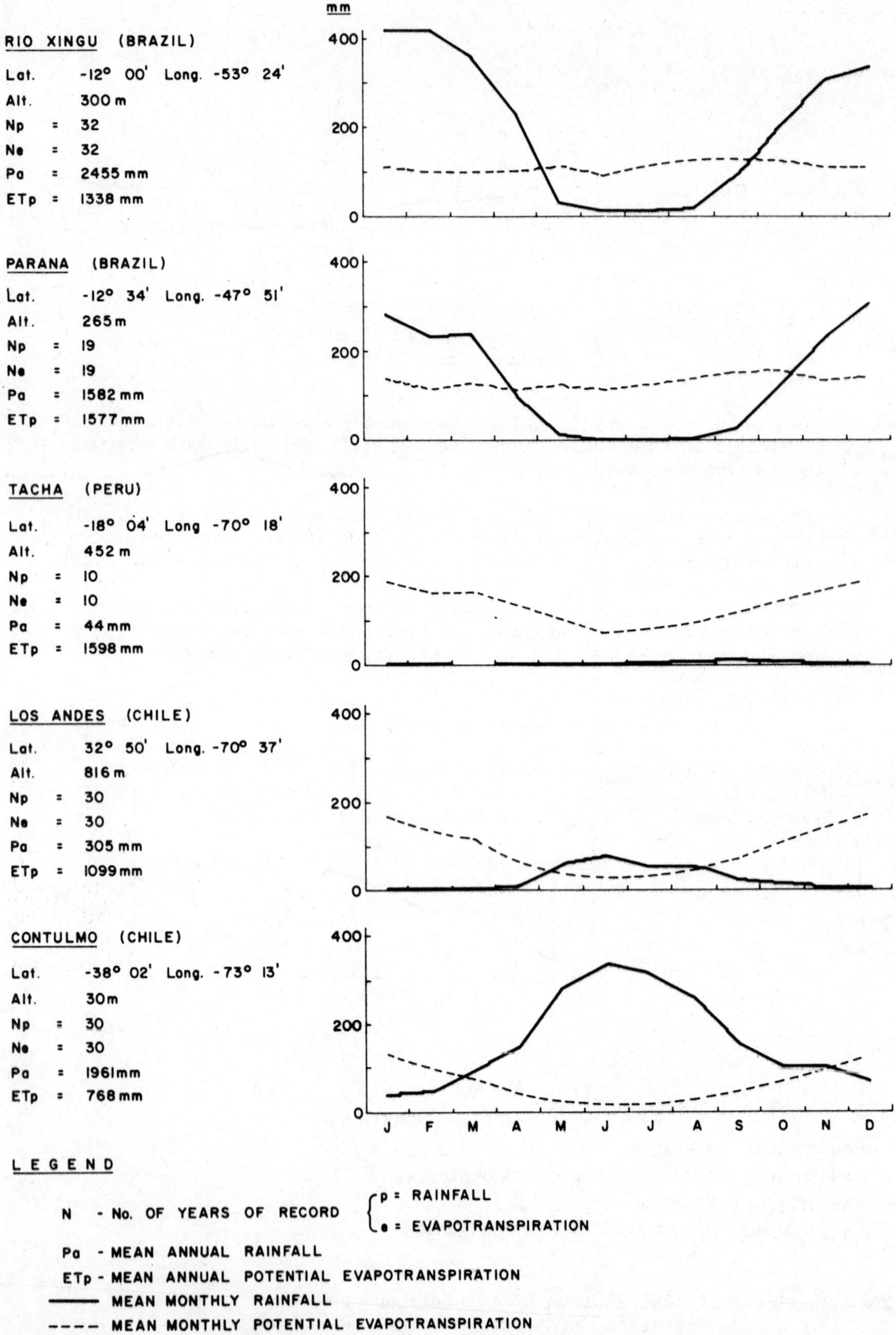

Fig. 7.2(b) Seasonal distribution of rainfall and potential evapotranspiration for a range of latitude in South America

FIGURES 7.2(a) and (b)

Seasonal Distribution of Rainfall and Potential Evapotranspiration for a Range of Latitude in Africa and South America

The graphs show the way in which the amount and seasonal distribution of rainfall P and potential evapotranspiration ETp vary with latitude in the two contenents.

In Africa near the equator, P is bi-seasonal, and is relatively high reflecting the seasonal migration of the ITCZ in low latitudes. ETp is relatively constant throughout the year and is approximately equal to P. With increasing latitude aridity increases, the duration of the ITCZ effect diminishes and rainfall amount decreases and is concentrated into a shorter time interval. ETp maxima becomes markedly summer seasonal and annual ETp increasingly exceeds P.

The difference between ETp and P, and thus maximum aridity, is greatest at the mean northern limit of ITCZ.

Further north a Mediterranean climate prevails, P is winter seasonal, the ITCZ effect is absent, ETp maxima occur in the summer and the difference between P and ETp reduces. Aridity is seasonal.

Approximately the same pattern is discernible in South America.

8. FIELD MEASUREMENT

8.1 MEASUREMENT OF PRINCIPAL CLIMATIC ELEMENTS

The measurement of climatic elements used for estimating evaporation, evapotranspiration and for agro-meteorological studies in general is covered by FAO and Meteorological Organisation (WMO) technical papers and manuals, e.g.

(a) Agro-meteorological Field Stations, Irrigation and Drainage Paper 27, 1976.

(b) Guide to Meteorological Instrument and Observing Practices, WMO No.8 TP3, 1969.

(c) Guide to Agricultural Meteorological Practices, WMO No. 134 TP61, 1963.

Instruments, station layouts and operational procedures are described in detail in these publications. This section deals only with those aspects of instrumentation, measurement, data processing and analysis procedures which require special consideration due to the arid environment.

8.1.1 Evaporimeters

The most widely used of these are the American USWB Class 'A' pan and the Russian CGl 3000 tank. The result of comparisons made by the World Meteorological Organisation between these pans and tanks with a $20m^2$ evaporating surface led to adoption of the Class 'A' pan as an international reference evaporation pan.

Production of reliable data, for the USWB Class 'A' pan presents many problems in arid zones. Lack of a readily available water supply often results in both pan and water becoming discoloured and a tendency for the water level to remain for long periods below the minimum required to remain constant energy storage and shade conditions. Ingress of blown sand and dust varies the pan reflectance and conductance properties. Birds drink from the pan unknown quantities of water. Data from pans screened to prevent this must be adjusted to account for the non-standard radiation and aerodynamic effects. The wooden pan platform, designed to allow free circulation of air beneath the pan is not easy to maintain due to deposits of wind-borne (and sometimes alluvial) sediments and the action of termites.

Sunken tanks of the CGI 3000, Colorado and other types suffer from the defect that they are more difficult to empty, clean and maintain in a standard condition. Their main advantage is the relative constancy of the pan factor (Doorenboos et.al. - op.cit.).

Standard methods of measuring the daily depth of evaporation from an evaporation pan or tank are well documented. Problems in obtaining reliable data cannot always be overcome but unreliable data can often be detected.

WMO Technical Note No. 83 discusses evaporimeters in some detail and gives the results of world wide comparisons between Class 'A' pan data and other evaporation measurements.

The Lake Hefner experiments show the importance of taking advected energy and heat storage effects into account when estimating the evaporation from deep water bodies in arid situations. The ratios of Lake Hefner evaporation ($\simeq$ 1400mm) to Class 'A' pan evaporation ranged from 0.13 to 1.32 during 1950-51 (Table 8.1). This result should be kept in mind when attempting to estimate lake evaporation on the basis of pan data.

A mean correction factor of 1.10 should be applied to screened pans to adjust to the unscreened condition. Significant differences between the colour and material of pans may result in comparable errors. If evaporation pan data are required, instruments should be standardised.

TABLE 8.1

USWB Class "A" Pan Ratios(1)

LOCATION	RATIO	YEAR	J	F	M	A	M	J	J	A	S	O	N	D	YEAR
Lake Hefner (USA)	Lake / Class "A" Pan	1950					0.35	0.60	0.71	0.82	0.93	0.90	1.32	1.06	
		1951	0.76	0.13	0.51	0.39	0.52	0.65	0.64	0.72					
Khartoum (Sudan)	12ft Tank / Class "A" Pan	1960	0.68	0.60	0.65	0.62	0.64	0.67	0.62	0.65	0.69	0.69	0.64	0.61	0.64
		1961	0.67	0.66	0.64	0.62	0.59	0.61	0.68	0.74	0.69	0.67	0.66	0.69	0.65
	12ft Tank(2) / Kohler Equ.	1960	1.07	0.99	1.02	0.97	0.99	1.03	0.94	1.00	1.01	1.05	0.99	0.98	1.00
		1961	1.06	1.01	1.00	0.98	0.97	0.95	0.98	1.03	1.04	1.01	1.00	1.03	1.00

(1) From WMO Technical Note No. 83

(2) Kohler's method of computing lake evaporation by correcting Class "A" pan evaporation for heat transfer through the sides and bottom.

For the reasons given above, much of the data produced by evaporation pans appear to be of questionable value. ETo, Eo or ETp values calculated from the readily obtainable temperature, radiation and wind-speed data are often much more reliable as indices and are normally required for agricultural studies other than those relating to crop water requirements.

The Piche evaporimeters, sometimes found in meteorological stations, does not respond to solar radiation in any way comparable to plants or open water since it is placed in the Stevenson screen. Piche evaporation is an index of advective energy but is very sensitive to the screen design. It is not recommended for hydrology in arid zones.

8.1.2 Temperature and Humidity Measurement

Measurement and recording of air temperature in thermometer screens presents no special technical problems provided that screen colour, material and state of repair meet WMO standards of air flow through the screen and reflection of radiation. Wooden screens should be painted glossy white and maintained in a clean condition to minimise heat storage capacity.

Humidity measurement is difficult in arid climates. Sensing units consisting of hair elements may produce consistently unreliable results in strong wind and high temperature. Laboratory calibration becomes meaningless when ideal conditions are replaced by exposure to hot, dust-laden winds. Cleaning the hair element by washing or brushing has no lasting effect. It is preferable to carry out field calibration by aspirated psychometer over the range of humidity experienced, with the hair element in its normal (far from standard but much more stable) condition.

In certain cases it may be possible to install electric resistance thermometers and even to record both wet and dry bulb temperatures automatically. It is certain that such sophisticated methods will provide more reliable humidity data. More often, reliance must be placed upon the normal procedure of observing temperatures at fixed times and care must be taken to ensure proper wetting of the wet bulb. The aspirated psychrometer is in this case to be preferred. If such methods are used, the relationship between the measured "point" time humidities and mean daily humidity must be established by well-controlled continuous wet and dry bulb recording.

8.1.3 Radiation Measurement

The only field problems normally encountered which are characteristic of the arid zone are those of dust accumulation on the outside of the hemispheric dome housing the sensors, "sand blasting" of the outside of the domes and, in the case of thermopile instruments, the occurrence of condensation within the glass domes. Dust effects can be controlled (but not eradicated) by frequent cleaning of the domes. Damaged domes can readily be changed in the workshop when damaged by sand and condensation can be prevented by re-sealing hemispheric domes to their mountings - again in the workshop.

8.1.4 Windspeed Measurement

Daily wind-run at 2m above ground level presents no problems other than occasional replacement of anemometer counters. This instrument should be read morning and evening in order to differentiate night wind and day wind.

Anemometers are sometimes installed to record the wind run over Class A evaporation pans for correction of Epan to Eo using Kohler's formula.

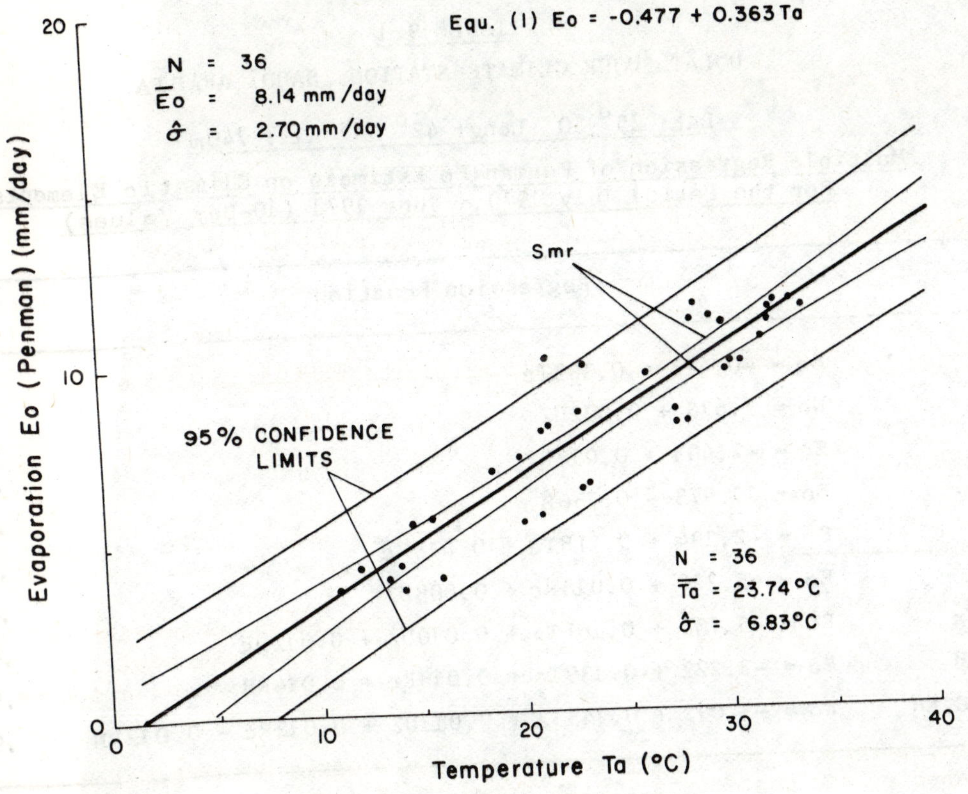

Fig. 9.1(a) Relationship between Penman Eo and mean daily temperature

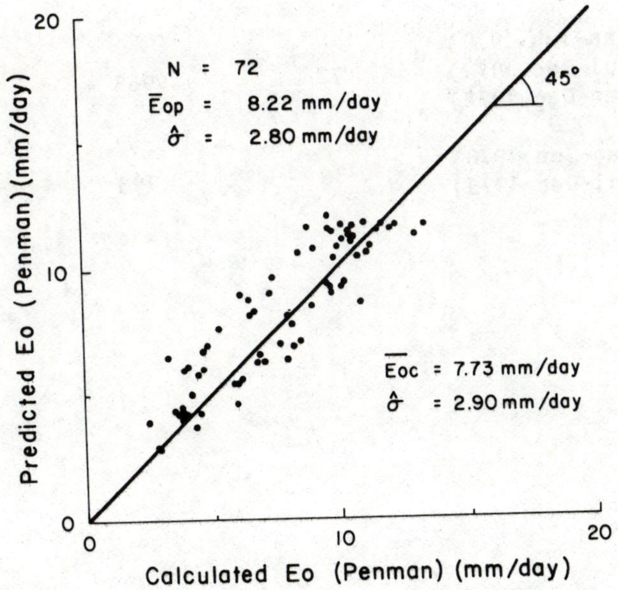

Fig. 9.1(b) Relationship between Eo calculated and Eo predicted from Equation (1)

8.2 EVAPORATION STATION NETWORKS

Network requirements are controlled by local climatic conditions and the spatial rate of change of climatic elements which determine potential evaporation. In general, relatively sparse networks of evaporation stations are required. These should be located at sites unaffected by micro-scale modifications of the general climatic pattern, in order to provide a measure of macro-scale climatic variability. For example, sites at which the local climate is influenced by irrigation should not be included in the minimum network.

When the standard error of estimate of mean seasonal and annual values of the more important climatic elements approach the tolerances on the measurements themselves the only additional benefit accrues from long-term sampling for periodicity. This can be achieved with fewer stations and saving in staff time.

Among the parameters which most affect evaporation, wind is usually the most variable in space except where there is no relief and the area is situated far from extensive water bodies. Air temperature and insolation are usually less variable in space and data from distant stations may well be satisfactory. Sometimes the density of synoptic stations is sufficient to meet the data requirements except for detailed investigations.

9. ESTIMATION OF EVAPORATION AND EVAPOTRANSPIRATION

Methods of estimating evaporation abound in the literature. Of these the most useful is the Penman equation for which the necessary climatic data are generally available.

9.1 PENMAN EQUATION

The reader is referred to Paper No. 24 of the present series for a complete discussion of the Penman equation including tables required to calculate a value of ETo based on a modified version of the method.

The value of Eo may also be calculated using the same tables after substitution of the value of the albedo from 0.25 to 0.05 in the appropriate tables.

The Penman estimate of evaporation Eo relates to a hypothetical body of water of infinite extent and shallow depth. These conditions are only approached in respect of water in shallow depressions following heavy rainfall when evaporation from water and land surfaces occurs at the potential rate. Values of Eo serve therefore only as indices of prevailing evaporative forces which may be modified to provide estimates of potential evapotranspiration losses from natural surfaces.

It may be shown that an adequate estimate of Penman's Eo may be calculated from regional multiple-regression models based on data of limited duration. Table 9.1 shows, for a station in Saudi Arabia, the relationships between each of the main elements used in the Penman equation and Eo. The regressions, based on 36 10-day mean temperature, windspeed, radiation and humidity data reveal the magnitude of the effect of each element on the result. Thus it is seen that a reasonably accurate estimate of Penman Eo for this station may be made using Equation No. 3 and radiation data only. The effect of introducing temperature and windspeed is to increase the precision of the estimate. Introduction of humidity data has little effect on the overall quality of the regression.

Bearing in mind the purposes for which an Eo or ETo value is required, estimates based on regionalised regression models may be adequate for most purposes. Such models may be developed from relatively sparse networks of fully-instrumented stations, with a greater density of stations at which radiation, temperature and wind-run only are measured automatically. Dense networks of fully-equipped climate stations may produce large quantities of unnecessary and imprecise data which provide little useful information and place a heavy demand on staff time.

Figures 9.1a, b, c and d show the relationship between Penman's Eo and Eo predicted from Equations 1 and 3 of Table 9.1.

Van Bavel (1966) showed that Penman's basic model is applicable to a wet evaporating surface in the presence of strong advected energy, provided that the transfer coefficient of the aerodynamic term is calculated to take account of surface roughness. Van Bavel proposed the substitution of Penman's Ea term by an aerodynamically - based simple wind function from which all empiricism, other than that relating to the surface roughness parameter, is removed.

9.2 ESTIMATION OF ACTUAL EVAPOTRANSPIRATION ETa

The problems of estimating actual evaporation and transpiration (ETa) losses for conditions in which the moisture supply to the evaporating surface is limited by shortage of water, by stomatal closure or by capillary conductivity, are severe. Actual losses for a well-gauged drainage basin may be estimated as the mean residual of the basin water balance, calculated for a period of several years to render storage changes insignificant. However, the results of such calculations in respect of large basins cannot readily be

TABLE 9.1

UQLAT SUQUR CLIMATE STATION, SAUDI ARAB

Lat: 25° 50' Long: 42° 11' Alt: 740m

Multiple Regression of Penman Eo Estimate on Climat
For the Period July 1970 - June 1971 (10-Day Va

Equ. No.		Regression Equation
1.	Ta	Eo = -0.477 + 0.363Ta
2.	U2	Eo = 7.673 + 0.002U2
3.	Rc	Eo = -2.403 + 0.018Rc
4.	RH	Eo = 11.973 - 0.156H
5.	Ta,Rc	Eo = -2.294 + 0.118Ta + 0.013Rc
6.	Rc,U2	Eo = -5.226 + 0.018Rc + 0.008U2
7.	Rc,U2,Ta	Eo = -5.735 + 0.161Ta + 0.010U2 + 0.012Rc
8.	Rc,Ta,RH	Eo = -3.222 + 0.139Ta + 0.013Rc + 0.014RH
9.	Ta,U2,RC,RH	Eo = -5.092 + 0.145Ta + 0.011U2 + 0.012Rc - 0.012RH

Test of Regression Equations

Equation No.	Test Period	Data No.	r^2	SE-mm
3.	Jan-Jun 1970) Jul-Dec 1971)	36	.92	0.7
9.	Jan-Jun 1970) Jul-Dec 1971) Jan-Dec 1972)	72	.968	0.5
1.	Jan-Jun 1970) Jul-Dec 1971)	36	.793	1.3

FIGURES 9.1(a), (b), (c), (d)

Relationships of Penman Eo with Temperature Ta and Radiation Rc

(a) and (c)

These are simple linear regressions of Eo on Ta and Eo on Rc using 72 sets of 10-day mean values.

The standard error of the mean relationships Smr is an indication of the mean error of the regression line with respect to its true position. If we calculated 72 10-day mean Eo values from the regression equation we could expect an overall error of the size indicated by Smr.

However, errors in individual 10-day Eo values would be within the range indicated by the confidence limits shown in 19 cases out of 20.

(b) and (d)

These show correlations between Eo predicted by the regression equations and the calculated Penman values for a further 72 sets of data.

The correlation tests the validity of using the regression equations developed as shown in (a) and (c) to predict Eo using temperature or radiation data only.

The results show the error about the 45% line through the data points to be greater in the case of Eo predicted from Ta than from Rc.

It is concluded that, for many purposes, measurement of Rc alone will produce sufficiently reliable Eo estimates.

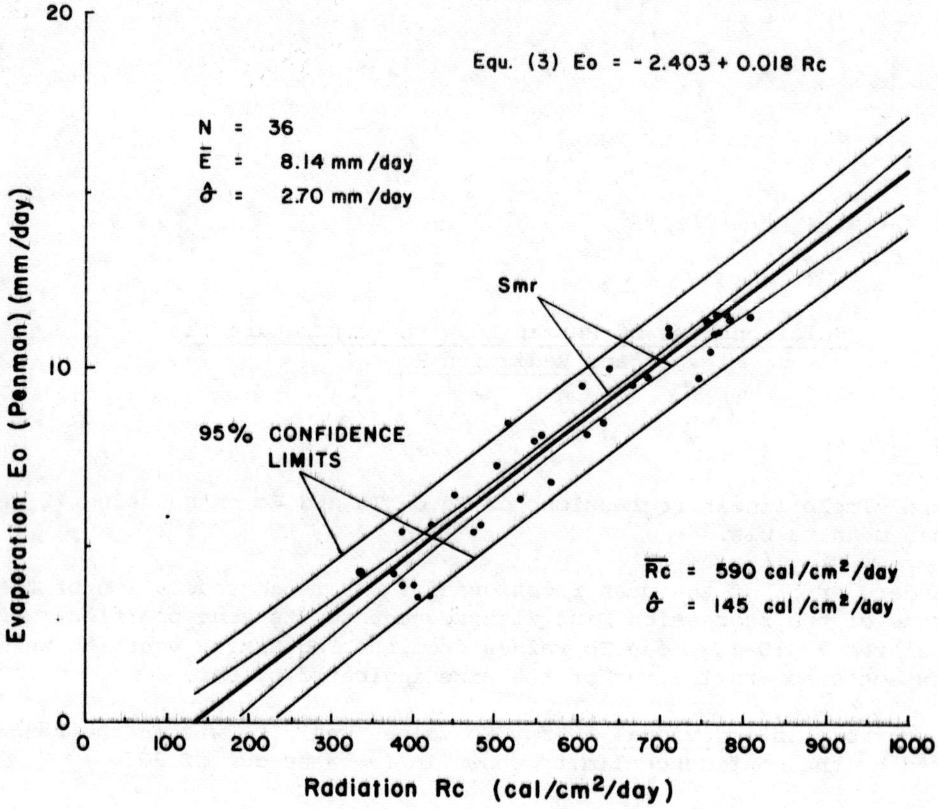

Fig. 9.1(c) Relationship between Penman Eo and Radiation

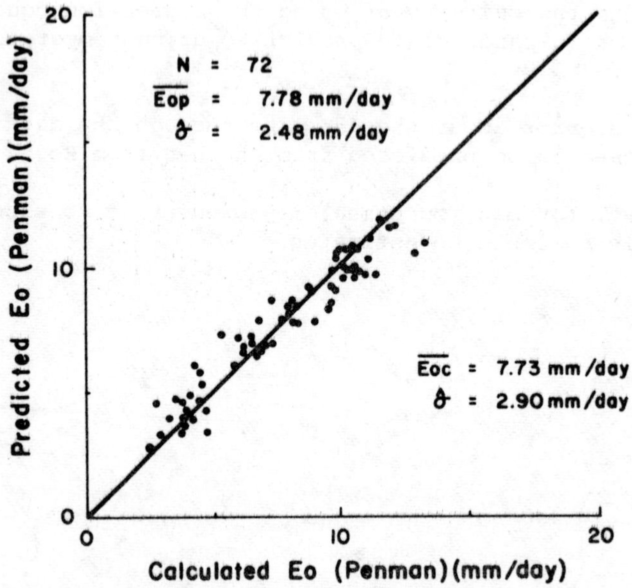

Fig. 9.1(d) Relationship between Eo calculated and Eo predicted from Equation (3)

fitted to ETa prediction models appropriate to small areas since ETa from large, unirrigated areas is governed almost entirely by the amount and distribution of rainfall in time and space.

The problem is largely of academic interest in basin resource studies which can only normally be undertaken on the basis of measurement of the water body to be developed; stream-flow or groundwater.

For some studies, however, it is of great importance that direct or indirect estimates should be made of ETa, there being no other way of apportioning the balance residual between two or more possible balance components.

An investigation carried out into the very large losses, as underflow to the Arabian Gulf or to evaporation sinks formed by coastal playas, was carried out in Saudi Arabia and has been described by Pike (1970). The objectives of the study were to evaluate evaporation losses from a shallow water table ($<$ 1m deep) beneath the extensive silty clay and sand flats overlying the Neogene aquifer formations in eastern Saudi Arabia.

Two basic techniques were employed:

1. The energy budget method (WMO Technical Note No. 83).
2. The combination formulas of Van Bavel (op.cit.) and Tanner and Fuchs (1968).

For the energy budget method, net radiation above ground level and just below the surface was measured by thermopile sensors and the temperature and humidity gradients between 0.5m and 2.0m above ground level were measured using thermocouple sensors. Recording was by means of portable micro-volt recorders utilising the galvanometer - photo-electric feedback principle with a simple transistorised amplifier. It was assumed that, in the homogeneous surroundings 10 km inland from the Gulf, energy advection across the site equalled that leaving it and that the measurement of the vertical energy component provided an accurate determination of evaporation.

Evaporation was calculated from the equations:

$$E = - \frac{Rn + S}{1 + H/E} \tag{9.1}$$

and

$$H = \frac{(Rn + S)\ H/E}{1 + H/E} \tag{9.2}$$

in which E is the evaporative heat, Rn the net radiation, S the soil heat flux and H the sensible heat flux to the air. The Bowen ratio calculated as appropriate to the site was

$$H/E = 0.49\ \Delta t/\ \Delta e \tag{9.3}$$

in which Δ t and Δ e are the temperature and vapour pressure gradients respectively between the two observation levels.

The results of two sets of measurements, made in the winter and spring seasons, suggested that the proportion of net radiation energy utilised for evaporation (one-tenth) was about the same on each occasion. The two sets of measurements were inadequate to assess the actual annual losses so the relationship of actual to potential evaporation, calculated using Van Bavel's modification of the Penman equation was assessed. This indicated ratios in the range of 0.2 to 0.8 from late summer to late winter and these values were used to estimate annual ETa losses.

The combination method developed by Tanner and Fuchs (op.cit.) was used to calculate hourly ETa rates and was found to agree well with values estimated by the energy balance method (Figure 9.2). This study shows that the two basic techniques provided a useful approach to the initial quantification of a large item in an important water balance.

An alternative equation which avoids direct estimation of the Bowen ratio is

$$E = \frac{Qn - G}{L} \left(1 - \frac{\gamma}{S + \gamma} \frac{\Delta Ta}{\Delta Tw}\right) \qquad (9.4)$$

as given in WMO Technical Note No.38.

In this equation Qn is the net radiation and G is heat storage, both in cal/cm^2; L is the latent heat of vapourisation (583 cal/cm^3); γ is the psychrometic constant $4.2 \times 10^{-4}/^{\circ}C$; $S \cong 0.63/P$ where P is pressure; Δ Ta and Δ Tw are the air temperature and wet bulb vertical temperature gradients between the two measurement levels.

The calculations involved in application of equation 9.4 in the energy balance method for a small grass plot in Saudi Arabia is given as Example 9.1 and illustrated by Figure 9.3.

The formula of Tanner and Fuchs may be written as:

$$ETa = \frac{\gamma + \Delta^*}{\Delta^*} ETp - \frac{\varepsilon \rho \; Cp}{\Delta^*} .h\,(e_o - e_z) \qquad (9.5)$$

in which Δ^* is the slope of the saturation vapour pressure curve at surface temperature Ts (bar/$^{\circ}$K), γ is the psychometric constant, ε the water air molecular weight ratio, ρ the density of air (kg/m^3), Cp the specific heat of air, h the transfer coefficient for heat and water vapour from the surface to height z (m/s).

This formula gives an estimate of evaporation from a drying soil with no restrictions or assumptions. If ETp is the value calculated using the Van Bavel combination method, the only additional measurement required is that of soil temperature Ts just below ground surface.

9.3 EVAPORATION FROM NATURAL SURFACES AND WATER USE BY NATURAL VEGETATION

9.3.1 Natural Surfaces

Evaporation from arid zone soil and rock surfaces represents a high proportion of rainfall.

Evaporation from rain-wetted soil surfaces is rapid without runoff generation, even locally, and accounts for 90% or more of rainfall. Evaporation from bare, rain-wetted rocks is much quicker and is usually quite small due to the low detention storage properties of rock surfaces. Evaporation from runoff reception areas of greater storage capacity, e.g. local depressions, rock interstices, sediment-filled gullies and major alluvial channels is more prolonged but is typically less in the long term than evaporation from the soil or rock surface receiving the rainfall.

Deep drainage of rainfall via infiltration at the point of fall, producing direct recharge to groundwater, may be a rare but important component in the hydrologic balance. Surface runoff which collects in local depressions can also give rise to groundwater recharge.

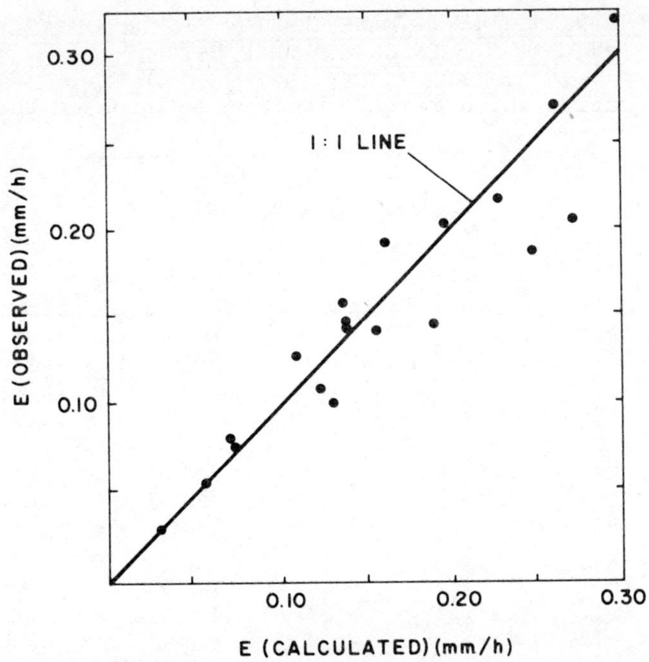

Fig. 9.2 Observed and calculated hourly value of E, Sabkhah an Nabiyah

Evaporation of Groundwater from Coastal Playas, Saudi Arabia

The figure shows the relationship between actual hourly evaporation (ETa) estimated using the energy budget method and values estimated using the combination method of Tanner and Fuchs (op. cit.).

For the former estimates heat fluxes and air temperature and humidity gradients are measured. For the latter, the only addition to the conventional (Ta, Rh, Rc, n and U_2) measurements made for Penman estimates is that of soil temperature immediately below the surface, which provides an indication of the moisture available for evaporation.

The relationship shown by the figure is not an indication of the quality of the estimate of ETa, but only of the results of estimating ETa using different methods.

Example 9.1

COMPUTATION OF ACTUAL EVAPOTRANSPIRATION FROM A SMALL GRASS PLOT BY THE ENERGY BALANCE METHOD
RIYADH - JULY 1972

TIME	Rn (Cal/cm²/min)	G (Cal/cm²/min)	Δ Ta °C	Δ Tw °C	Δ Ta / Δ Tw	Rn -G / L mm/min	1-γ/S+γ . Ta/Tw	ETa mm/min
2300	- .04	- .07	-1.0	+2.5	- .40	.00051	1.156	.00059
2400	- .04	- .08	-1.0	+2.3	- .43	.00068	1.168	.00079
0100	- .05	- .08	-1.0	+2.1	- .48	.00051	1.184	.00070
0200	- .05	- .08	-1.0	+2.0	- .50	.00051	1.195	.00061
0300	- .05	- .08	-0.5	+2.0	- .50	.00051	1.109	.00057
0400	- .05	- .07	-1.0	+1.6	- .60	.00034	1.234	.00042
0500	- .04	- .06	-1.0	+1.5	- .67	.00034	1.262	.00043
0600	- .02	- .05	-1.0	+1.3	- .77	.00051	1.300	.00066
0700	+ .04	- .04	-1.0	+1.1	- .91	.00136	1.356	.00184
0800	+ .06	- .02	-1.0	+1.0	-1.00	.00136	1.390	.00189
0850	+ .08	0	-1.0	+1.2	- .83	.00136	1.324	.00180
0855	+ .74	0	-1.0	+1.5	- .67	.01260	1.262	.01590
0920	+ .56	+ .10	-1.0	+1.8	- .56	.00780	1.218	.00960
1000	+ .64	+ .10	0	+2.0	0	.00915	1.000	.00915
1100	+ .67	+ .10	-0.5	+2.3	- .22	.00964	1.086	.01048
1200	+ .70	+ .09	-0.5	+2.5	- .20	.01035	1.078	.01113
1300	+ .71	+ .09	-0.3	+2.7	- .11	.01052	1.043	.01098
1400	+ .69	+ .07	-0.1	+2.8	- .04	.01052	1.016	.01068
1500	+ .60	+ .06	-0.5	+3.0	- .17	.00916	1.066	.00977
1600	+ .44	+ .01	-0.5	+3.0	- .17	.00720	1.066	.00775
1700	+ .04	- .06	-1.2	+2.9	- .41	.00170	1.160	.00197
1800	+ .00	- .07	-1.3	+2.8	- .46	.00119	1.180	.00140
1900	- .03	- .06	-1.5	+2.7	- .56	.00051	1.218	.00062
2000	- .04	- .06	-1.3	+2.6	- .50	.00034	1.195	.00041
2100	- .01	- .05	-1.2	+2.5	- .48	.00068	1.187	.00081
2200	- .02	- .05	-1.2	+2.5	- .48	.00051	1.187	.00060
2300	- .01	- .04	-1.5	+2.5	- .60	.00051	1.234	.00063
Total	+5.51	- .39	-24.1	+58.5		1.0037		.11218
Mean	+0.20	- .014	- .89	+2.17	-0.46	.0377 (= 5.35 mm/day)		.00415 (= 5.98 mm/day)

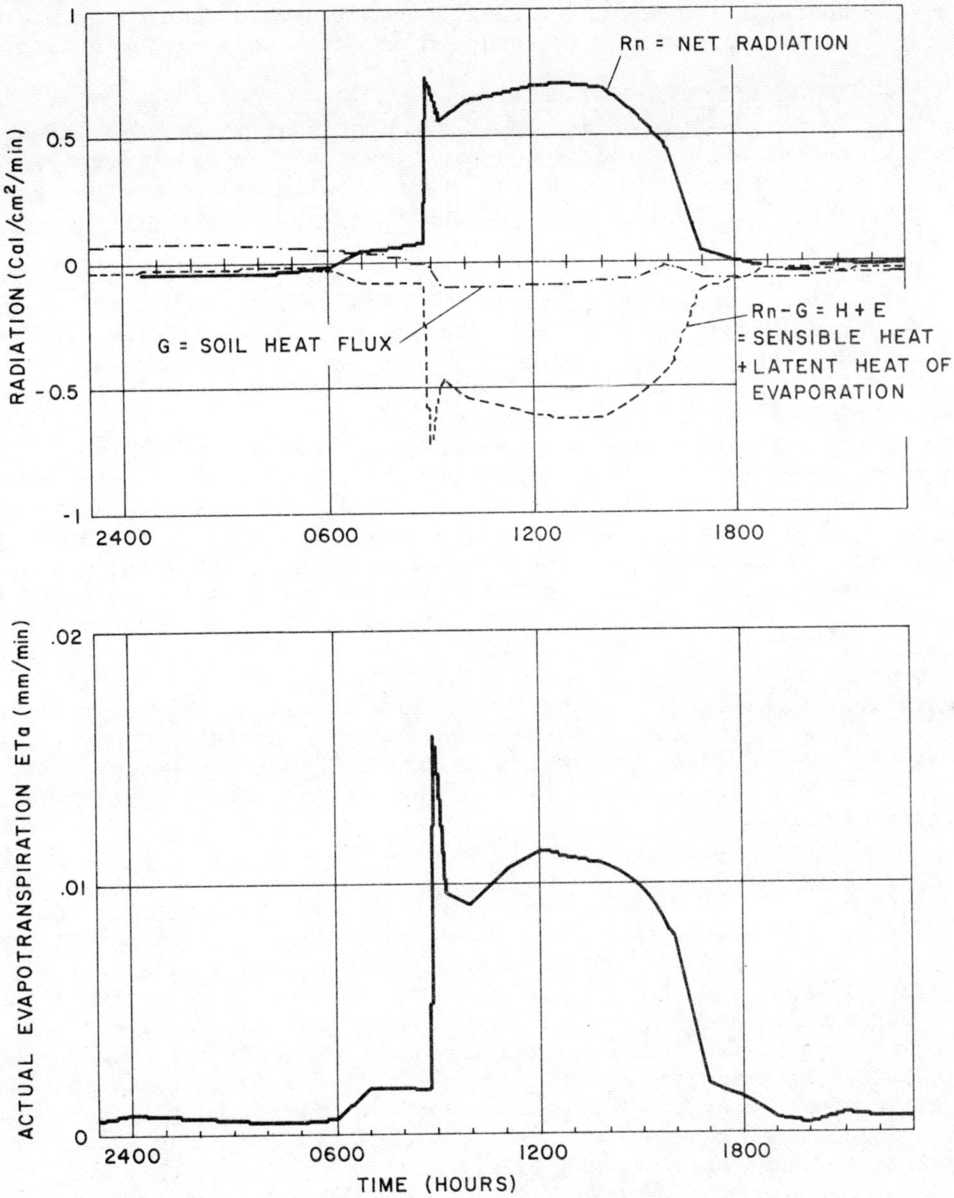

Fig. 9.3 Energy balance for a small grass plot - Riyadh, Saudi Arabia

The figure shows the results of applying the energy budget method as illustrated by Example 9.1. It is similar to Figure 7.1 which illustrates the same method used to estimate ETa from a stand of phreatophytes. In this case the symbols and terminology are consistent with those of the Example.

The net radiation flux is measured above the evaporating surface. The soil heat flux deducted from net radiation represents the total incoming energy used for heating the air (sensible heat energy) and for evaporation (latent heat of evaporation). The apportionment of the net incoming energy between sensible and evaporative heat is calculated from air and humidity gradients measured between 0.5 m and 2 m above ground level.

The vertical portion of the graphs is due to shading of the site during the morning.

Both infiltration to and evaporation from bare soils are controlled largely by temperature gradients within the soil profile. For the same above-surface meteorological situation, more evaporation occurs under conditions of upward than of downward soil heat flux. Thus the rate of evaporation from a depth of moist soil may be greater in autumn than spring.

A notable feature of the arid zone is the fact that in gravels, dune sands and sand river beds rates of evaporation from depths greater than about 30 cm is very low. Moisture may usually be found at this depth beneath wadi beds and the upwind slope of dunes many months after the last rainfall or runoff event, in the absence of vegetation to exploit this moisture storage.

The implications for direct (rainfall) and indirect (via water courses and depressions) recharge through surface sands and gravels are considerable. Important sources of fresh water supplies are found adjacent to or within sand desert dunes where mean annual rainfall is only about 50 mm. This is due to the occurrence of recharge as a result of storm rainfall depths greater than the field capacity of the recharge medium for the depth of dry material. This may amount to some 30 mm for a 30 cm dry depth, assuming 10% moisture content at field capacity. In the case of alluvial water courses the natural efficiency of the indirect recharge mechanism is due largely to limited evaporation losses from wetted surfaces and the consequent need to replenish relatively small soil moisture storage before indirect recharge takes place. For example in Wadi Najran, Saudi Arabia, the evaporation losses from the wetted surface area of the wadi following runoff over the principal recharge zone was estimated as about 2% of total runoff.

Evaporation from groundwater overlain by bare soil, is controlled mainly by the capillary conductivity of the soil and the prevailing temperature gradients. It may be found that despite the existence of a moist or damp soil surface the evaporation rate is due much more to the capacity of the soil to transmit moisture by capillarity than by the available evaporative energy.

9.3.2 Natural Vegetation

Transpiration of arid zone vegetation is controlled by both climatic and plant physiological factors. Annual vegetation species develop and subsist briefly on occasional rainfall events and the limited soil moisture capacity of the soil or rock crevices. Perennial vegetation develops and subsists in locations where soil moisture storage capacities are high or where moisture replenishment is relatively frequent or both. For example acacia species typically occur in the alluvial outwash fans at the foot of the hills, phreatophytes occur where they are able to root in the capillary fringe of alluvial channel aquifers and salt-tolerant grasses occur where the groundwater table is perennially near the ground surface.

Many perennial species survive by controlling their transpiration rates through adjustment of stomatal openings of their leaves. Stomatal closure can be induced by the plant's internal water deficit. When stomata are closed due to this deficit, transpiration virtually ceases. Slatyer and Mabbutt (1964) report on the decline of ETa from acacia groves (plant cover ≃ 40%) with increasing soil moisture stress, in Alice Springs, Australia. Here, the ratio of ETa to ETp (Penman) declined from 0.6 to 0.1 over an available soil moisture range of 50 mm to zero. By contrast, in the intergrove areas (plant cover 15%) ETa/ETp ranged from 1.0 to 0.

9.3.3 Phreatophytes

Of all the many phreatophytes species, tamarix (salt cedar) appears to be the most wide-spread and the most pernicious in the arid zone, world-wide.

Robinson (1965) estimated that in the western part of the United States over 3700 km^2 were covered by the plant in 1961 and predicted that this figure would reach 4000 km^2 by 1970. He estimated that the corresponding consumptive use in 1970 would be 6 x 10^9 m^3 per annum, equivalent to 1.5m depth of evapotranspiration and enough water to irrigate 200,000 hectares.

In South Africa, Hellwig (1973) estimated consumptive use by phreatophytes from a sandy bed reach of the Swakop River covering 8300 hectares, to be 76 Mm^3 per annum. Of this amount, ETa from 3140 Ha was estimated to contribute 68% (52 Mm^3). Evaporation from permanently wet areas (820 Ha) was 19.5% (15 Mm^3) and from temporarily wet areas (4340 Ha) 12.5% (9 Mm^3). Thus ETa from the phreatophytic vegetation was equivalent to 1.66m and E from permanently wet areas, 1.83m. Hellwig concluded that removal of the phreatophytes would result in an ETa reduction of 16 Mm^3 per annum.

In Saudi Arabia studies of two alluvial aquifers showed mean annual phreatophyte ETa losses to represent 26 Mm^3 (55% of the total groundwater balance) in one case and 17 Mm^3 (18%) in the other.

9.3.4 Estimation of Phreatophyte Evapotranspiration

Hellwig (op.cit.) found good agreement between measured total losses (from wet and moist sand and from phreatophyte vegetation) and Penman estimates.

The energy budget method was used to estimate ETa from a tamarix stand in Saudi Arabia. The result of a single day's measurements suggested that at this location, a very humid cultivated area near the Red Sea coast, ETa is approximately 0.7 Eo. If applied to annual Eo the annual ETa value would be 0.7 x 2.5 = 1.75m. Assuming no corrections for volume density of foliage, depth to groundwater (generally less than 4m), or salinity (2-6 mmhos/cm), the mean annual consumptive waste represented by this ETa rate over the 15 km^2 area would be 26 Mm^3.

In the USGS Gila River Phreatophyte Project in Arizona, colour - infrared photography was used as a relative radiometer to obtain data from 38 photographic missions flown between 1967 and 1971 (Jones, 1972). Remote sensing measurements of evapotranspiration from 13 missions flown during 1968 were related to water-budget measurements for a 680 ha area cleared of vegetation and an 880 ha phreatophyte-covered area. Correlation coefficients between water-budget and remote sensing measurements were 0.88 for the cleared area and 0.86 for the vegetated area. Photographic data were also correlated with depth to groundwater, soil moisture, foliar cover and volume of canopy. It was determined that colour-infrared photography and computer analysis cost about one-tenth of conventional species classification and canopy-measurement techniques.

For the USGS Buckeye Project (on the Gila River 40 km west of Phoenix, Arizona, USA) six 9m x 9m x 3.5-4m deep tanks (evapotranspirometers) were constructed in 1959 for the long-term study of water use by salt-cedar (Van Hylckama, 1970). After six years of study it was found that:

- Water use is little affected by thinning of the stands. Cutting one-half the branches in each clump (weighted by diameter) in two tanks and complete cutting in two others resulted in relatively small reductions in ETa.

- It was shown that reduction of foliage volume density by 50% reduced ETa by 10-15%.

- After complete cutting vegetation quickly regenerated (daily increases in twig length up to 50mm/day) and resumed transpiration.

- It is common practice to relate ETa directly to volume density of foliage. If the tank experiments may be taken to represent natural conditions, the assumption that ETa will be reduced in direct proportion to reduction in foliage volume density could result in overestimation of water savings. Equally, the ETa estimate for natural vegetation will be overestimated if it is assumed that a natural increase in volume density will produce a corresponding increase in ETa.

- Depth to groundwater was found to be an important factor in ETa rates. The fact that tamarix appears to thrive with a low groundwater level is not necessarily an indication that it is transpiring at the potential rate.

- ETa was found to vary greatly with the conductivity of the water supply. In the tank experiments water use was found to range from 500mm/year at EC = 40 mmhos/cm to 3000 mm/year at EC = 10 mmhos/cm.

Under natural conditions, it is difficult to take account of depth to water table, salinity of groundwater and volume density of foliage since these are variables. However, it is important to know that these are constraints and that the common assumption that ETa is at least as great as ETp for 100% volume density, may not be valid in all cases.

Similar evapotranspiration tank studies were carried out by the USGS in Nevada (Robinson, 1970). Here the phreatophytes studied were greasewood, rabbitbrush, willow and wildrose. Mean annual rainfall in the study zones ranges from 125 to 200mm and Epan is about 1450mm. The tanks ranged in size from 9m x 9m x 3m deep to 3m x 3m x 2m deep. One of the tanks was left bare of vegetation.

It was found that:

- More than two thirds of the annual water use by greasewood and rabbitbrush occurred during June-August,

- Measured evapotranspiration, expressed on an areal basis as depth over unit area gives no indication of the growth conditions for which the information was obtained and may result in serious error if transposed to areas of dissimilar growth conditions. Some of this uncertainty may be reduced by expressing ETa as volume of water per unit of foliage volume.

- ETa varied widely among species; the mean 1963-1967 measured tank values being:

  Greasewood 428mm
  Rabbitbush 503mm
  Willow 698mm
  Wildrose 484mm

  During the same period Epan was 1450mm.

Some of the methods commonly used to estimate ETa are discussed below:

a. Evapotranspirometers

Evapotranspirometers were used to measure water use by salt cedar (tamarix gallica, sometimes called tamarix pentranda) during studies on the Pencos River in New Mexico. Here, the estimated average annual use was 1.83m including 0.3m rainfall. Tank experiments to determine the use of water by salt cedar at 100% volume density in the Safford Valley of the Gila River, Arizona during 1943 and 1944 gave results ranging from 2.80m, for an average depth to water level of 1.22m, to 2.13m for a depth of 2.44m.

Phreatophyte water use appears to be highly sensitive to temperature and is controlled more by the volume density of its foliage than by its areal density. Areal density is the numerical density of plants per unit of area with respect to the maximum. Vertical density relates to the vertical depth of the foliage with respect to the maximum possible.

b. Extrapolation of ETa rates from other Areas

If the rate of evapotranspiration is known for a unit area of phreatophyte species at a particular volume density of foliage it can be extrapolated to other areas having the same species, volume density and climate. The total water used by each phreatophyte species was computed as the product of its area at 100% volume density and the unit rate of water use for that density.

c. Inflow-Outflow Method

If the surface and groundwater inflow to and outflow from an area in which phreatophytes occur and the changes in soil moisture and groundwater storage volume in the reach can be measured, the total loss of water by evapotranspiration is given by the residual of the water balance.

Usually the measurement of balance components other than ETa is difficult and highly inaccurate and the residual is thus a poor estimate of ETa. In a large alluvial aquifer the observation well network required to produce reliable estimates of groundwater storage change could be prohibitively costly.

In the Acme-Artesia studies on the Pecos River it was found possible to estimate ETa by this method despite the complications of irrigation diversions, local tributary flows, irrigation drainage water inflows, sewage effluent etc. The mean annual groundwater consumed by phreatophytes over the 164 km^2 area for the three-year period 1956-1958 was 58Mm^3 corresponding with an average annual ETa of 0.35m.

d. Pumping-well Analogy Method

ETa may be estimated by assuming phreatophytes to behave as a well-field and by simulating the drawdown due to phreatophytes by extraction from "ideal" wells with a variable pumping rate. If the aquifer, transmissibility and storage are known, a table can be prepared showing the proportion of water diverted to a well for various durations and rates of pumping and for various distances from the recharge source (in the case of a flowing stream).

In the case of groundwater contours normal to the recharge line-source, as with an ephemeral stream some time after flow has ceased, distance from this source is less relevant. Groundwater levels and fluctuations recorded in observation wells are assumed to represent the pumping-well effect of the phreatophytes and ETa is calculated using aquifer test formulas of the Theis non-equilibrium type.

This method was applied in the Acme-Artesia studies and produced a three-year mean ETa estimate of 29Mm^3 corresponding with an average annual ETa of 0.18m.

e. Transpiration Well Methods

In this method fluctuations in observation wells located in phreatophyte stands are assumed to reflect the variations in consumptive use of the vegetation. The difference between maximum and minimum levels in the well multiplied by the specific yield of the aquifer material is an indication of ETa. However, since this drawdown effect is highly variable areally, a number of such wells would be required to be representative and capable of being extrapolated.

Water level fluctuations in phreatic aquifers may be used to estimate ETa as originally conceived by White (1932). The diurnal groundwater level cycle approximately represents a cumulative curve of the rate of transpiration.

The number of piezometers used must depend upon variations in the amplitude of diurnal groundwater level fluctuations in the area of interest. Amplitude is greatest near the centre of the phreatophyte stand and reduces with distance from it. Hence, two piezometers located within and 50m from the stand would produce data from which estimates of ETa would be quite different due to differences in rate of change of water table level.

9.3.5 Comparison of Estimation Methods

Table 9.2 summarises the results of four of the methods discussed above as applied to the Acme-Artesia reach of the Pecos River in 1956-58. It may be seen that computed ETa varies considerably.

Of the four methods the pumping-well analogy method was expected to produce the lowest estimate for this particular study.

Estimates based on climatic elements, e.g. Penman, Blaney-Criddle, Thornthwaite etc., do not take account of water table depth, groundwater quality or phreatophyte species and must therefore rely for their precision on correlation with measured ETa values by one or more of the methods discussed. The quality of correlations may show that temperature-sensitive estimates such as the Thornthwaite or Blaney-Criddle method are most reliable.

9.3.6 Conclusions

Where evapotranspiration by phreatophytes represents a major component of a water balance it is essential that methods other than those based on climatic and foliage measurements should be used. Possibly the simplest of these is the transpiration-well method, using a representative network of small-diameter piezometers fitted with maximum and minimum water level indicators and with reliable estimates of specific yield.

The inflow-outflow method should only be attempted where groundwater storage changes represent a small item in the balance and where evapotranspiration of phreatophytes can be separated from that of other vegetation and together represent a large proportion of the balance. Climatic methods may produce misleading results. Investigations in the Karakum Canal desert areas showed that transpiration from cane undergrowth could be five times that of open water evaporation and amounts to 9m/year.

The radiation energy balance of the phreatophyte stand cannot account for such rates but the "oasis effect" by which the advected energy represented by dry, hot air traversing the area most certainly can, provided that depth to groundwater and salinity are both low.

The estimation of phreatophyte evapotranspiration losses remains an area of considerable uncertainty and is worthy of continued experimentation.

TABLE 9.2

USE OF WATER IN THE PHREATOPHYTE AREA IN THE ACME-ARTESIA REACH OF THE PECOS RIVER, NEW MEXICO, 1956-8

Method of Computation	Use of Water (acre-feet)								
	1956			1957			1958		
	Ground-water	Precipitation	Total	Ground-water	Precipitation	Total	Ground-water	Precipitation	Total
Extrapolation of water use rates from other areas	44,000	18,000	62,000	38,000	29,000	67,000	32,000	41,000	73,000
Inflow-outflow	44,000	18,000	62,000	50,000	29,000	79,000	47,000	41,000	88,000
Pumping-well analogy	36,000	18,000	54,000	23,000	29,000	52,000	14,000	41,000	55,000
Transpiration well	70,000	18,000	88,000	72,000	29,000	101,000	66,000	41,000	107,000
Average			66,000			75,000			81,000

10. CHARACTERISTICS OF SURFACE RUNOFF IN ARID ZONES

10.1 GENERAL

Significant surface runoff in arid lands originates from well-defined runoff-generating zones; it generally contains relatively small amounts of dissolved solids - electrical conductivity being typically less than 500 micromhos/cm - and occurs sporadically as short, isolated "flow periods" separated by longer periods of low or zero flow. Sustained flow is rare, and prolonged base flows occur only in channels fed from outside the region or by discharging groundwater.

Peak flow rates generated by sporadic, high-intensity rainfall may occur within a matter of minutes of the start of a rise in discharge. The highest of the series of flood peaks generally occurs within a few hours of the start of a rise. Their abruptness results from the nature of arid zone rainfall and sometimes from the steepness of the channels draining the runoff-generating zones. The latter feature, together with the natural formation of an efficient hydraulic shape and the channel-forming process, often causes near-translatory flood wave types to develop. In these circumstances tributary inflows to the main channel occurring during the passage of the initial flood wave are propagated downstream with a velocity greater than that of the original peak. Hence, peaks have a tendency to over-run each other with the result that, at any fixed location on the main stream, a "wave-train" is formed producing a hydrograph made up of a series of sharp, rises. This hydrograph feature is also due to wide variations in lag-times between randomly distributed multi-cell convective rainfall occurrences within the catchment and the runoff response at the point of observation.

Flood peaks and volumes in the ephemeral streams typical of arid lands usually diminish during their passage downstream infiltration into the channel bed and overbank spill. Figure 10.1 shows a flood event along the course of a fifty kilometre channel reach traversing an alluvial plain. No significant diversions for agriculture took place, and the losses are due to bed infiltration. On this occasion, about one percent of the runoff emerging from the hard-rock areas arrived at the end of this reach.

Transmission losses represent the single most important recharge mechanism for alluvial aquifers common to arid lands.

It should therefore be expected that total discharge volume expressed as depth over the basin reduces with basin area. The relationship between progressive basin area and absolute runoff volume within the same basin may be expected to be non-linear, with a typically upward convex shape (see Figure 10.2).

The potential for infiltration of flood waters over any channel reach is controlled by:

- the position of the water table below the channel bed and the storage coefficient of the unsaturated alluvium;
- the area of channel bed occupied by the flood-waters;
- the surface infiltration capacity and the vertical permeability of the channel alluvium.
- the water temperature.
- the moisture content of the unsaturated zone.

- the nature of the sediment load.

Recharge by infiltration and percolation from floods to the groundwater table below alluvial channels reduces with rise in groundwater level. This is typical of deeply incised intermontane drainage systems where relatively shallow depths of alluvium are found. Here, water tables are usually near the surface and recharge amount is effectively controlled by the limited volume of aquifer storage available beneath the channel bed. However, in alluvial fans and plains groundwater levels are usually too low to influence infiltration processes and recharge amounts are controlled essentially by the availability of flood waters.

The width of flow over an alluvial channel during the passage of a flood varies with instantaneous discharge and sediment load and is a function of the hydraulic parameters of the channel section and bed material, the flow and its rate of change. The variability of alluvial channel characteristics in space and time and their sensitivity to changes in discharge make the wetted width at any section difficult to determine. Figure 10.3 shows the relationship between width of flow area and discharge for a South Arabian wadi. A typical example of bed movement with time in an alluvial channel is shown by Figure 10.4.

The infiltration capacity of a channel bed is determined by its surface condition. In general, it varies as an exponentially decreasing function of the time since the onset of infiltration. It may not be greatly influenced by the depth of water standing above the infiltration surface. It would appear that the coarse sand to gravelly channel beds typical of ephemeral streams in arid zones may have initial surface infiltration rates in excess of 15cm/h, gradually falling off after periods of prolonged flow to relatively constant terminal rates determined by the shape and proximity of the recharge mound or by the thickness and permeability of the saturated recharge medium.

10.2 CHARACTERISTICS OF FLOOD EVENTS

Individual flood events are generated by discrete storms which may take the form of single cells or multi-cell clusters which move as part of a more or less homogeneous air mass or along preferred alignments. The form of the discharge hydrograph is an integration of the physical characteristics of the drainage basin, which control the lag times between rainfall and runoff, and the time and location of separate flood-generating storm events. These are generally sporadic due to the normal cell development, decay and regenerative features of convective storm activity.

The magnitude of floods in the main drainage stem is determined by the areal characteristics of the storms, their orientation and movement relative to individual sub-catchments and the coincidence or otherwise of flood peaks generated from contributing sub-catchments. Experience in a wide range of arid zones suggests that there is no general relationship between the magnitude of the events and the catchment area except for small basins, in contrast with more temperate zones.

Within hard-rock, mountainous drainage systems the translatory nature of larger flood waves in ephemeral streams may result in peak discharges of individual floods remaining relatively constant within long channel reaches despite some runoff volume reduction by transmission losses. Example A.3.4 given in Annex 3 illustrates the transposition of an actual, major wide-spread ($> 5000\ km^2$) storm to the most favourable flood-generation location within a particular basin. This example shows that, even with severe assumptions as to the proportion of rainfall becoming runoff at the basin outlet, a discharge peak in excess of 5000 m^3/s would be unlikely. There is an infinite range of depth-area-duration-frequency and initial and transmission loss conditions which could, in theory, produce a peak discharge in excess of $5000m^3/s$. There is, however, little possibility of modelling the processes capable of producing such a discharge with any confidence in arid zones. We are left, therefore, with the alternative of statistical analysis of the records of peak discharges, without regard to the conditions which caused them.

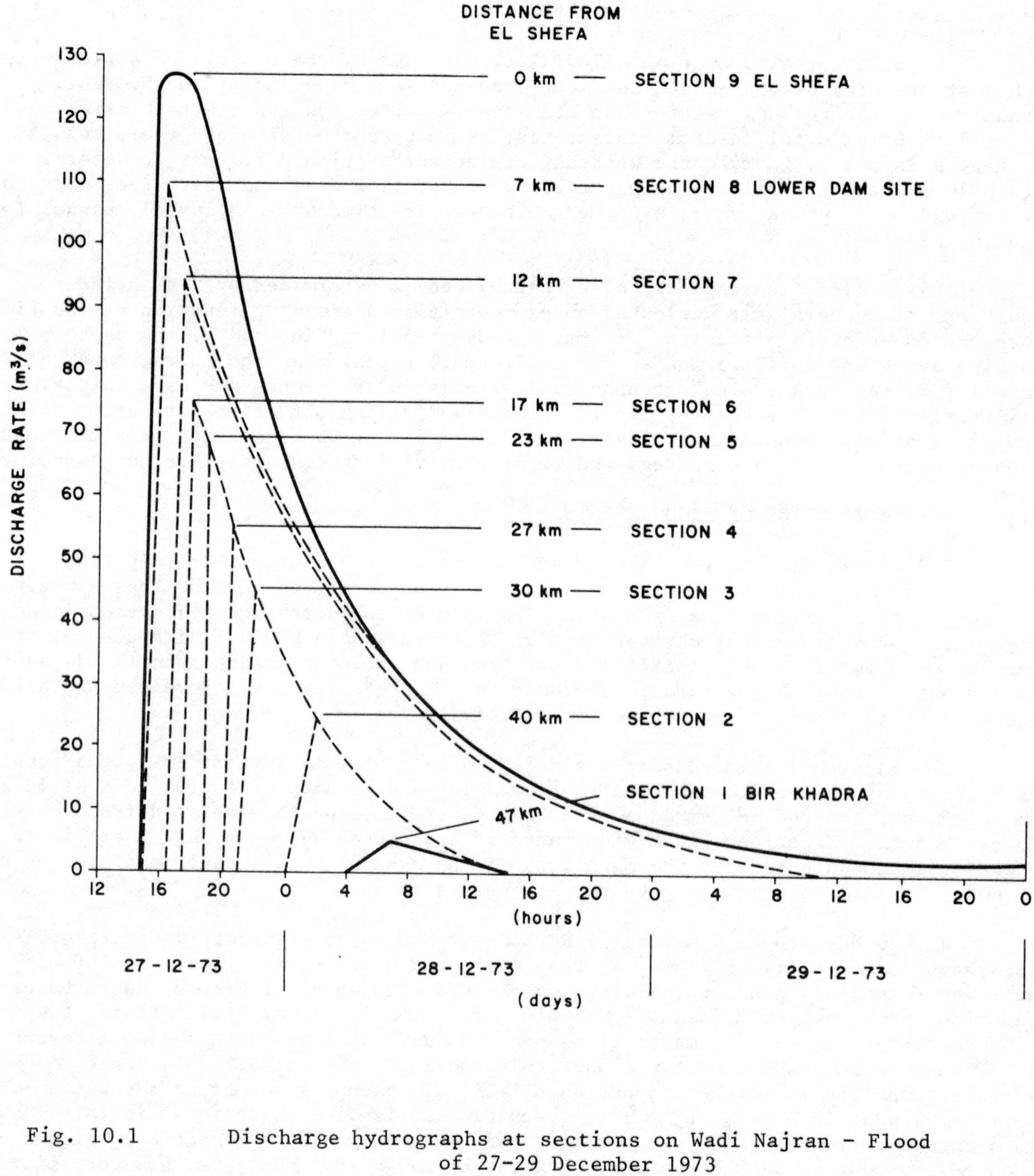

Fig. 10.1 Discharge hydrographs at sections on Wadi Najran - Flood of 27-29 December 1973

Figure 10.1

Discharge Hydrographs at Sections on Wadi Najran - Flood of 27-29 December 1973

The hydrographs shown on the figure were observed at various cross sections of Wadi Najran alluvial channel at the positions indicated. They illustrate well the transmission losses from a single-peaked hydrograph to a dry alluvial wadi bed in a situation where depth to groundwater increases downstream.

For the purpose of the illustration, the hydrographs have been idealised to some extent but retain their main characteristics. These are:

- the flood peak and volume diminishes with increasing distance downstream as flow is absorbed by the wadi bed alluvium;
- the flood wave has barely reached the lowest section, some 40 km downstream;
- the timing of the commencement of hydrograph rises and peaks indicates an average wave celerity of 1.3 m/s;
- increasing storage capacity in the wadi alluvium downstream results in cessation of flow earliest downstream and latest upstream where baseflow issues from saturated alluvial storage.

This situation is generally typical of arid region drainage systems at their emergence from hard-rock runoff-generating zones.

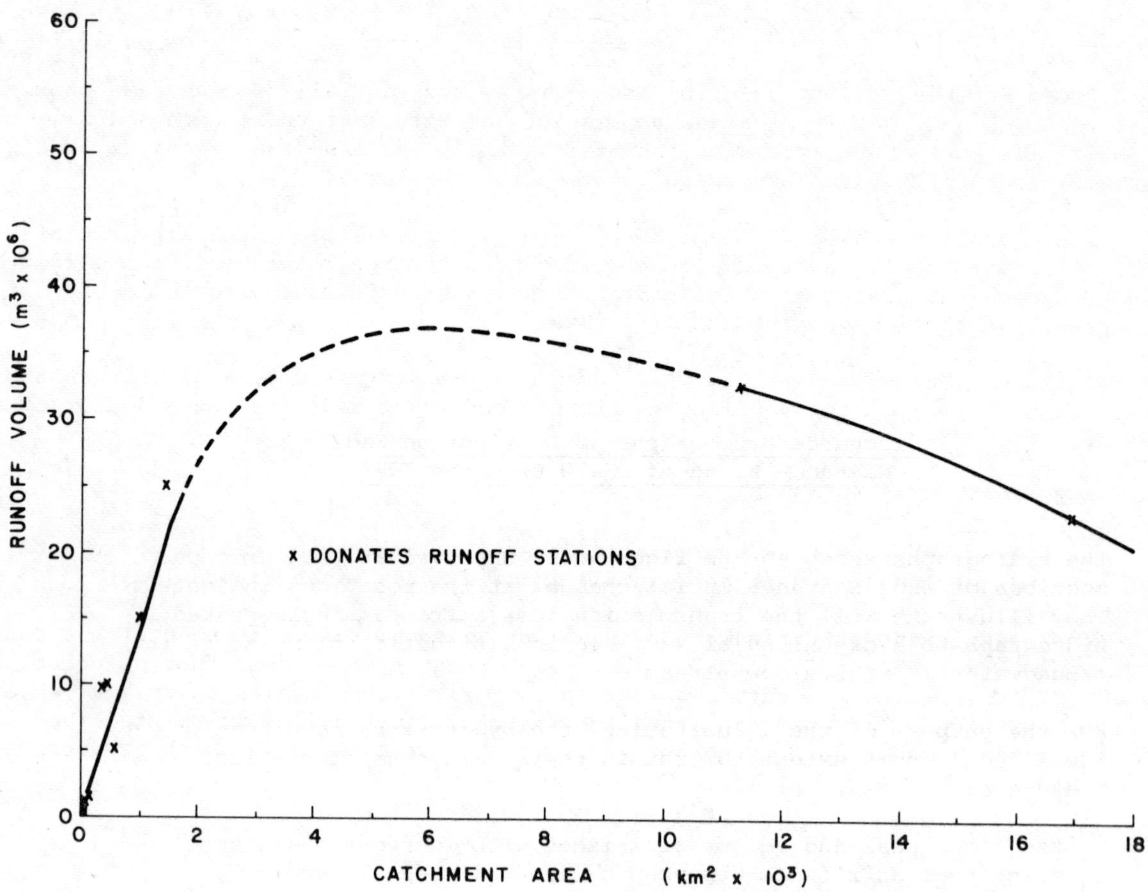

Fig. 10.2 Relationship between catchment area and runoff volume - Wadi Bishah Drainage Basin, Saudi Arabia - 1971

In the upper part of the Basin, runoff exceeds infiltration rate and alluvial channel storage capacity.

Further downstream, rainfall and runoff are less, alluvial channel storage capacities are greater and channel runoff decreases. This latter process is also illustrated by Figure 10.1.

Permanent base runoff station should ideally be sited at or near points of maximum runoff.

It can be seen that the centre part of the curve is not well defined and an additional field gauging station is desirable.

Storm periods of long duration are unlikely substantially to increase peak discharges in upland, hard-rock areas because of the nature of the growth and movement of storm cell clusters. Such storms do, however, greatly increase the discharge peak, volume and duration of flood events in remote downstream areas.

Flow-duration and flow-volume curves are generally steep but a significant proportion of the total flow volume occurs at high discharge rates (see Figure 18.4(c)). The instantaneous discharge-duration characteristics of arid zone streams is of importance for both flow gauging and water use.

Although the major part of runoff occurs at relatively modest discharge rates, by far the greatest part of the sediment is transported at or near the peak discharge as shown in Section V.

10.3 ANNUAL AND SEASONAL RUNOFF AMOUNTS

It will be obvious that no reliable general relationships can be derived to relate seasonal or annual runoff to rainfall in arid zones. (see Figure 18.1)

The coefficient of variation of annual or seasonal runoff in ephemeral streams typically exceeds 100% and is normally 30 to 50 percent greater than that of point rainfall. This is in marked contrast to temperate climates in which basin carry-over storage effects tend to smooth out annual or seasonal fluctuations in rainfall and in which runoff has a lower coefficient of variation than the rainfall which produces it.

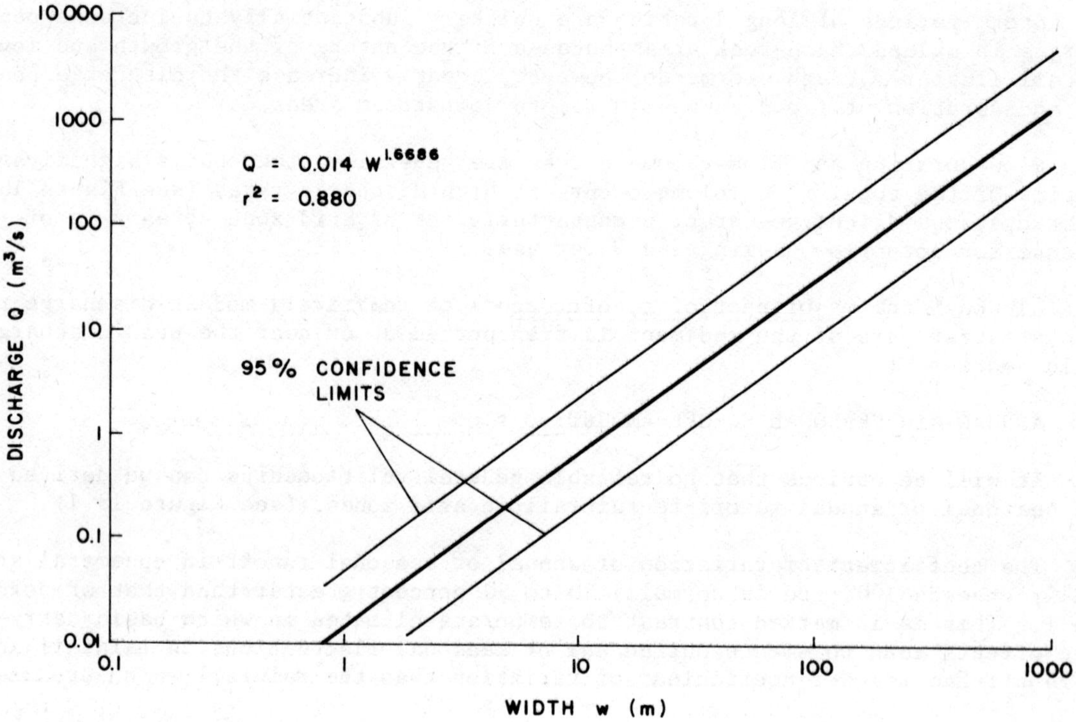

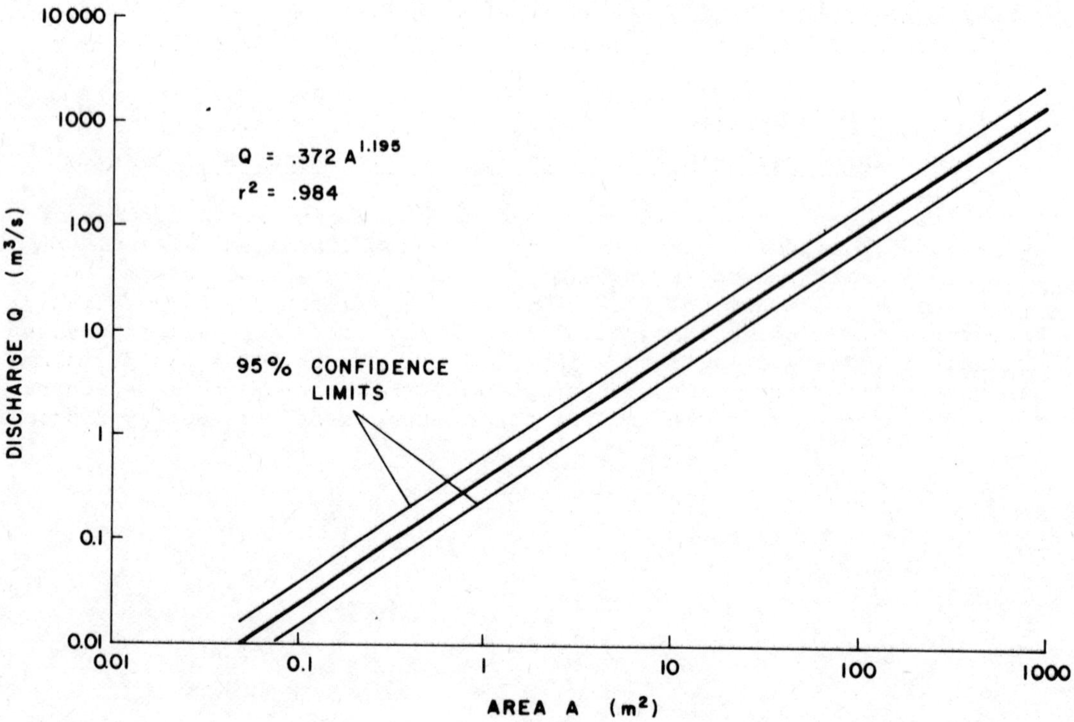

Fig. 10.3 Discharge vs. area and width of flow in an alluvial channel
(individual points are too numerous to be shown usefully)

FIGURE 10.3

Discharge vs. Area and Width of Flow in an Alluvial Channel

a. Discharge vs. width excluding slope - area measurements

The relationship between discharge and width of flow in a self-forming alluvial channel is important for two reasons:

- The scatter about the regression line is a measure of the variability of the indirect recharge opportunity presented by a given discharge flowing over an alluvial aquifer. The mean relationship makes it possible to estimate the mean recharge per unit of alluvial channel length. This may be achieved by producing an infiltration-duration curve from an instantaneous discharge duration curve, the width-discharge relationship and an estimate of infiltration capacity of the recharge medium.
- The characteristic variability of the flow width for a given discharge provides a useful illustration of flow gauging problems. For example, the 95% confidence limits suggest a width range of 45 to 220 m for $Q = 100\ m^3/s$.

b. Discharge vs. area excluding slope - area measurements

The relationship between discharge and area is useful for discharge estimation in alluvial channels. In the case illustrated, it would be possible to make a good estimate of discharge over a flood period by measuring only the area of flow. By omitting velocity measurement an observer could measure the area of flow at much more frequent intervals during the passage of a flood. Thus, an estimate of total runoff volume (though not the instantaneous hydrograph) would possibly be more accurate than one based on more precise but fewer measurements of rapidly fluctuating flow.

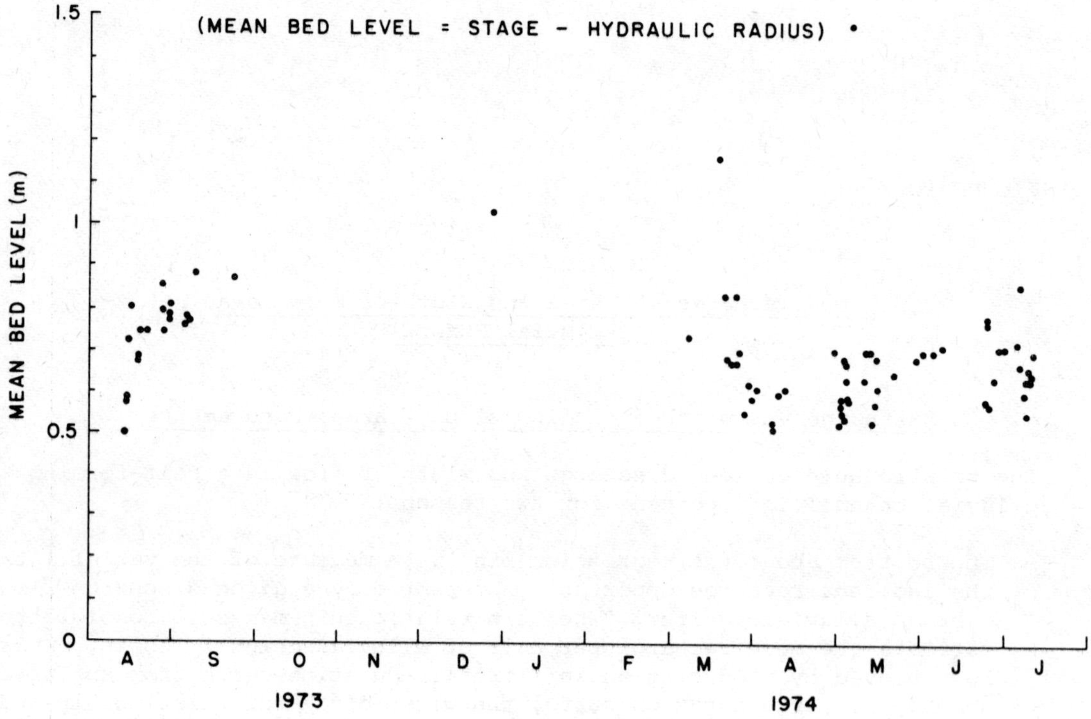

Fig. 10.4(a) Variation in mean bed level with time in an alluvial channel - Wadi Najran, Saudi Arabia

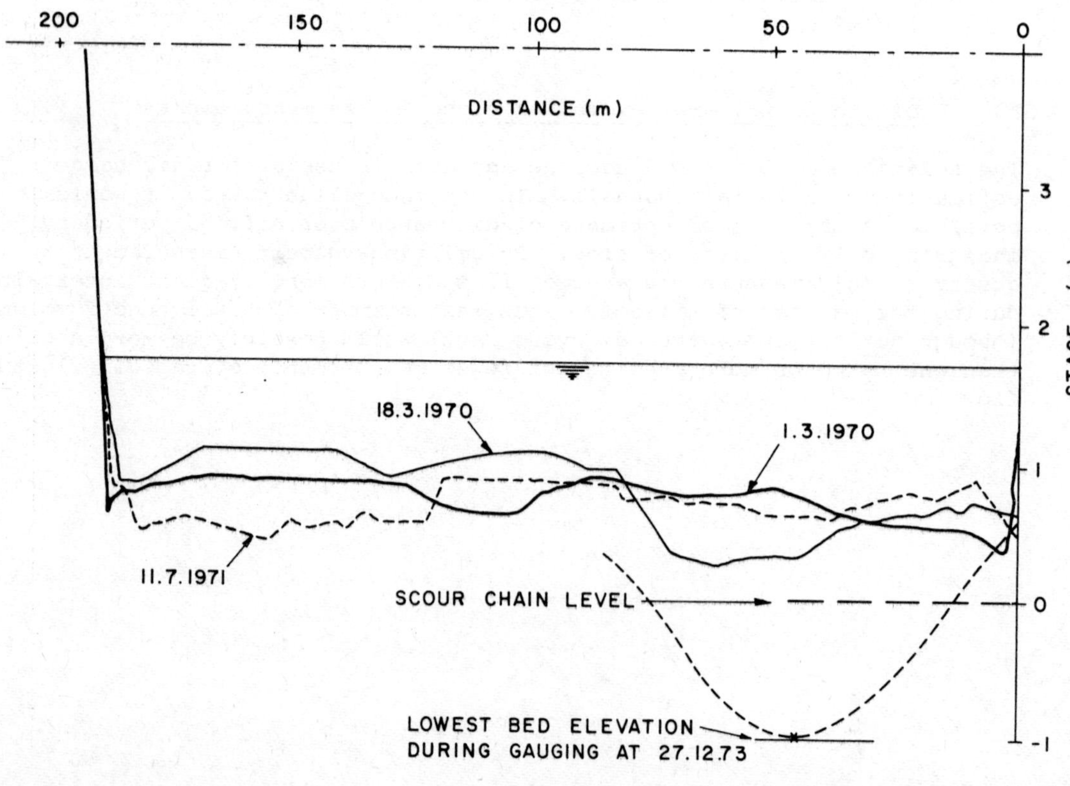

Fig. 10.4(b) Cross sections at recorder house showing bed movement caused by floods - Wadi Bishah, Saudi Arabia

FIGURE 10.4(a)

Variation in Mean Bed Level with Time in an Alluvial Channel - Wadi Najran, Saudi Arabia

The figure shows the mean bed level, at a fixed gauging station current meter section, based on current meter measurements made during flow periods in 1973 and 1974.

The range of mean bed level displayed by the figure (about one metre) is less than actually occurred, as the plotted points do not include high discharges when greater scour depths would have been attained.

FIGURE 10.4(b)

Cross Sections at Recorder House Showing Bed Movement Caused by Floods - Wadi Bishah, Saudi Arabia

This figure shows cross sections of the alluvial channel at the recorder house surveyed as part of post-flood surveys for slope-area peak discharge estimation. The range of minimum bed elevations shown by the figure is about one metre. The actual range of minimum bed elevation is much greater as is indicated by the lowest bed level of -1 m stage measured during a cableway measurement on 27.12.73.

11. FIELD MEASUREMENT

11.1 STREAMGAUGING NETWORK REQUIREMENTS

The World Meteorological Organisation (1974) set out, in general terms, the ma: objectives of minimum networks as, "... to obtain an appreciation of the availability c surface water resources, their geographical distribution, and their variability in time The magnitude and frequency of floods and drought periods are of particular importance this regard." The WMO guide then goes on to suggest the size of the area to be covered by each station for minimum networks. Network stations are classified into those on streams draining basins larger than 10,000 km^2 and those gauging smaller basins. It is suggested by WMO that an equal number of stations should be in each category.

For the purposes of this Paper arid zones water-courses are classified into two main groups:

- Those whose catchment area lies wholly within an arid or semi-arid zone (as earlier defined). These are endogeneous water courses.
- Those whose catchment area lies partly within a precipitation zone other than arid and whose flow regime is characterised to a greater or lesser degree by the greater precipitation of that zone. These are exogenous rivers e.g. the Nile, Colorado and Gila (US), Rufiji (Tanzania), Okovango (Botswana), Niger, Senegal and Volta (W. Africa).

Within these broad groups lies an infinite variety of runoff regimes. Exogenous rivers may for instance include those whose regime and importance as a water source, as they traverse arid lands, derives mainly from relatively high groundwater-derived base-flows, the attenuation of flood flows generated in the high-precipitation sub-catchments or snow-melt. In other exogenous water courses the sporadic runoff component from the arid sub-catchment is relatively high. Endogenous arid zone basins may range from those whose relatively stable regimes are governed largely by a high groundwater base-flow component (e.g. the Yarmouk in Jordan) or by large basin storage effects (e.g. the Wami in Tanzania) to those whose regime is characterised by ephemeral flows consisting of isolated, sporadic events only.

The regimes described in Section 10 make it difficult to adhere to general guidelines as to minimum networks, particularly as local conditions vary widely. Because of the lack of correlation between runoff at points within and between basins, the minimum network must sometimes be regarded as a number of unrelated sampling points whose individual purpose and results are specific and whose aggregate data output may not (as raingauge networks do) add much to knowledge of spatial runoff occurrence.

Runoff stations may be classified as follows:

1. Permanent base stations with good controls capable of being calibrated over the greater part of the discharge range, located at or near points at which runoff is near maximum.

2. Representative basin stations with artificial control structures. These should be located at the outlet of small basins with distinctive geological and geophysical characteristics which may be classified systematically in order that the results of rainfall and runoff measurements may be transposed to similar areas.

3. Project stations whose sole purpose is to produce data which, together with data from permanent and representative basin stations, and from groundwater and rainfall studies, form the basis for project feasibility assessment.

Clearly, sites for permanent stations located in the ideal conditions described above are not readily found but they should be sought. It may be found that, in terms of unit basin yield, the criteria for (1) also satisfy (3), since project areas are usually those in which agriculture is already extensively practiced or where considerable potential exists. Stations regarded as permanent, but sited almost entirely for convenience of access and the prospects of employing a permanent observer may produce little usable runoff information, other than an event record, unless the other conditions are met. The most important factor is control stability; access can usually be provided and permanent observers are not always essential. Stations must be capable of measuring the normal discharge range by current meter from a bridge or cableway by pre-rated control or by dilution techniques.

Representative basins must be small if deterministic modelling of rainfall-runoff processes is to be achieved. However, this may not be necessary in all cases and such basins may with good reason be instrumented only for runoff measurement. The aim may be to produce a long-term time-series of runoff processes within basins at locations selected for the typicality of the runoff response which can be transposed to other, similar areas and to which stochastic modelling techniques can eventually be applied. Full use should be made of the opportunity afforded by small storage reservoirs to provide hydrologic information.

Measurement of natural spring flows can be an important aspect of hydrological activity in arid zones. No special problems attach to clearly defined emissions which can be canalised and measured by current meter, "V"-notch or flume. However, in some areas spring flow is the result of upward leakage from deep-seated confined aquifers producing artesian flow. Such springs commonly produce an oasis of dense, aquatic and/or phreatophytic vegetation and are usually developed for agricultural purposes.

Project stations must, of necessity, be located at or near the point of interest whatever the control conditions. If control conditions are poor the station may consist of conventional equipment with a resident observer team whose job it is to measure all runoff by continuous current-meter definition of the runoff hydrographs at several locations during flow periods.

The difficulty of obtaining accurate runoff data in arid and semi-arid zones cannot be over-stated. The problems are:

- Rapid rise in discharge after the onset of storms.
- Debris and sediment load impeding current meter measurement.
- Changes in cross section shape during current meter measurements.
- Lack of effective natural controls due to high mobility of alluvial channels and periodic changes in cross sections upstream of permanent section controls.
- High velocities.
- Difficulty in maintaining good quality, continuous water level records due to environmental factors such as waves and high sediment loads.
- Poor accessibility in wet weather.

11.2 GAUGING STATION EQUIPMENT AND OPERATION

Stream-gauging practice is well described in U.S. Geological Survey Water Supply Paper 888 (1945) and many other text books. Gauging stations are equipped with with some or all of the following:-

1. Water level recorder.
2. Primary staff gauges incorporating crest gauges.
3. Stilling wells and gauges.
4. Slope gauges.
5. Cableway.
6. Current meter equipment.
7. Sediment sampling equipment.
8. Float-velocity measurement markers.
9. Wading measurement section markers.
10. Scour indicators (where appropriate).
11. Control.
12. Bench-mark.

The following discussion covers only those aspects which are of special relevance or present special problems in the arid zone.

There has been a significant advance, in recent years, towards automation of data collection systems using electronic sensing and logging methods and data transmission by radio. These developments have aimed to produce more accurate recording and computer-compatible records. In some cases they are used for flood warning purposes to ensure that occurrences of flood-producing rainfall or of upstream flooding are communicated to operations centres or flood-control dams for operational forecasting. In other cases, for instance densely-instrumented representative or experimental basins, sophisticated sensing, transmission and central recording systems may be essential to the purpose of the studies which would otherwise suffer from inaccuracies in the timing of events. Water level recorders are being installed which automatically transmit data through satellite repeaters.

None of these systems reduce the need for regular, field trips by experienced hydrologists and hydrometrists. In arid zones it is more important than elsewhere that such visits should ensure the efficient recording of infrequent flood events whenever they occur.

11.2.1 Water Level Recorders

Both float-operated and gas-purge, pressure-type recorders have been found by experience to be capable of producing reliable results in arid zone conditions. An essential requirement of all autographic recorders is an open time-scale strip-chart system on which the rapid rise and fall of water levels normally experienced is accurately reproduced.

It is sometimes necessary to use two recorders to measure the range of discharge in a channel due to the coming into operation of low and high discharge controls.

(a) Float-Operated Recorders

Autonomy is normally one month but may be as long as six months.

Hydrologists should have regard to experience in arid zones before procurement of new equipment types.

Flat floats, though more sensitive than long narrow ones have the advantage that they follow the water level to a lower level as stilling-well siltation advances. They are less likely to become embedded in the silt as it settles at the bottom of the well.

The need for frequent stilling-well desilting operations renders float-operated recorder mechanisms vulnerable to maintenance by inexperienced observers.

It is often difficult, to discern the direction of movement of the pen trace on the chart due to frequent stage changes at or near the reversal stage and failure of the reversal indicator to produce a clear indication. For this reason observers, who should, where possible, attend runoff stations during floods, should make occasional clear calibration marks on the chart trace, encircling the pen position and writing the stage, time and direction of pen movement.

Stilling-wells of recorder installations on ephemeral streams are usually dry. For this reason there is usually no water level gauge reading against which to check the pen setting. With some instruments the float tape is graduated numerically. The tape reading, corrected by a positive or negative constant, gives the stage reading represented by the float elevation. Where the float cable or tape is not graduated, an index mark should be provided on the float cable and a reference point specified.

In arid zones the dusty conditions make instrument maintenance difficult. It is only by high maintenance standards in difficult conditions that reliable stage records can be obtained. Instruments must be meticulously checked during field visits. This of course applies to all instruments but the dearth of runoff events could mean that water level recorder failure for one month may result in loss of record of the greater part (or the whole) of the runoff of an entire year.

(b) Pressure Recorders

Two basic types of pressure recording instruments have been used with success in arid zone hydrometry. These incorporate the gas-purge, "bubbler" system and the enclosed pressure-bulb system. The former is normally considered to be the more accurate though requiring a higher degree of skill and effort in operation and maintenance.

The clarity and precision of the hydrograph trace produced by these instruments is variable and depends not only on the instrument design but on such factors as:

- bubble rate, length of tubing and air outlet point design in the case of gas-purge systems.
- length of tubing, moisture content of air in the tubing and quality of sealing in the case of pressure-bulb system.

The pen-trace response of gas-purge system recorders to water level fluctuations may be made more sensitive by increase of bubble rate and by increasing the gas outlet diameter by the addition of a piece of piping to form a vertical, bell-like tapping point. It should be noted that evidence of sensitivity to rapid level fluctuations is not necessarily an indication of precision in stage recording unless system components and adjustments are matched to the site; fluctuations can be well-recorded at the wrong level.

11.2.2 Primary Staff Gauge

In arid zones, floods in water courses bring with them large quantities of detritus including branches and whole trees which, with high discharge velocities, renders staff gauges highly vulnerable to damage. It is desirable though often difficult, due to the rocky banks at gauging stations (where rock is highly desirable), to recess gauges into the bank in order to avoid damage, and create relatively quiescent conditions for accurate stage observations.

The importance of well-maintained crest gauges cannot be overstressed. They should be cleaned and painted with fresh indicating solution after each flood. The most suitable indicating material is a thick solution of cleansing powder. Alternatively white paint which emulsifies in water may be used. The flood mark is usually clear on any clean light-coloured surface; it is the cleaning which matters most. Flood debris should be removed from gauges after each flood.

11.2.3 Stilling-Wells

Siltation can be reduced by minimising the inlet pipe size but care must be taken not to introduce lag effects. Storage space for silt accumulation must be provided below normal minimum bed level and must be desilted frequently.

A number of small diameter pipes set flush with the stilling-well walls is preferable to long, protruding, large-diameter pipes in conditions where the lower pipe(s) may be submerged beneath bed sediments. Flush-fitting pipes are also more likely to reduce well drawdown effects due to turbulence and eddies at the entrance to pipes protruding into the flow. Static tubes are not encouraged due to difficulty in desilting them.

Pipe stilling-wells with cone-shaped ("hopper") bottom sections having a small aperture at the apex have been found effective. In principle, water velocity accelerates past the aperture and prevents blockage; the velocity of the flow of water-sediment mixture to and from the well via the aperture is rapid; water level oscillations due to waves and surges maintain a variable differential head on the aperture, preventing a sediment pack from forming. A cone-shaped plug with a small aperture rests in a larger aperture at the bottom of the "hopper" and is attached by a chain to the top of the well. Periodic cleaning is performed by raising and lowering the plug while water is in the well, thus creating a surging motion.

The aperture should be set not too low, to avoid sedimentation problems, not too high, to avoid losing record of the arrival and cessation of runoff. The "hopper bottom" is preferable to the use of numerous small-diameter holes which, while damping out oscillations, usually result in failure to record flood recessions and subsequent peaks due to blockage.

Where long pipes are necessary to reach low flow courses, it is sometimes useful to thread a loop of small diameter cable through the lowest inlet pipe and a higher one. Pulling the cable through the pipe during low flow can speed up the pipe cleaning process where there is a water level difference to induce a flow in the pipe.

In cases of stilling-wells fabricated from sheet metal or steel tubes, desilting doors or panels are required. Lack of adequate silt storage in this type of well makes them generally unsatisfactory unless constructed with cone-shaped bottom sections with self-cleansing aperture as discussed above.

11.2.4 Slope Gauges

These are installed to measure water surface slope during floods. They should incorporate crest gauges in order to indicate the water surface slope at the flood peak. The distance apart of the slope gauges is determined by channel bed slope, by likely observation errors and the degree of surge and wave action anticipated. For a bed slope of 0.0025 and an estimated stage error of $\pm$ 5cm the distance apart should be about 100m. Slope-gauges are often combined with float-reach markers for convenience.

11.2.5 Scour Indicators

Quantitative data on variations in the stream cross sectional area during flood flows in alluvial streams, cannot be obtained reliably without direct measurement. However, an indication of the existence and possible extent of bed scour can be gained by installation of scour indicators in the channel bed.

Indicators may be of many different types, ranging from simple markers to irradiated ("labelled") natural or artificial bed material tracers (IAEA, 1968). Some of the simpler devices commonly used are:

1. Chains buried vertically in the channel bed and anchored by means of heavy concrete blocks below the expected scour depth. In theory, the depth of scour is the depth to that part of the chain subsequently found to be lying horizontally.

2. Tiles or other expendable materials may be buried at specified vertical intervals. The depth of scour is assumed to be between the depth to the first tile subsequently located and that previously above it.

All these methods suffer from the defect that evidence of depth to indicators is not necessarily an indication of the area of flow at the flood peak. Nor does it mean that the maximum cross-sectional area indicated occurred at any particular time during the passage of the flood, since part of the section may have been scouring while another was filling. All that can be said is that scour to something approximating the indicated depth probably did occur (though it may have been aided by the presence of the indicator itself) and that, if significant scour and redeposition is revealed, the station should be located elsewhere. Scour depths greater than 10m have been observed in sandy river beds.

11.2.6 Controls

To obtain reliable runoff data with minimum effort stations should be located where the stage-discharge relationship is relatively stable. Although this is obvious, there are many instances of runoff stations sited where there is no prospect of control stability nor the likelihood of producing data by continuous current meter measurement. It is infinitely preferable to install runoff stations at points where controls exist or can economically be constructed, even if such sites are not readily accessible or are less relevant to the problem in hand, than to do so where the data may be meaningless or dangerously misleading.

Bridges sometimes appear to provide possibilities of both partial control and the means with which to measure high discharges. Usually, however, the cross sectional area of flow under the bridge is variable for a given stage. In addition to this the bed level downstream of the bridge is variable as elsewhere in alluvial channels. However, bridges, do provide the means of rapid gauging of floods using cranes and gauging reels. Truck-mounted gauging equipment may be appropriate in some situations.

Culverts may provide complete control and should be considered whenever possible. It may be necessary to record both upstream and downstream levels to show whether the control is complete or partial, or when control passes from the culvert to the downstream channel and vice versa.

Rating curves may be developed as for channels with back-water effects by developing relationships between stage, discharge and difference in upstream and downstream water level. Culverts are seldom designed to accommodate major floods and, in such conditions, roads are overtopped (and often washed out). However, a reasonably good estimate of peak discharge may be obtainable by post flood surveys upstream and downstream of culverts using the methods described in WMO TN 90 (1968). This type of estimation requires careful selection of the appropriate equation and discharge coefficients and is aided by the availability of photographs taken near the peak of the flood. Such photographs are useful in all hydrology but in arid zone ephemeral flow conditions they are especially valuable in recording flow conditions.

Current meter measurement in culverts presents problems where velocities exceed the capacity of the count or where debris load is high. Ultrasonic velocity measuring systems (Ellis and Green; 1977, 1978: Herschy and Loosemore, 1974) allied to culvert controls are a practical (though expensive) possibility, well worth considering in relation to experimental or representative basin programmes.

Despite the doubtful value of data produced by stations on alluvial channels, it is sometimes necessary to operate such stations for project purposes. For permanent stations, controls are essential and difficulty of access to sites where controls exist should not be allowed to be a major deterrent. With well trained and equipped field teams, a natural control may be possible to rate over a single flood period. Stage data collected prior to rating are always relevant.

11.2.7 Chemical Gauging

Chemical (or dilution) gauging may be applied successfully in arid zone conditions for rating runoff stations with permanent or semi-permanent natural controls or for check-calibration of artificial, pre-rated controls where current meter measurement is not possible or is subject to intolerable inaccuracy. The main difficulty with dilution gauging is the requirement for well-organised, trained and equipped teams the success of whose operations is governed by their team-work and mutual dependence to a much greater degree than for current-meter teams. The need for skill in site selection and for coordination of effort at the injection and sampling sites, which may be remote from each other, is usually difficult to fulfill. Whereas current meter work can be successful with only one proficient team member - dilution gauging cannot.

The prospect of continuous dilution gauging in unstable alluvial channels, for which techniques are available, is attractive. Experimentation is required to assess whether or not the technique is capable of being developed for routine application in ephemeral streams.

11.2.8 Runoff Estimates from Reservoir Records

Storage reservoirs are potentially valuable gauging structures capable of producing good quality runoff data with little effort in comparison to that required for the effective operation of conventional runoff stations.

The reservoir storage change is estimated from water level records and the elevation - storage relationship established by survey.

Rainfall on the reservoir surface is included in the storage change term of the water budget equation and may usually be ignored. Evaporation must be estimated using methods referred to in Section III. For practical purposes, the measured reservoir hydrograph rise (which, in arid zones, occupies a relatively short time period) and estimated outflow produce adequate inflow estimates, without evaporation estimation, where prolonged surface runoff recession or base flow represent a small proportion of runoff.

The effect of sedimentation on the precision of the inflow estimate depends on its magnitude in relation to the flow. The precision of spillway discharge estimation and draw-off for water supply or irrigation should be high (say ± 10%). That of the evaporation term is low unless heat storage in the reservoir can be accounted for (say ± 20%). The precision of the storage change estimate can only be expressed as an absolute value (because the percentage error will normally depend greatly on reservoir contents) whose effect is dependent on the size of the other balance items. Seepage is not usually measured.

12. DATA PROCESSING AND ANALYSIS

12.1 GENERAL

In arid zones it is the extreme unsteadiness of runoff rates which creates the main problems. In arid conditions the total flood flow can pass a gauging station in one hour of the day while virtually no flow occurs in the remainder of the day or possibly of the month or even the year.

In order to develop a resource which manifests itself in so unpredictable a manner we must ensure that nothing in processing and analysis masks these characteristics, indeed we must stress them.

Procedures for processing data at stations in arid zones are basically no different from those applicable elsewhere. The extra requirement on the part of the data processor is an awareness of those characteristics of arid zone watercourses and runoff processes likely to produce unacceptable or unrecognised errors.

Natural water courses in arid zones may be divided roughly into three main types:

(a) Stable, rocky, steep, deeply-incised, irregular channels which almost wholly control the flow characteristics of the water/sediment mixture they occasionally contain and are changed relatively slowly by it.

(b) Unstable, disorganised, braided alluvial channels whose slope, depth, shape, sinuosity and bed-form are controlled by the water-sediment flow characteristics and are in a state of dynamic equilibrium with them.

(c) Minor watercourses in flat alluvial plains or approaching the terminal reaches of major basins which flows seldom reach. In these watercourses vegetal growth dominates the depth discharge relationships. There may be no obvious single preferred channel, water finding its way among the large hummocks created by the growth.

Type (a) are degrading channels while types (b) and (c) are aggrading. In many countries, most agricultural development is concentrated in the vicinity of type (b) water courses.

Discharge measurement in type (a) channels may be subject to large errors due to high velocity, high turbulence and high debris load. If the control is stable, increasing the number of measurements containing large random errors, increases the precision of the station rating and thus the overall precision of runoff measurement.

Measurements in types (b) and (c) channels are subject to large errors due to the configuration (sometimes changing during measurement) of the water course. Each measurement is a "spot" sample which relates only to the particular flow event and can seldom be used for the construction of a rating curve. Increasing the number of measurements improves the quality of the runoff record but this high frequency of measurement must continue for as long as the station is operated.

The more common sources of error are discussed in the following subsections. Sample computations are given for illustration of the probable error for different circumstances and stages of computation.

12.2 DISCHARGE MEASUREMENT DATA PROCESSING

12.2.1 Current Meter Measurement

The total discharge error XQ is composed of random and systematic errors and the method of its estimation is described by the International Standards Organisation (1978). The partial errors associated with each component of current meter measurements should be assessed in relation to the measurement site, flow conditions and measurement method. For steady flow in the channel conditions described above XQ is especially sensitive:

- For type (a) channels; to the complex configuration of rocky, non-uniform channels necessitating a large number of verticals and point-velocity measurements in each vertical to define the flow boundary and the spatial and temporal variability in the velocity distribution normal to the section;

- For type (b) channels ; to the formation, movement and decay of bars, dunes and ripples. These conditions ideally require near-instantaneous measurements to define the flow boundary with a large number of verticals and the velocity distribution at each vertical by multiple-point measurements.

For unsteady flow conditions the problems are increased and accuracy is virtually impossible. Where stable controls are available the inevitable lack of precision of point-measurements made to calibrate them may be improved by increasing the number of measurements. Where controls are lacking, the question of permanent rating does not arise and the precision of the hydrograph is only as good as the individual measurements and the interpretation of the stage data used to define it. It is suggested that current meter measurements made in steady flow conditions in typical arid zone conditions may be in error by up to ± 20% to which the error attributable to non-steady flow should be added.

12.2.2 Slope-Area Methods for Flood Gauging

12.2.2.1 General Considerations

In the absence of a cableway or a bridge, slope area methods can furnish at least the order of magnitude of flood flows; however the difficulty of estimating the wetted area during various flood stages decreases the accuracy of the method and may give rise to large systematic errors.

12.2.2.2 Observational Requirements

Both water surface slope and float velocity measurement must be of a high order of accuracy and, because of the inaccuracies inherent in measuring water levels in highly turbulent conditions and in the selection and identification of pieces of floating debris for use as floats, lack of quality must be compensated by an adequate quantity of observations.

(a) Roughness Factor

The roughness factor n may be estimated from discharge measurements made under similar conditions of depth, velocity and bed material at other stations where cableways have been installed and where water surface slopes along a uniform reach of channel have been measured.

These values are not necessarily transposable to other channels but will provide a good indication of the likely range under similar conditions. Care should be exercised in selection of n values by extrapolation from values determined for low stages. Changes in configuration of the channel bed during rising and falling stages may cause considerable variations in n which are not necessarily valid beyond the range of flow for which they were determined. Values of n also vary over a fairly wide range for similar stages because of variations in discharge caused by changes in the bed configuration, energy gradient and hydraulic radius. Variations for the same discharge are likely to be smaller.

Dawdy (1961) showed the discontinuity in the relationship between hydraulic radius and velocity due to the rapid transition from lower to upper regime bed configuration. This regime change results in a very marked decrease in bed-form roughness as dunes are removed during increasing discharge and are re-formed during flow recession (though not necessarily at the same discharge).

The variations of roughness with flow regime in sand channels is discussed in detail in WMO Technical Note 90. Numerous examples of roughness coefficients will be found in text books on open channel hydraulics (eg. Chow, 1959).

(b) Water Surface Slope

Slope measurements are made by observations of primary and slope gauges during float measurements. Observations should be made at the upstream and downstream gauges simultaneously with observation of the middle gauge made as a check on gross error.

(c) Float Velocity Measurement

This is carried out by using standard procedures but giving attention to the following important points:-

- For wide, sand-bed channels the only practicable method of float velocity measurement during floods is by timing floating debris over the measured reach.

- As many float measurements as possible, at varying distances across the channel, should be made on the rising and falling flood. The more variable the surface velocity across the channel, the more float measurements are required to define the variation.

- Estimation of the distance of the float across the channel is difficult and likely to be in considerable error. It is more satisfactory to estimate the proportion of the total width from the water's edge. Observers should measure the width of flow at the peak from high water marks on each bank as soon as wading is possible. The distances should be measured from selected "initial points" for each set of gauges.

12.2.2.3 Discharge Computations

The discharge is computed using the Manning equation if the velocity is not measured:-

Manning's formula is:-

$$Q = \frac{A\, r^{2/3} S^{1/2}}{n} \qquad (12.1)$$

Where:

Q = Discharge (m^3/s)

A = Cross section area of flow (m^2)

r = Hydraulic radius (m)

= $\frac{A}{P}$ where P = Wetted perimeter (m)

S = Slope of the energy gradient in a channel reach of uniform shape and roughness. This may be taken to equal the water surface slope in uniform channels.

n = Roughness factor

For wide, relatively shallow alluvial channels the discharge equation may be written as:-

$$Q = VPd \qquad (12.2)$$

Where:

d = Mean depth (m) $\approx$ r

V = Mean velocity (m/s)

$$= \frac{r^{2/3} S^{1/2}}{n} \qquad (12.3)$$

Rearranging (12.3)

$$r = \left[\frac{Vn}{S^{1/2}}\right]^{3/2} \qquad (12.4)$$

and

$$Q \approx VP\left[\frac{Vn}{S^{1/2}}\right]^{3/2} \qquad (12.5)$$

$P\left[\frac{Vn}{S^{1/2}}\right]^{3/2}$ can be seen to represent the area A.

If float measurements have been made, the mean velocity V may be estimated from Vs (surface velocity) and will normally lie in the range of 0.8 - 0.9 Vs. For flood conditions in an alluvial channel the value will normally be closer to 0.9. The factor to be used in individual cases should be estimated after consideration of the vertical velocity distribution in similar channels for the same depth, if observational data are available.

The computation may be carried out in either of two ways depending on the quality of observation and the distribution of velocity across the channel. The method using "panels" of discharge, as for normal current meter measurements, is necessary where considerable variations in velocity occur across the section and where the number of float observations is adequate to define this variation. Where the surface velocity distribution across the section is relatively uniform, the mean of the individual float measurements may be used to estimate $\bar{V}s$ and thence V. An illustration of the shorter computation method is given as Example 12.1.

Example 12.1

Wadi Bishah at Heifa - Flood of 2nd March, 1970
Application of Modified Manning Formula

The station was visited by a hydrologist on 2nd March 1970, one hour after the commencement of water level rise in response to heavy rainfall when the stage was 1.75 m on the primary gauge.

The mean surface velocity, estimated from five float measurements was 3.1 m/s. The estimated mean section velocity was 3.1 x 0.85 = 2.63, say 2.6 m/s. The factor of 0.85 used to reduce surface to mean velocity would be high if flow had been in the "lower regime" but it is shown later that this was not the case.

The slope of the energy gradient was estimated by taking the normal bed slope (.002) plus an allowance for decreasing depth and velocity head in a downstream direction due to the rapid rise in water level. It is assumed that, at the front of the flood wave, depth and velocity head were zero.

$$S = S_o + \frac{D}{L} + \frac{V^2}{2gL}$$

Where:

S_o = Wadi bed gradient ≃ .002

D = Stage minus bed level (before flood) ≃ 1.3 m

L = Distance from station to flood wave front 3600m (estimated from the time from commencement of the flood rise multiplied by an estimated mean velocity of 1 m/s).

$$S = .002 + \frac{1.3}{3600} + \frac{2.6^2}{19.6 \times 3600}$$

$$= .00246$$

W.M.O. Technical Note No. 90, gives Manning's basic "n" value as 0.020 for a median bed material size of 0.4mm as determined from sieve analyses of samples from eleven test pits across the wadi at the recording section (assuming that channel irregularities, sinuosity, vegetation etc. are negligible and do not add to the basic value).

The estimated discharge from equation 12.5 is:-

$$Q = 2.6 \times 185 \left(\frac{2.6 \times 0.020}{.00246^{1/2}}\right)^{3/2}$$

$$= 516 \text{ m}^3/\text{s}$$

Extrapolation of the station rating curve through this point to the peak flood stage of 2.40 m. gives an estimated peak discharge of 2100 m^3/s.

The estimate of the flood peak using Manning's formula in the normal way and based on a slope-area survey made after the flood had receded would have been

n	V	s	$s^{1/2}$	$d = \left(\frac{Vn}{s^{1/2}}\right)^{3/2}$	$Q = VP\bar{d}$	$F = \frac{V}{\sqrt{g\bar{d}}}$	Chezy C
0.015	2.6	0.00246	0.0496	0.69	334	0.995	63
0.020	"	"	"	1.07	519	0.804	51
0.025	"	"	"	1.50	721	0.678	43
0.030	"	"	"	1.97	947	0.593	37

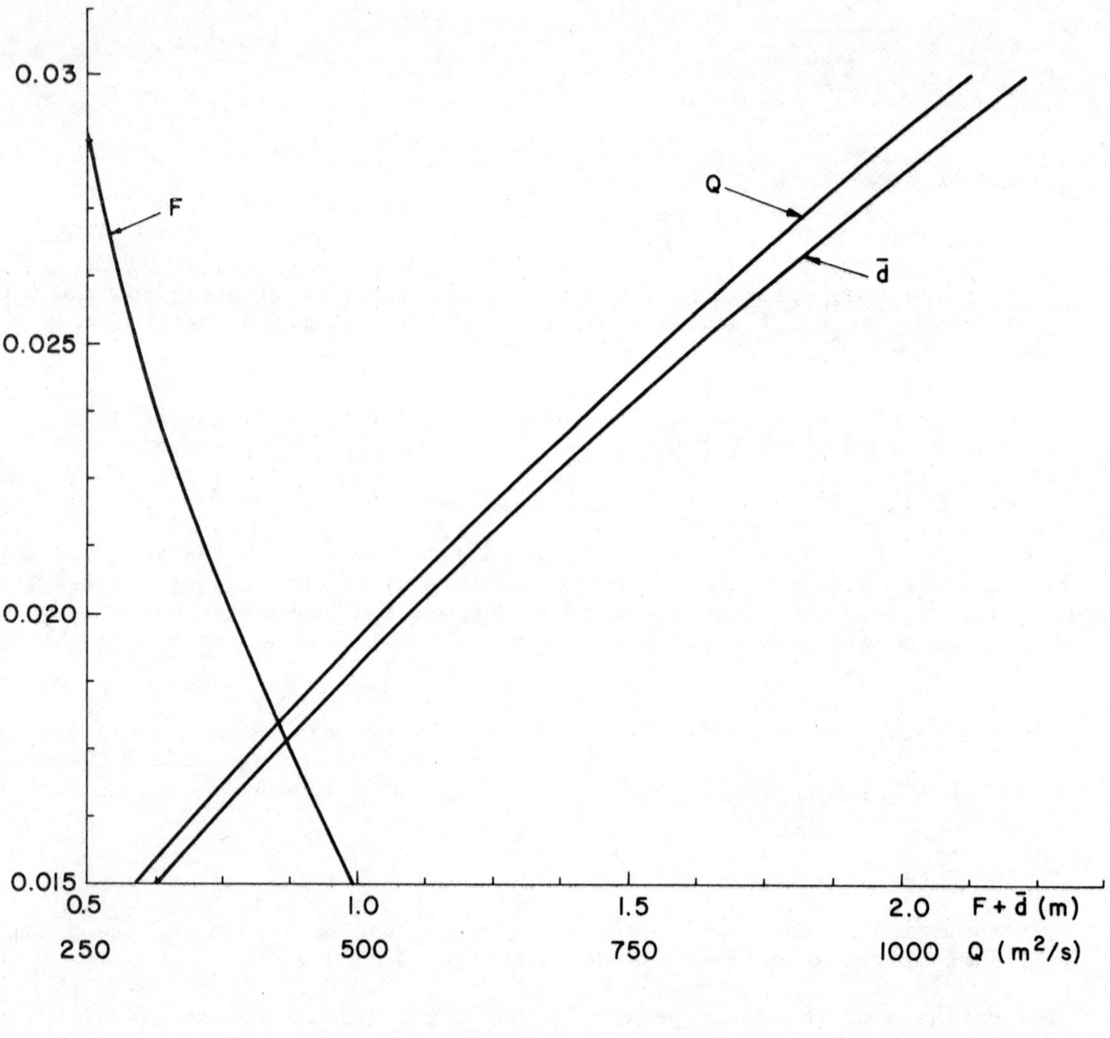

Fig. 12.1 Froude number (F), discharge (Q) and mean depth ($\bar{d}$) for different values of Manning's 'n' used in estimating discharge in Wadi Bishah at Heifa at 17.00 hours on 2 March 1970 P = 185 m

FIGURE 12.1

B-402 Wadi Nishah at Heifa - Saudi Arabia

The figure shows the relationships:

Manning's n vs. discharge Q
Manning's n vs. depth $\bar{d}$
Manning's n vs. Froude number F

for an alluvial channel during a flood when surface velocity was observed. Values of Q corresponding with a range of n values and an estimated water surface slope were calculated using the modified Manning formula (Section 12.2.2) as shown on the figure.

Assuming mean velocity to have been accurately estimated from surface float measurements the likely range of discharge can be assessed by reference to F and $\bar{d}$. Thus an n value of 0.03 would result in Q = 950 m^3/s, $\bar{d}$ = 2.0 m and F = 0.6. Since standing waves were observed (and photographed) n must be lower than 0.03.

An n value of 0.015 would result in Q = 350 m^3/s, $\bar{d}$ = 0.7 and F = 1. For mean depth to be critical, as implied by F = 1, deep standing waves would have formed; these were not observed.

The visual evidence supports the WMO recommended value of n = .020 for median bed material diameter of 0.4 mm and produces a discharge estimate of 520 m^3/s. The discharge range for an error of ±20% in n (0.016 to 0.024) is 375 to 675 m^3/s.

$$Q = \frac{Ar^{2/3} S^{1/2}}{n} = \frac{270 \times 1.42^{2/3} \times .002^{1/2}}{.020}$$

$$= 763 \ m^3/s$$

which corresponds to a mean velocity of 2.8 m/s.

The velocity derived from the conventional use of Manning's formula at peak stage is thus only slightly higher than that derived from velocity observations well below the peak. The rating curve derived from equation 12.5 was preferred for estimating run-off volume for this event.

It is interesting to consider the effect of varying the value of n for the above calculations. Figure 12.1 shows variations in discharge, mean depth and Froude number for different n values and illustrates the sensitivity of the roughness factor. For the estimate of 516 m^3/s at the time of observation, the Froude number value is about 0.8 which suggests a bed configuration varying across the section between "plane regime" and "antidune standing waves". The visual evidence, including photographs taken at the time, fits the theoretical assumptions remarkably well. High Froude numbers provided by lower n values would necessitate the formation of deep standing waves, which were not apparent, and the correspondingly smaller mean depth would indicate a higher bed level than that subsequently found, which is most unlikely. Lower Froude numbers would result from higher n values which would indicate larger bed material size, deeper scour and possibly "lower regime" flow.

The range of median bed material diameter size is 0.2 to 1.0mm and the median diameter is 0.5mm. The range of basic Manning's n values corresponding to these sizes is .012 to .026.

12.2.3 Alam and Kennedy Method

Discharge estimates based on slope-area survey methods and channel parameters may also be made using depth-discharge predictors of the type discussed in the ASCE Sedimentation Engineering Manual (1975). Among these is the method proposed by Alam and Kennedy (1969).

For this method, channel roughness is apportioned between friction factors associated with grain roughness f_f and bed-form roughness f''_b which together comprise the overall Darcy-Weisbach friction factor f_b.

The derived functional relationships are:

$$f_f = f_f \ \frac{Vr_b}{\nu}, \frac{r_b}{D_{50}}$$

$$f'' = f'' \ \frac{V}{g \ d_{50}}, \frac{d_{50}}{r_b}$$

Where V = mean velocity (ft/s)

g = gravitation constant (ft/s^2)

d_{50} = median grain diameter of bed sediment (ft)

r_b = hydraulic radius of bed (ft)

ν = kinematic viscosity of fluid

$\frac{Vr_b}{\nu}$ = Reynolds number R_e

$\frac{V}{g\ d_{50}}$ = Froude number based on mean bed material diameter.

The functional relationships are shown in Figures 12.2(a) and (b). It should be noted that, as with the Manning equation, scour and deposition during flow are not accounted for. The cross-sectional area of flow and hydraulic mean radius are assumed to be constant as found from the post-flood survey.

Simplifying assumptions are as follows:

1. The channel shape which controls the slope of the depth-discharge relationship is defined by the parameters, hydraulic mean radius of the bed r_b and flow width W.

2. The mean relationship between r_b and W may be used to estimate W from r_b as in Example 12.2.

3. W is free to increase as r_b increases, i.e. there are no lateral constraints and regime alluvial channel conditions obtain.

4. The energy gradient S is constant.

The calculation is carried out as follows (using fps units). Symbols are as given in the ASCE Sedimentation Engineering Manual (1975).

1. Select a value for V.

2. Assume a value for r_b.

3. Calculate $V/\sqrt{gd_{50}}$, r_b/d_{50} and Vr_b/ν .

4. Obtain f_f from Figure 12.5(a) and f" from Figure 12.2(b).

5. Calculate $f = f_f + f''$.

6. Calculate $r_b = f \frac{V^2}{8gS}$

7. Repeat the calculation until assumed and calculated values of r_b are about equal.

8. Calculate q (discharge per metre of width) from $q = 0.093Vr_b$ $m^3/s/m$ (0.093 converts to metric).

9. Calculate $Q = Wq$ m^3/s - where W is the alluvial bed width in metres.

Calculations relating to a range of velocities for a South Arabian wadi are shown as Example 12.2. In this case the calculations were used to estimate the slope for extrapolating the rating curve. The relationship between hydraulic mean radius r and flow width W was estimated from the Q-W and Q-A relationships shown by Figure 10.3. The high degree of scatter about this relationship is shown by Figure 12.3a. For the purposes of Example 12.2, two cases are considered:

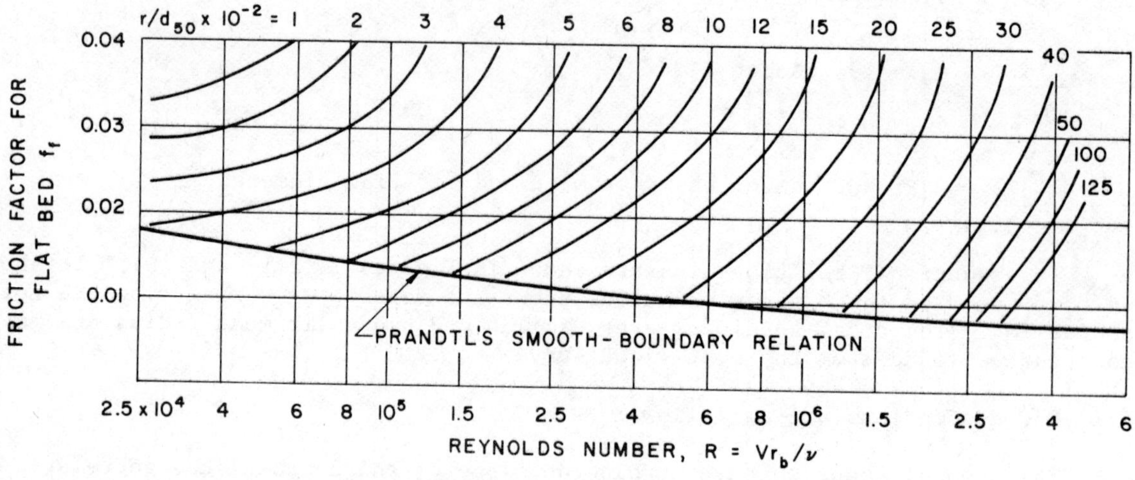

Fig. 12.2(a) Flat bed friction factor diagram for alluvial streams (after Lovera and Kennedy, 1969, Fig. 1)

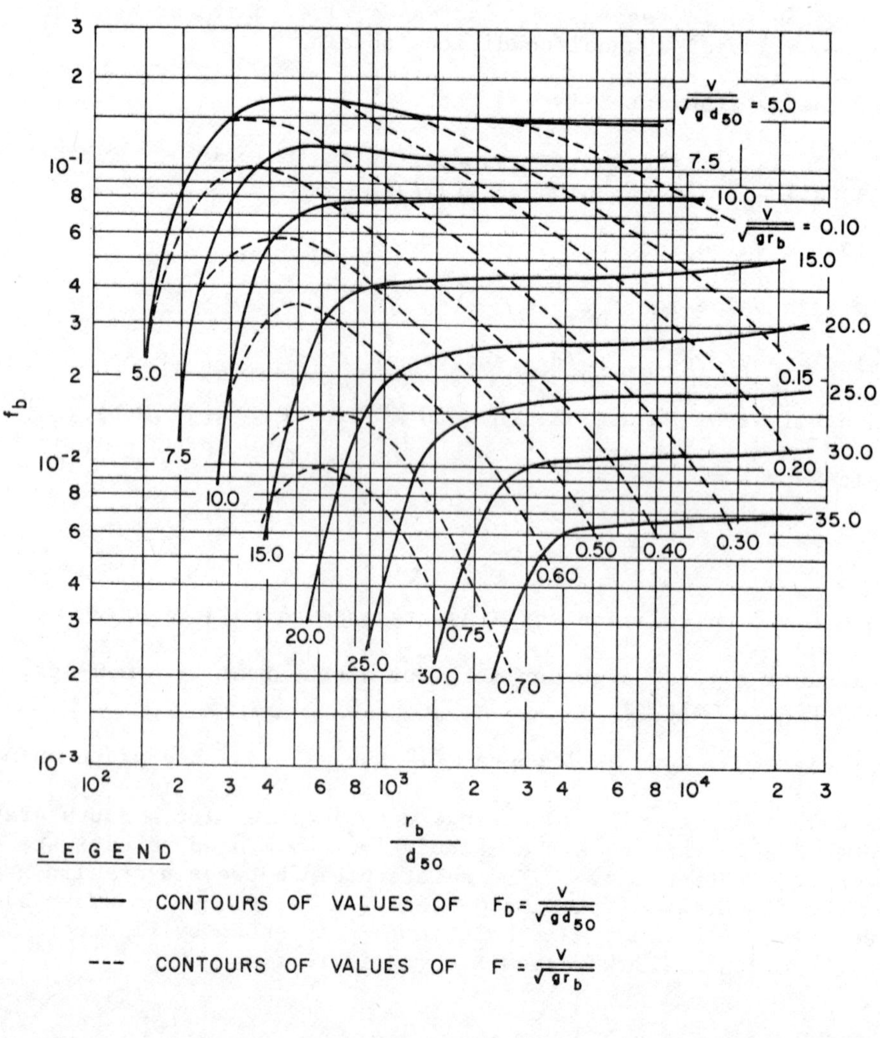

Fig. 12.2(b) Graphical expression of f as function of Froude number and r_b/d_{50} (after Alam and Kennady, 1964, Fig. 3)

Example 12.2

ESTIMATION OF MEAN-DEPTH DISCHARGE RELATIONSHIP FOR A WIDE ALLUVIAL CHANNEL, USING THE METHOD OF ALAM AND KENNEDY

V		$V/\sqrt{gd_{50}}$ (1)	r_b	r_b/d_{50}	re	V/ grb	f"	f_f	f(3)	$r'_b = f\,V^2/8gs$ (2)		$q = V_b$	W (4)	Q
ft/s	m/s		ft							ft	m	$m^3/s/m$	m	m^3/s
2	0.61	8.31	2.0	1110	0.4	0.25	.09	.024	.114	0.71				
			1.0	555	0.2	0.35	.10	.026	.126	0.78				
			0.8	444	0.16	0.39	.08	.030	.110	0.68				
			0.7	389	0.14	0.42	.08	.032	.112	0.70	0.21	0.13	18.7	2.4
5	1.52	20.8	3.0	1670	1.50	0.5	.022	.040	.062	2.4				
			2.5	1390	1.22	0.57	.020	.040	.060	2.3	0.70	1.07	321	343
10	3.05	41.6	7.0	3890	7.0	0.7	.002	.040	.042	6.6	2.01	6.13	3860 (730)	23660 (4470)

Note: CHANNEL PARAMETERS

(1) d_{50} = 0.55mm = 0.0018 ft.

(2) S = 0.0025

8gS = 0.645

$\sqrt{gd_{50}}$ = 0.241

(3) The resultant f values correspond with Manning's n from 0.026 to 0.023 for increasing q.

(4) $W = ar^b$, in which r is taken to equal r_b and a and b are derived from the relationship shown on Figure 12.3(a) to produce the equation $W = (16.50\ r_b)^{2.3585}$. Bracketted W and Q values relate to the maximum channel width (W max) at the location used for the example.

Example 12.2 Continued

ESTIMATION OF MEAN VELOCITY AND DISCHARGE FROM POST-FLOOD SURVEY DATA USING ALAM-KENNEDY METHOD

DATE	STAGE	A	W	r_b		V		S	d_{50}		$V/\sqrt{gd_{50}}$	r_b/d_{50}	Re	f"	f_f	f	r_b	Q
	m	m^2	m	m	ft	m/s	f/s		mm	ft			$x10^6$					m^3/s
28.7.67	1.66	212	164	1.29	4.24	2.76	9.05(1)	.0025	0.55	.0018	37.6	2360	3.84	.001	.04	.041	5.78	
						2.13	7.0	"	"	"	29.1	"	2.96	.010	.04	.050	3.81	
						2.29	7.5	"	"	"	31.2	"	3.18	.006	.04	.046	4.02	485
21.1.70	1.62	105	200	0.52	1.71	0.85	2.8(1)	"	"	"	11.6	950	0.48	.065	.032	.097	1.00	
						1.22	4.0	"	"	"	16.6	"	0.68	.030	.040	.070	1.74	128
21.7.70	1.23	80	161	0.495	1.62	1.10	3.61(1)	"	"	"	15.0	900	0.58	.040	.040	.080	1.62	88
31.8.70	1.64	127	230	0.55	1.80	1.56	5.10(1)	"	"	"	21.1	1000	0.92	.016	.040	.056	2.26	
						1.22	4.0	"	"	"	16.6	"	0.72	.032	.040	.072	1.80	155
6.4.71	1.20	57	136	0.42	1.38	1.13	3.7(1)	"	"	"	15.4	767	0.51	.038	.040	.078	1.66	
						0.91	3.0	"	"	"	12.4	"	0.41	.055	.035	.090	1.26	
						0.98	3.2	"	"	"	13.3	"	0.44	.050	.038	.088	1.40	56
9.4.71	1.60	138	174	0.79	2.59	1.82	5.97(1)	"	"	"	24.8	1440	1.54	.010	.040	.050	2.76	251
13.3.72	1.70	166	232	0.74	2.43	1.63	5.35(1)	"	"	"	22.2	1350	1.30	.017	.040	.057	2.54	
						1.58	5.2	"	"	"	21.6	"	1.26	.017	.040	.057	2.40	262
29.4.72	1.18	60	151	0.40	1.31	1.09	3.58(1)	"	"	"	14.9	730	0.47	.035	.040	.075	1.49	
						0.91	3.00	"	"	"	12.4	"	0.39	.055	.034	.089	1.25	55
5.4.69	3.05	500	195	2.56	8.40	3.74	12.3	"	"	"	51.0	4670	10.30	-	.040	.040	9.4	
						3.28	10.0	"	"	"	41.5	"	8.40	-	.040	.040	6.2	1770
						3.54(1)												

(1) As f is constant at 0.04 for this situation,

$$V = r_b 8gs/f = \frac{8.4 \times 8 \times 32.2 \times .0025}{.04} = 11.63 \text{ f/s} = 3.54 \text{ m/s.}$$

TABLE 12.1

COMPARISON OF THE RESULTS OF MANNING AND ALAM-KENNEDY METHODS APPLIED TO THE SAME POST-FLOOD SURVEY DATA

No.	GH m	r m	A m^2	W m	S	n	f	V m/s	Q m^3/s	REMARKS
1	1.66	1.29	212	164	.0025	.022		2.69	571	M
							.046	2.29	485	AK
2	3.05	2.56	500 (1)	195 (1)	"	"		4.26 (1)	2128 (1)	M
							.040	3.54 (1)	1770 (1)	AK
3	1.62	0.52	105	200	"	"		1.47	154	M
							.070	1.22	128	AK
4	1.23	0.50	80	161	"	"		1.30	104	M
							.080	1.10	88	AK
5	1.64	0.55	127	230	"	"		1.53	194	M
							.072	1.22	155	AK
6	1.20	0.42	57	136	"	"		1.28	73	M
							.088	0.98	56	AK
7	1.60	0.79	138	174	"	"		2.06	284	M
							.050	1.82	251	AK
8	1.18	0.40	60	151	"	"		1.23	74	M
							.089	0.91	55	AK
9	1.70	0.70	163	232	"	"		1.80	293	M
							.057	1.58	262	AK
10	1.55	0.52	105	202	"	"		1.47	154	M
11	2.15	0.99	303	345	"	"		2.26	684	M
12	2.28	0.80	229	284	"	"		1.96	449	M

M - Manning, AK - Alam-Kennedy

(1) This estimate relates to partial cross section in alluvial channel only.

TABLE 12.2

RELATIONSHIPS BETWEEN Q, V, R, C, AND n IN AN ALLUVIAL CHANNEL (1)

RANGE OF Q FROM m^3/s	TO m^3/s	$\bar{Q}$ m^3/s	$\hat{\sigma}_Q$ m^3/s	$\bar{V}$ m/s	$\hat{\sigma}_V$ m/s	$\bar{R}$ m	$\hat{\sigma}_R$ m	$\bar{C}$	$\hat{\sigma}_C$	$\bar{n}$	$\hat{\sigma}_n$	No.
20.7	44.0	28.0	6.0	0.89	0.15	0.38	.09	30.0	7.1	.030	.009	17
10.4	18.2	14.2	2.6	0.71	0.14	0.41	.11	22.9	4.8	.040	.010	25
5.2	5.8	5.5	0.2	0.56	0.08	0.30	.10	21.5	4.7	.040	.013	14

(1) It is assumed that Slope S = .0025 is constant.

(2) It should be noted that, although there is no correlation between Q and C or n, there does appear to be a systematic difference between mean values of both C and n for discharges above and below 20 m^3/s. This is to be expected as discharge changes from lower towards upper regime.

a) That there are no lateral constraints and that W is free to increase with discharge. In some arid zone alluvial channels this situation may occur up to very high discharges.

b) That lateral constraints on W exist, with discharge per unit width of channel being applied to the maximum width of the section, Wmax, when the corresponding value of r in the equation $W = ar^b$ (Example 12.2) is exceeded.

The relationships $r(\simeq r_b)$ and H(m) vs Q (m^3/s) are shown by Figure 12.3(b). Estimates of discharge peaks calculated from post-flood survey data using the Manning ($n = 0.025$) and Alam-Kennedy methods are also shown on Figure 12.3(b) and Table 12.1.

The relationship between measured flow depth d and discharge per unit width q (= dv) is shown by Figure 12.3(c) for a Saudi Arabia wadi. The relationship calculated by the Alam-Kennedy method (Example 12.2) is shown on the same figure for comparison and is seen to be a reasonably good fit to the data points.

It may be concluded therefore that the Alam-Kennedy method provides a relatively accurate estimate of the d vs q relationship. It is suggested that it may also be used to derive estimates of the much less stable r vs Q relationship (Figure 12.3(b)), provided that lateral constraints on W are taken into account.

To convert this relationship to one of stage vs discharge is virtually impossible as is shown by Figure 12.3(d) and by the plot of mean bed level against time on Figure 10.4.

The practical implications of the validity of the Alam-Kennedy method for estimating d vs q and r vs Q relationships in alluvial channels reside in the possibility it presents of estimating q or Q from knowledge of the d or r only. For example, by applying the relationship $q = 1.90\ d^{1.671}$ (derived from Example 12.2) to the depths measured for the gaugings of 27.12.73 and 1.4.75 the discharges are estimated as 338 m^3/s and 448 m^3/s. The errors of these estimates are +8% and -13% respectively with respect to the field measurements.

Applying the equation $Q = 1.90\ Wr^{1.671}$ to current meter measurements made in Wadi Najran produces the result illustrated on Figure 12.3(e). The high correlation coefficient and low standard error of this relationship show that the Alam-Kennedy equation could be used to estimate runoff from cross sectional measurements only. In arid zones, where the precision of runoff estimation depends to a greater extent on the number of discharge determinations made to define the hydrograph shape than on the precision of individual measurements, this is an important factor. The observers at the stations illustrated would utilise their time most effectively by continuous cross sectioning of the stream, with intermittent use of current meters.

It should be noted that comparable results are obtainable by application of the regression of Q on A (Figure 10.3(b)). However, in this case the Q vs A relationship has been determined empirically from field measurements, whereas the q vs r relationship has been calculated using the Alam-Kennedy procedure.

12.2.4 Chezy's Equation

In this method mean velocity V is a function of gravity g, hydraulic radius R, slope S and a dimensionless number (a) which may depend on boundary roughness, Reynolds number and cross-sectional shape.

The Chezy equation may be written as $V = C\sqrt{RS}$ in which the coefficient C ($=\sqrt{g/a}$) may be estimated as $\sqrt{8g/f}$ where f is the Darcy Weisbach friction factor (Henderson, 1966). The relationship between Chezy C and Manning n may be written $C = R^{1/6}/n$.

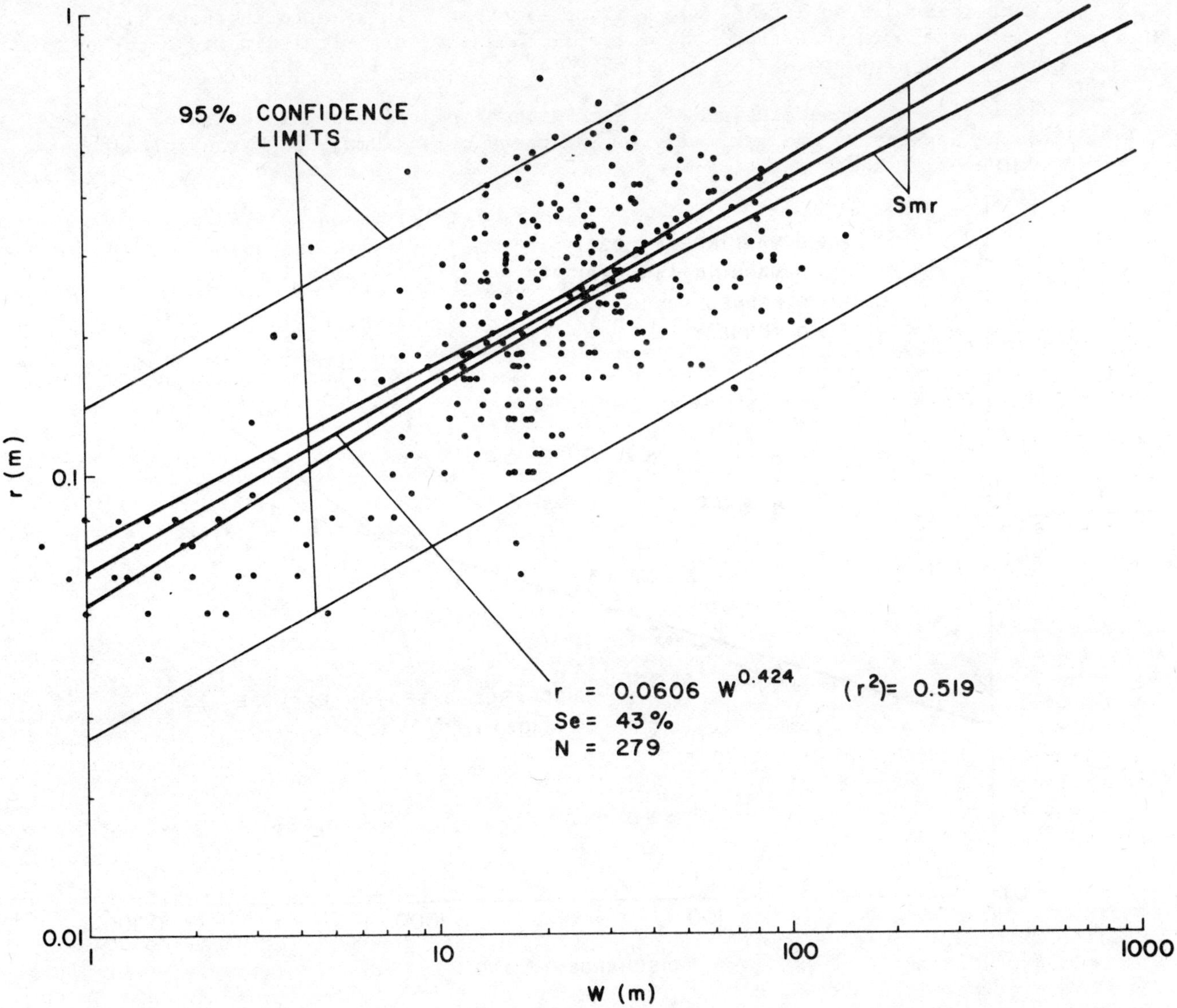

Fig. 12.3(a) Relationship between hydraulic radius r and surface flow width W in an alluvial channel - Saudi Arabia

The data points are for current meter measurements in a self-forming alluvial channel with no lateral constraints to surface flow width.

The 95% confidence limits are calculated by standard statistical formula and show the range within which 95% of measured r values lie for specified W. If we increase the number of measurements we would be unlikely materially to reduce these confidence limits.

The Smr confidence limits are those within which the relationship of r - W population values would be expected to lie at 95% confidence. They give no indication of the reliability of an r estimate for given W, or vice versa, as would be the case with a fixed functional relationship where the scatter represented measurement error only.

Thus, the 95% confidence limits on prediction of r from a single W observation are shown by the outer limits and the confidence interval on prediction of the mean value of r for given W is given by the inner limits.

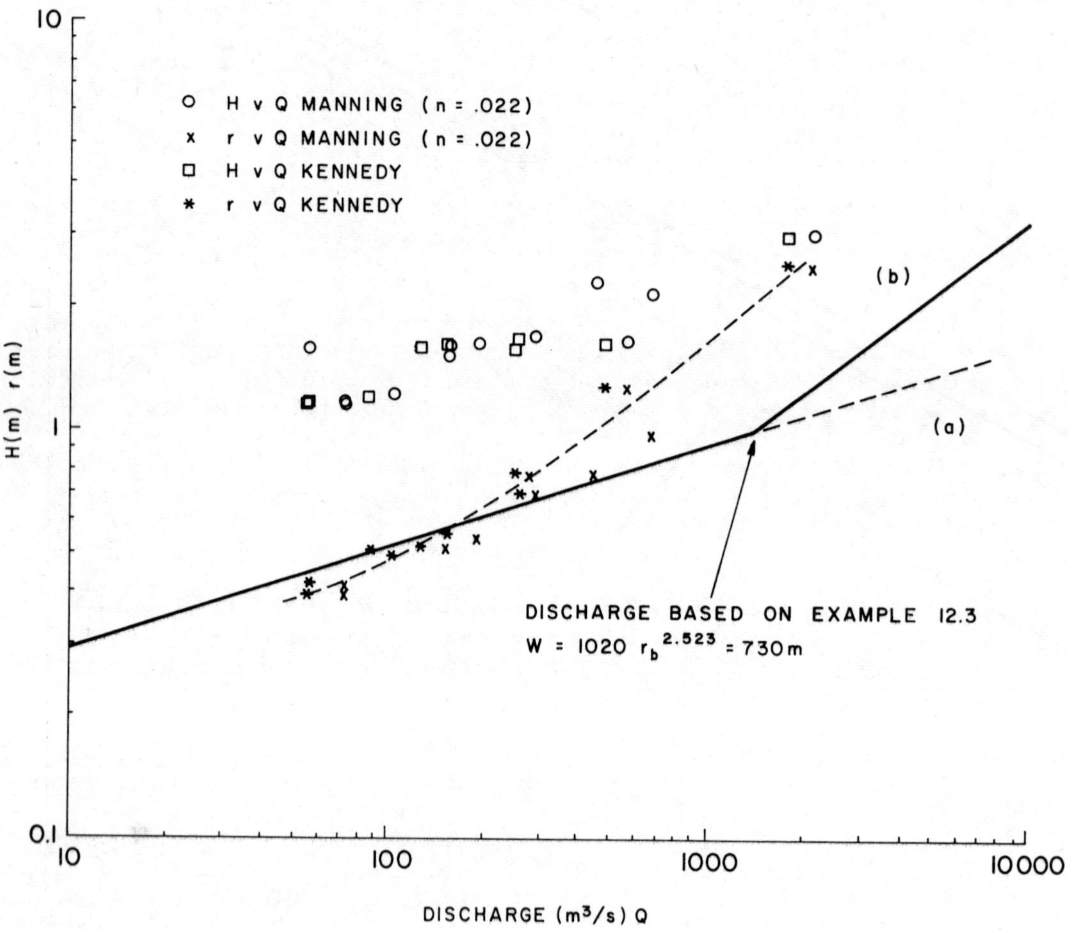

NOTES:

(a) RATING CURVE ASSUMING LATERAL W CONSTRAINTS ABSENT
(b) RATING CURVE ASSUMING W MAX = 730 m

Fig. 12.3(b) Stage and hydraulic radius vs. discharge based on post-flood slope-area survey

FIGURE 12.3(b)

Stage and Hydraulic Radius vs. Discharge Based on Post-Flood Slope-Area Surveys

The figure shows the results of applying the Manning and Alam-Kennedy methods to post-flood survey data for an alluvial channel. Calculated Q values and plotted against stage H and hydraulic radius r. The two methods are shown by the figure to produce comparable results.

Also shown on the figure as a segmental linear relationship is the result of the conjunctive use of the Alam-Kennedy method and the relationship between r and W derived from Figure 12.3(a). Thus, the discharge per unit width of channel based on an assumed value of the bed hydraulic radius r_b, is multiplied by the corresponding value of W to produce an estimate of total discharge. At W = 730 m lateral constraints came into effect, W is fixed at this value and the rate of increase of Q reduces to take account of this; hence the break in slope of the r-Q relationship.

The line of relationship is seen to plot to the right of the Manning/ Alam-Kennedy data points for post-flood surveys at discharges greater than about 200 m^3/s. It may thus be regarded as a more realistic relationship since indirectly (through the empirical r-W relationship) it takes account of bed movement. However, since it is based on extrapolation of a very poor r-W relationship the r-Q relationship should be used with caution.

Since the relationship between H and r is unstable the r-Q curve cannot be applied in calculating discharge hydrographs from stage hydrographs.

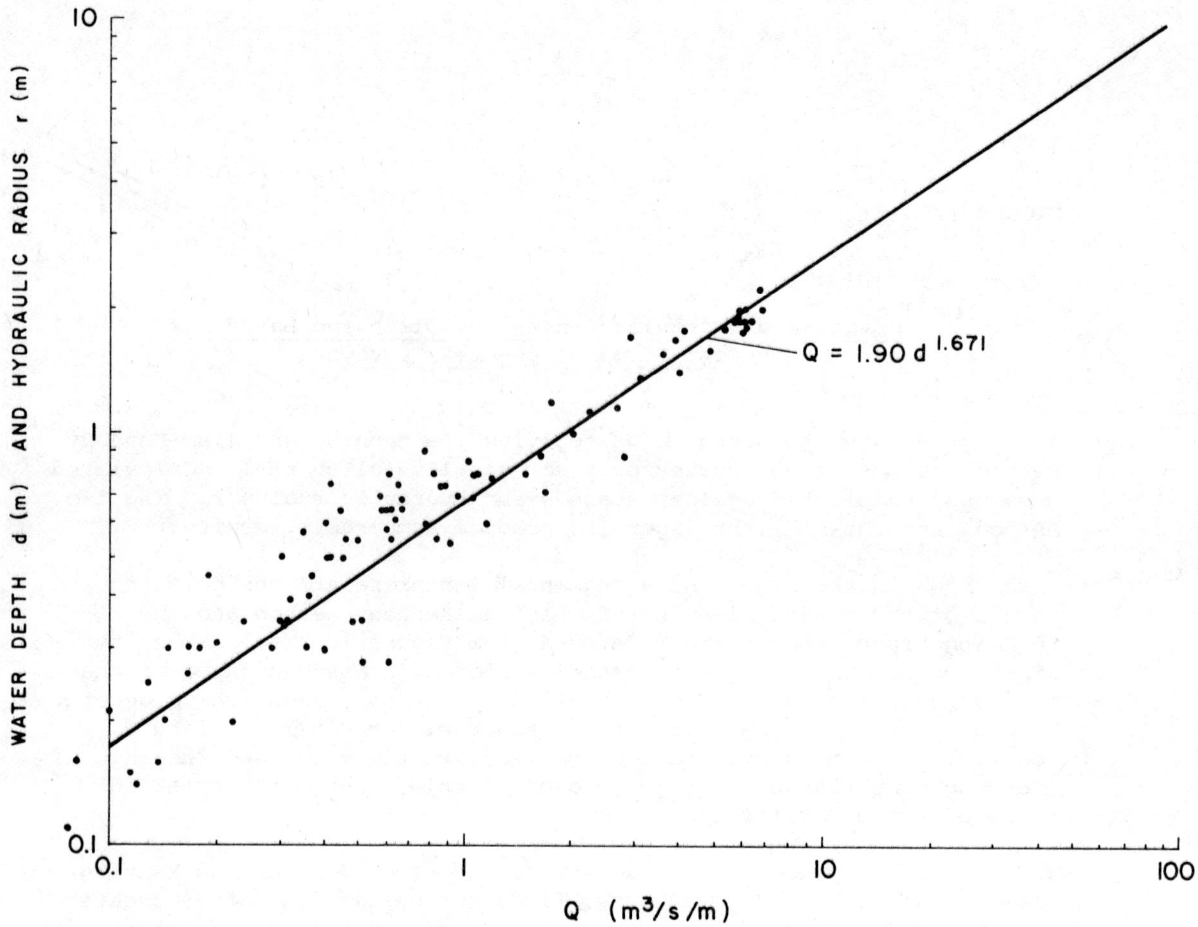

Fig. 12.3(c) Relationship between d and Q in alluvial channel - Saudi Arabian wadis

Note: The equation $Q = 1.90\ d^{1.671}$ has been inferred from Example 12.2 by assuming that d = r.

The figure is a plot of channel hydraulic radius, or water depth at a single point on a cross section of flow, against discharge per unit width of the stream based on current meter measurements in Saudi Arabian wadis.

The line of relationship shown is that calculated using the Alam-Kennedy method as shown by Example 12.2. It may be seen that the calculated relationship is a good fit to the data points.

The implication of the results shown on the figure is that the Alam-Kennedy method could have been used to predict the d-Q relationship reasonably well and that Q could be estimated by measuring d or r. This conclusion is similar to that illustrated in Figure 10.3 which shows the relationship between Q and area A. In the case of Figure 12.3(c), however, the relationship is based wholly on calculation and suggests that, in ungauged alluvial channels, measurement of the cross sectional area of flow is capable of producing a reasonably accurate measure of total runoff.

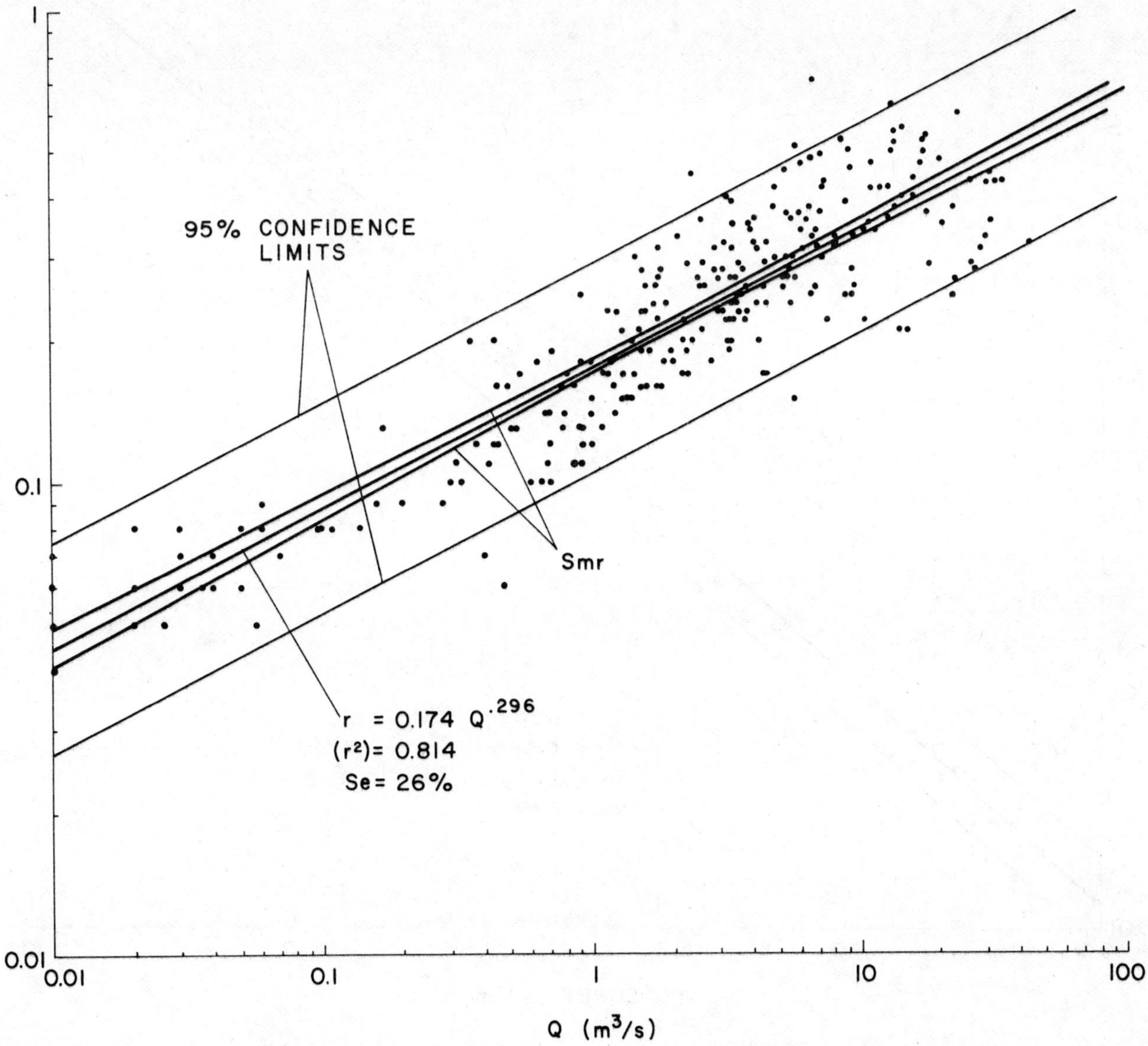

Fig. 12.3(d) Relationship between hydraulic radius r and discharge Q in an alluvial channel - Saudi Arabia

The scatter of the data points shown on the figure may be compared with that of Figure 12.3(b) for the d and r - Q relationship. The data are from current meter measurements and the range of Q within 95% confidence limits for specified r is seen to be very large. This reflects the uncertainty in the r-W relationship shown by Figure 12.3(a).

Thus it would be of little benefit to measure r alone; it is the combination of r <u>and</u> w or A which provides the best means of estimating Q.

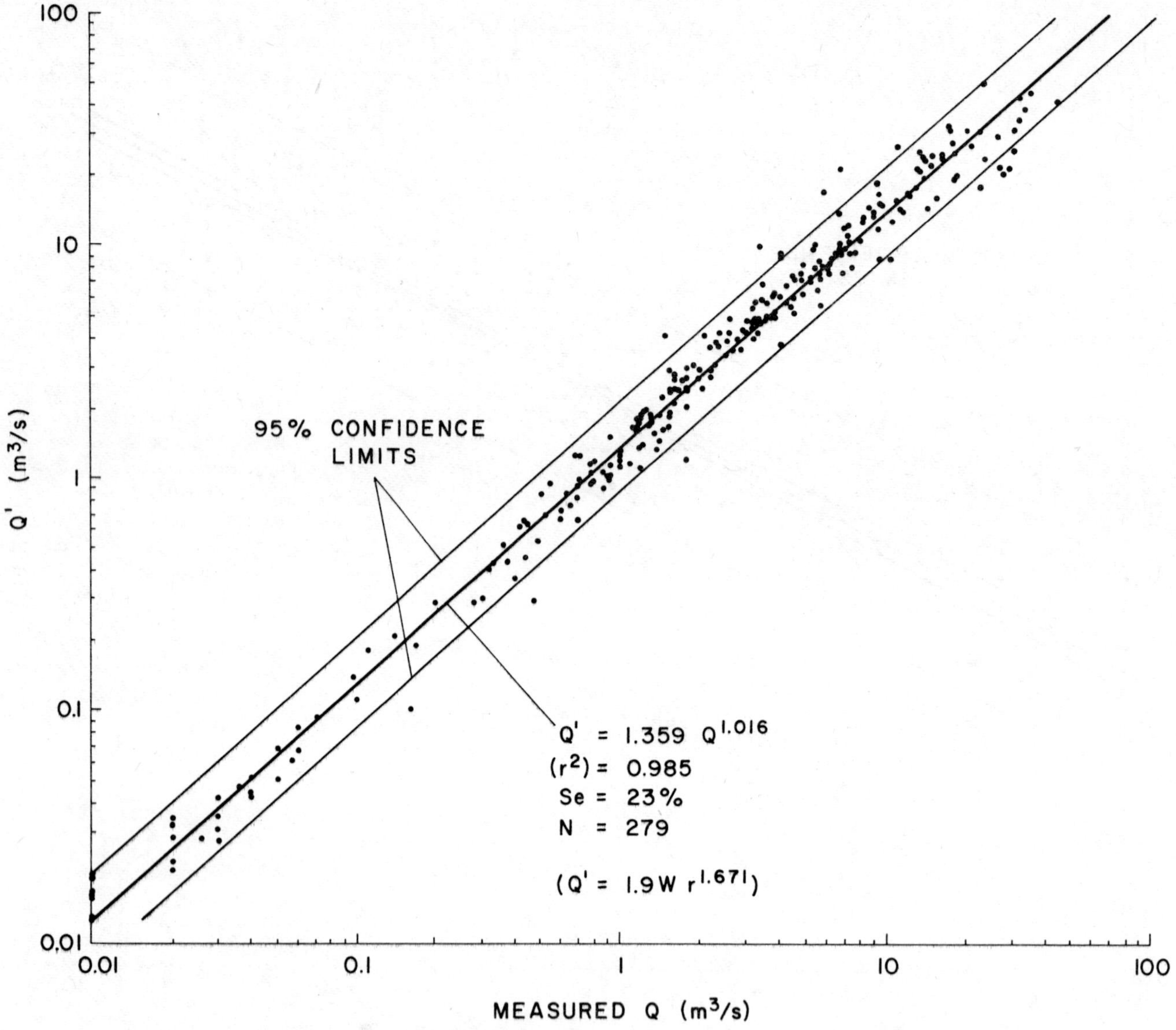

Fig. 12.3(e) Relationship between measured discharge and discharge estimated by Alam-Kennedy

The figure summarises the conclusions of the Figure 12.3 series. This conclusion is that, if Q cannot be measured by current meter, a reasonably accurate estimate may be obtained by measuring the cross section of flow to determine r, using the Alam-Kennedy method to estimate q and measured W to estimate Q.

The reliability of the resultant estimate of runoff volume would depend on the number of cross section measurements made during high discharge.

Calculation of C from a large number of current meter measurements made in an alluvial channel in Saudi Arabia produces the results shown in Table 12.2 which also shows calculated Manning's n for comparison.

C and n were calculated assuming S to be constant at .0025, which is a reasonable approximation, having regard to the fact that all current meter measurements were made during recessions from flow peaks. There was found to be no correlation between C and r nor between V and R. The extreme values of the range, C = 9 to C = 58, represent those conditions where small discharge measurements were made where the true water surface slope would have been important.

Table 12.2 shows C to be rather less variable than n and, since it embodies the shape parameter R, the Chezy equation may be more useful for estimation of Q when C values, have been derived from Q, R and S measurements by current meter under flood conditions.

Comparison of flood peaks at a reasonably well-rated upstream station in Saudi Arabia, where a good control exists, with those estimated by means of slope-area methods at a downstream station provides an example of the possibilities for estimating a notional value of Chezy's coefficient, C , indirectly. In this case it is reasonably certain that waves of translation travel between the two stations when the channel alluvium is fully saturated. Table 12.3 shows, for three floods the discharges estimated from the upstream rating curve and the downstream slope-area survey. Also shown are the C values required to produce similar flood peaks. C is not Chezy's C because it is not based on a measured cross-sectional area. It simply indicates the size of C required to compensate for channel scour at flood peak discharge for this particular alluvial channel, assuming the peak discharges at the two stations to be equal.

TABLE 12.3

COMPARISON OF PEAK DISCHARGES ESTIMATED AT TWO STATIONS 9KM APART ON THE SAME WADI IN SAUDI ARABIA

Date of Flood	Upstream	Downstream				
	Qpu m^3/s	Qpd(1) m^3/s	A(2) m^2	r(2) m	S	C (3)
27.12.73	290	150	105	0.52	.0025	77
19. 3.74	830	680	303	0.99	.0025	56
7. 5.74	900	449	229	0.80	.0025	88

Notes: (1) Qp estimated from Manning's equation with S=.0025 and n = .022.
(2) From post-flooded survey.
(3) Calculated from $C = Qpu / (A \sqrt{RS})$

For the maximum measured discharge of 44 m^3/s at the downstream station, C was 32.5.

If C is calculated from the equation $C = \sqrt{8g/f}$ using f values derived as shown in Section 12.2.3 using the Alam and Kennedy method, the results would be as shown in Table 12.4.

TABLE 12.4

Depth-Discharge Relationship - Chezy Equation

r m	S	f	C	$V = C\sqrt{rS}$ m/s	q $m^3/s/m$	W m	Q m^3/s
.21	.0025	0.112	26.5	0.61	0.13	21	2.7
.70	.0025	0.060	36.1	1.51	1.06	103	109
2.01	.0025	0.042	43.2	3.06	6.15	420	2583

The resultant Q values are very similar to the results shown in Example 12.2, as would be expected. As the width of flow W corresponding with R is derived from extrapolation beyond the range of experience Q may only to be estimated up to about 2500 m^3/s in this particular case. The relationships should, strictly, not be used for channels other than that for which the relationships and coefficients were derived. However, the R/W and C/S relationships could be applied to other channels with similar bed material grain size distribution and slope where lateral flow constraints are absent.

12.2.5 Conclusions and Recommendations

Slope-area methods applied to mobile sand-bed alluvial channels cannot be expected to produce accurate results. This is due to the lack of fixed relationships (between hydraulic mean depth R, cross-sectional area A, flow width W, mean velocity V and discharge Q) such as exist for channels with fixed boundaries of constant roughness.

There is no known, reliable method of taking account of changes in cross-sectional shape which occur during the passage of a flood, when calculating discharge peaks from slope-area surveys.

There is little to choose between the Manning, Chezy and Alam-Kennedy formulas. They all produce roughly similar results when applied to a channel of known shape and slope. It is known that a discontinuity exists in depth-discharge relationships as bed-form regime changes during increasing and decreasing discharge in alluvial channels. For flood flows, upper regime conditions may be assumed to obtain and roughness coefficients, whether Manning's n, Chezy's C or Darcy-Weisbach f must take account of this. If the roughness value is well-chosen, the remaining error results from unknown cross-sectional area and shape.

The overall effect of the latter uncertainty is usually to underestimate discharge.

12.3 STAGE DATA PROCESSING

The main problems here are definition of the true water level by interpolation within the band-width of fluctuation shown on the recorder chart in some cases and establishment of the relationship between this level and the primary gauge level where systematic drawdown effects cause these to be different.

The former problem can only be dealt with by interpolation. The latter can be dealt with in one of two ways:

- by establishing the relationship between stilling-well and primary gauge from observations of each over the entire stage range;
- or by establishing rating curves of discharge against stilling-well gauge readings.

In many respects the latter approach is preferable.

An additional problem arises from the occasional need to reconstruct the stage-hydrograph trace when the stilling well is silted or the pressure tapping point blocked.

The stage hydrograph depicted by the recorder trace is typically irregular for all the reasons previously given. It is seldom worth the effort of attempting to produce a discharge hydrograph which faithfully reproduces these irregularities. Some generalisation is justifiable despite the fact that the quantities represented by the irregularities, if real, are appreciable. "Real" hydrographs should not be smoothed if the error in so doing is likely to be appreciable. This can only be judged on the basis of the rating curve shape and quality. If the possibility exists of stochastic generation of runoff sequences, every natural hydrograph rise must be quantified.

12.4 RATING CURVES

For stations with stable controls, standard procedures for rating curve construction are appropriate and require no elaboration here. It is recommended that, in the case of unstable controls, logarithmic paper should be used. This is because the rating problem can usually be "seen" in a more manageable perspective when the data are plotted in this form. The cease-to-flow level should be deducted from measured stage prior to plotting the data points for particular flood periods in an attempt to linearize the relationship.

The manner in which a rating curve, or set of curves, is constructed in the case of unstable controls cannot be prescribed with any certainty that the result achieved will be of tolerable accuracy. All that can be done is to mention some of the problems and possible methods of dealing with them, as follows:

1. A set of discharge measurements made during a flood recession may produce a rating for that particular part of the runoff hydrograph but would be completely inapplicable to any other. Similar ratings for subsequent flood recessions may have similar slopes and lie nearly parallel to the first. Shifts calculated from individual measurements or from the relative positions of successive ratings cannot reliably be applied to intervening flood flow periods, because rating shift varies indeterminately with time.

   In the absence of discharge measurements covering the entire period and range of flow, shift corrections cannot be related to stage and must therefore be assumed to vary only with time between each curve or between consecutive measurements.

   The result of adopting a time-based method of shift correction is likely to be a more random time-distribution of errors than arbitrary selection of a stage-shift relationship. But the fact that shift may have been very much greater between rating periods is not accounted for.

2. Rating curves produced from measurements made on flow recessions in alluvial streams are commonly flat and become progressively flatter as the channel bed rises due to post-flood aggradation processes. If no fluctuations in stage occur

during the recession period the best estimate of the discharge recession is a hydrograph segment drawn through the data points. In the event of minor fluctuations generally within the range of discharge measurement, interpolation and limited extrapolation is warranted provided that:

(a) the resultant series of hydrograph recessions is reasonably consistent with expectations based on measured discharges covering a similar range of recession flow,

(b) the intervening peak discharges are consistent with expectations of depth and channel roughness based on experience or on relationships such as those of Simons and Richardson (1966).

Study of data producing flat rating curves in the Middle East shows that the flatness results mainly from the gradual rise of the cease-to-flow level rather than from the shape of alluvial channel surveyed after a flood flow period. In these circumstances, straight-line extrapolation from flood recession measurements may produce a "criss-cross" pattern of contradictory (and often physically impossible) hydraulic relationships (Figure 12.4(a)). Since the shape of alluvial channels is a function of the momentary water-sediment discharge rate, the slope of rating segments on a logarithmic plot can only be guessed at in the absence of control data.

3. One method of dealing with this type of rating problem is to estimate the discharge corresponding with the maximum stage experienced or conceivable from the available geomorphological evidence and to "fix" the upper limit of individual rating segments by assuming them to converge on this point from points located separately on the lower segments, however defined. Figure 12.4(b) illustrates this concept.

Another method is to asume a minimum stage (corresponding with an assumed maximum scour depth) and to consider a single straight-line rating segment joining this notional point to that corresponding with the estimated maximum stage and discharge as a "master curve". All other derived rating segments are then drawn to meet or converge asymptotically on this line (Figure 12.4(c)).

The results obtained by either method will be in error to a degree which will normally be unknown. However, there is some merit in the implicit assumption of each method that, at the peak of the maximum flood, differential scour and deposition in a wide, long alluvial channel reach become relatively unimportant. The controlling reach is much longer than at lesser discharges and the scour and deposition effects more likely to be compensatory.

4. The type of problem discussed above is an extreme example which assumes a wide alluvial channel composed of non-cohesive, sand-silt particles with no vertical control and limited lateral control features. Permanent stations should not be located in such situations if they can be avoided. However, project requirements sometimes dictate the need for runoff-recharge estimates in respect of alluvial aquifers which cannot be obtained in any other way.

The problem of estimating the maximum discharge usually entails use of the slope-area survey method for which an estimate is required of the maximum scour depth in the mobile-bed element of the channel reach. Several trial-and-error assumptions may have to be made. It would have to be assumed in the case of a sand bed that upper regime flow prevails, for which a value of Manning's n can be selected. The resultant main channel velocity could be assumed not to exceed critical velocity, though it could be expected to be at or very close to it in some conditions. Thus for a wide channel of median bed material diameter $\simeq 0.5$mm and slope $\simeq 0.002$, upper regime flow could be expected at the range of depths exceeding about 2m at which Manning's n would be about 0.02. For depths greater than 2m, n would have to increase in order to maintain sub-critical flow conditions. This it would do through bed form change from plane bed to dune formation. In this way an arbitrary upper discharge value may be set to define the position of the "master" point or curve.

An example of the error involved in use of the "master curve" method is provided by comparison of the discharge hydrographs at two locations on a wadi in southern Saudi Arabia. The difference in catchment area and alluvial aquifer storage capacity between the two stations was such that significant differences in the hydrographs for the two stations were considered unlikely. The standard error of estimate of a regression of 3-day running total runoff volumes for the two stations was about 50% but there was little systematic bias.

12.5 CALCULATION OF DAILY RUNOFF VOLUME

Arid region flow regimes are typically unsteady, with very rapid changes in stage which may or may not be represented accurately by the stage recorder.

Daily runoff volumes should be calculated from discharges and time differences corresponding with points defining the shape of the discharge hydrograph, and not by the method of calculating the discharge corresponding with the mean stage for discrete time intervals.

Defining the typically non-linear discharge hydrograph shape may in theory only be possible to achieve by use of a large number of points on what may appear on the stage hydrograph to be a straight line. In selecting data points for manual processing, the reliability of the stage-discharge relationship is an important factor.

In the absence of stage records between two flow measurements made during a recession, interpolation between measured values can usefully be made assuming exponential decrease of the flow with the help of semi-logarithmic paper.

A detailed analysis of the errors involved in the determination of daily runoff volume taking into account all main sources of error will be found in Annex 2.

12.6 FLOODS

An extensive literature is available concerning the estimation of flood peaks and volumes of specified return period. Much of the discussion necessarily concentrates on fitting statistical distribution types to observed data. The development of empirical methods for converting storm rainfall to runoff (e.g. the unit hydrograph approach) continues to find favour in regions where basin rainfall input may be quantified by applying areal reduction factors to point rainfall distributions.

For large basins in arid regions there is little prospect of producing reliable flood estimates from rainfall estimates. The flood might result from high-intensity convective rainfall over part of the catchment area, or from relatively uniform, low-intensity rainfall over the whole area, or from any combination of circumstances between these extremes. For this reason, statistical methods coupled with geomorphological and flood level evidence are generally preferred to rainfall-runoff models. However, the latter may be useful in setting a physical upper limit on flood values estimated from a frequency distribution.

For full discussion of the theory underlying the selection of statistical distributions relevant to general flood hydrology, the reader is referred to WMO Technical Notes 81 and 98 and to FAO Irrigation and Drainage Paper No. 19 in addition to standard hydrological text-books. This Paper discusses only the problem of estimating flood frequencies in endogenous arid or semi-arid zones for which good quality data are usually lacking.

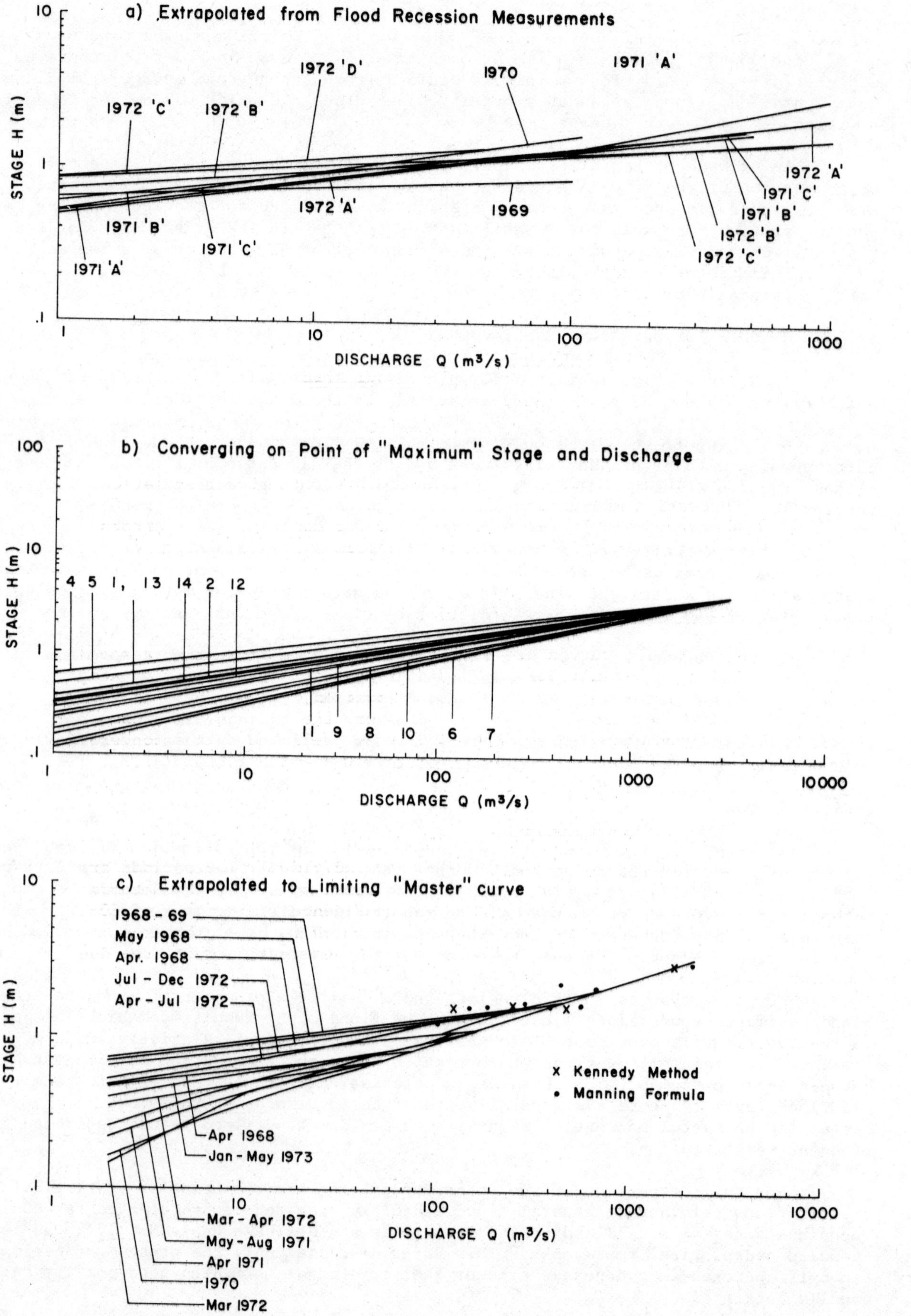

Fig. 12.4 Rating curves for a gauging station on an alluvial channel, Saudi Arabia

FIGURE 12.4

Rating Curves for a Gauging Station on an Alluvial Channel - Saudi Arabia

a. Extrapolation of flood recession measurements

The figure shows the futility of this form of extrapolation. Recession measurements produce typically flat ratings with correspondingly large errors in Q for moderate stage errors. The relationship beyond the data points is unknown.

b. Converging rating curves

The rating curves are shown to converge on a point corresponding with an estimate of the notional physical upper limit of stage and discharge. It should be recognised that there is no such fixed location in reality. However, the average difference between actual discharges and those predicted by the converging rating curves is likely to be less than with method (a).

c. Master rating curve

The rating curves established for individual flow periods are seen to convert on a curve established by slope-area methods (see Figure 12.3(b)). The 95% confidence limit curve of 60% would be a better master curve on which to base discharge estimates since it allows for a 33% underestimate of area due to scour.

In these conditions the problem usually resolves itself into that of estimating a mean annual flood peak and applying "growth factors" to this value according to the preferred distribution. In many cases it is necessary to apply all realistic methods of flood estimation in order to determine the sensitivity of the estimates to the underlying assumptions. In many cases, especially where spillway design is involved, the upper physical limit flood or that defined by the envelope curve of all regional or even worldwide floods may be resorted to by the engineer.

Methods discussed herein are as follows:

1. Statistical Methods:

(a) Single-station analysis.
(b) Regionalised flood frequency analysis.
(c) Partial-duration series.
(d) Stochastic generation of discharge sequences.

2. Deterministic Methods:

(a) Storm transposition and flood routing.
(b) Unit hydrographs.
(c) The Rational method.

3. Empirical Methods

Creagar's envelope curves.

12.6.1 Statistical Methods

12.6.1.1. Annual Series

The method is straightforward and consists in the adjustment of normal (Gauss), log-normal (Galton), extremal (Gumbel) log-extremal (Frêchet) or Pearson Type III distributions to the annual series by plotting on suitable probability paper or by using the general equation for hydrologic frequency analysis (Sokolov et al, 1976; WMO, 1974; Water Resources Council, 1967; Chow, 1964):

$$Q_T = \bar{Q} + \hat{\sigma} K_T$$

Where Q_T is the peak (Qp) or volume (Qv) for specified duration

T is the recurrence interval in years.

K is a frequency factor

and $\hat{\sigma}$ is the sample standard deviation.

The frequency factor K is a function of the recurrence interval T and the type of distribution. The technique is standard and applicable to data from all zones whether arid or humid.

An example of the estimation of the 100-year flood assuming an extremal distribution (Gumbel) is given in Example A.3.1 in Annex 3.

The same procedure may be followed to adjust a lognormal distribution by using appropriate frequency factors related to the skewness of the sample.

12.6.1.2. Regional Flood Frequency Analysis - USGS Method

The method, described by Dalrymple (1960), is applicable to the arid zone and, in incorporating all elements of regional hydrology and a homogeneity test, is a more powerful approach than the single-station methods described above.

The USGS method of regional flood frequency analysis is as follows:

- Select all stations within a hydrologically homogeneous area having at least five years of concurrent flood peak record.

- For each station, rank the flood peak values giving rank m=1 to the highest value.

- Calculate the return period T (recurrence interval) for each annual peak Qp using the equation $T = \frac{(n+1)}{m}$

- Plot Qp against T on Gumbel or Log-Gumbel (Frêchet) probability paper and draw best fit line visually.

- Read Qp corresponding with T=2.33 (mean annual flood) and T=10 and calculate the ratio Qp10/Qp2.33 for each station.

- Carry out homogeneity test to ensure station data are hydrologically similar, as shown in the example.

- Tabulate ratios $Qp_T/Qp2.33$ for several T values for each station.

- Determine the median ratio for each T.

- Plot median ratios against T on Gumbel or Log-Gumbel probability paper to derive regional $Qp_T/Qp2.33$ curve.

- To estimate Qp_T at any point within the region estimate $Qp2.33=\bar{Q}p$ and apply the ratio corresponding with the required return period.

This procedure is illustrated by Example A.3.2 and Figure A.3.1 (Annex 3).

In the USGS method $\bar{Q}p$ is estimated from regressions of this parameter on basin physiographic characteristics, eg area, slope, drainage density etc. In arid zones there is little possibility of establishing a relationship of $\bar{Q}p$ with basin area; generally there is an increase of $\bar{Q}p$ up to a particular area for individual basins. This area depends on rainfall characteristics, especially storm size in relation to basin area, and the infiltration and storage capacities of channel alluvium. It is more profitable to estimate $\bar{Q}p$ from channel morphology, local information and visual evidence of past floods, and use of slope-area methods, than to attempt to derive $\bar{Q}p$ values from basin physiography.

12.6.1.3. Regional Flood Ratios - Station Year Method

For a hydrologically homogeneous region within which annual flood events are not significantly correlated, the "station-year" approach may be applied, whereby the ratios $Qp/\bar{Q}p$ for each station are treated as a sample from a single distribution. The Gumbel or log-normal curve of best fit to the plotted values is used to derive the T-year ratio which is then applied to $\bar{Q}p$ as shown in Example A.3.3 and illustrated by Figure A.3.2. (Annex 3).

The validity of the result of fitting a log-normal or Gumbel distribution to the station flood ratios is debatable, especially where cross-correlation between stations requires that ratios applicable to floods occurring on a common day should be averaged. However the results appeared to be reasonable in the example given. The effect of a short record period has not been obviated, but the station-year concept inherent in the method, and supported by the evidence of the rainfall-depth-duration frequency analysis is believed to be a valid and logical use of data in the arid zone.

12.6.1.4. Partial-Duration Series

In many cases the use of partial-duration series analysis (in which all independent peak discharges above a specified threshold are analysed makes better use of the available data since all significant discharge peaks are utilised. Where long runoff records are available, partial-duration and annual series have been shown to be convergent for return periods greater than 10 years. This seems an attractive reason for using the method in arid regions where only a relatively short record is available but where flood peaks tend to be independent.

The results of applying the partial-duration series approach where only a few years of data are available, are sensitive to the method of calculating plotting positions P and to the assumed probability distribution.

12.6.1.5. Stochastic Generation

Stochastic runoff generation models may be used to synthesize a long instantaneous hydrograph record from which the annual maxima may be abstracted for analysis. The distribution of discharge peak Qp and discharge volume Qv is very strongly influenced by the QP (hydrograph rise) distribution inferred from the sample data. Model studies for a wadi in Saudi Arabia have shown that the T-year volume Qv for any duration derives from the occurrence of the T-year Qp and that, for practical purposes, the probability distribution of QP and application of a deterministic flow recession function are adequate for estimation of Q_{vT}.

12.6.1.6. Discussion of Statistical Methods

The results obtained in estimating Qp_{100} by using different statistical methods are given in Annex 3. Differences in results obtained by the regionalisation methods are due to the fact that the station-year method assumes the data samples to be "in series", whereas the USGS method assumes them to be "in parallel".

Other differences are due to sample size and treatment. The stochastic generation model can only be considered useful for Qp_T estimates where the data quality is good, the record is of relatively long duration (5 - 10 years) and contains a wide range of flow period durations and QP magnitudes. The degree to which the generated sequence represents the true runoff regime is affected more by the distribution type than are samples of actual or synthetic (regionalised) annual series.

Attour and Ibbitt (1966) concluded that the results obtained by fitting the Gumbel and log-normal distributions to the single-station data sets and fitting the Gumbel distribution to the basin ratio series were comparable. Higher Qp_T estimates were derived from fitting the log-normal distribution to the regional ratio series.

For approximate estimates the mean annual flood may be taken to be about 1 to 1.5 times its standard deviation. Flood peak magnitude may thus be assessed, very roughly, from an estimate of mean annual flood, based on channel hydraulics, multiplied by the relevant frequency factor plus one.

$$Qp_T \simeq (K_T+1)\ \bar{Q}p$$

Where K is taken from Table 5.3 of WMO Guide 168.

12.6.1.7. Non-Linearity of Flood Frequency Curves

Where long records of peak discharges are available, a sharp upward curvature may appear in the frequency curve. In desert areas this may be due to the fact that peak discharges from general storms of moderate intensity may produce relatively low annual peaks but infrequent storms of high intensity, centred over the basin, may cause an exceptionally high flood. In other words the lower portion of our empirical curves may be no indication of catastrophic events.

Beard (op.cit.) suggests that it may be best to omit zero Qps from flood-frequency calculations, to compute a tentative frequency curve for the remaining years and to adjust the exceedence frequency by the ratio of the number of years of record to the number of years with runoff. Beard also suggested that, in the cases of upward curvature, only the upper half of annual flood events should be fitted with a theoretical frequency curve.

In other cases it may be found that, on the contrary, the magnitude of flood peaks tends to increase very slowly in the upper region of the curve. This is common in semi-arid conditions. The rarer peak floods then tend to spill with the result that peak flows may be little greater than the bankfull discharge.

12.6.2 <u>Deterministic Methods</u>

12.6.2.1 Storm Transposition and Flood Routing

It is commonly found that major floods result from large scale weather systems which produce similar rainfall depths over large areas although the temporal and areal intensity variations within the storm period may be considerable, due to cell growth and decay.

In order to set an upper limit on the response to a rainstorm of this type, it is valid to define the isohyetal pattern of a historic storm and to transpose this to the area of interest. A time-area diagram based on isochrones of travel time from within the basin to the outlet is plotted and a storm profile based on a mass-curve of rainfall derived from one or a number of storm hyetographs is derived. Although unlikely to resemble the true rainfall pattern, the design storm is likely to represent a more severe situation the response to which is of interest. The storm profile is modified to account for surface retention losses and infiltration rate, if appropriate and is then convoluted with the derivative of the time-area diagram which forms a synthetic unit hydrograph. The resultant output is a prediction of the basin outlet hydrograph. Example A.3.4 illustrates the method (Annex 3).

12.6.2.2 Unit Hydrograph Method

This method usually assumes areally uniform percentage rainfall depth as a percentage of storm duration. Its usual application requires that an average initial loss and infiltration rate are assumed or derived empirically. Alternatively it may be assumed that an average percentage of rainfall will run off a given surface type for a given severity of rainfall and antecedent basin wetness. These assumptions may be reasonably valid in humid and temperate climates for design purposes but, except for very small basins, they are not valid for the truly arid zones and are difficult to substantiate in semi-arid regions.

Firstly, there are usually few data on which to base the estimate of an areal reduction factor, to be applied to point rainfall depth for specific return period and duration.

Secondly the surface response in arid zones may range from almost immediate runoff from hard rock areas to zero runoff from adjacent sand and gravel plains.

Thirdly, the surface and subsurface storage properties of drainage basins and channels are such that runoff generated on hard-rock areas may quickly infiltrate or be propagated downstream depending on the state of basin storage, as was discussed at the beginning of this Section.

The South African Hydrological Research Unit has calculated synthetic unitgraphs from 600 flood event records of 96 basins including the arid west (Midgley, 1972). The duration of effective rainfall was estimated by employing conventional S-hydrograph techniques where catchment rainfall data were lacking or poor. Unitgraphs of convenient unit duration were constructed in respect of each of which, basin lag T_L (defined as the time lapse between centres of area of input hyetograph and output hydrograph) was calculated. Dimensionless one-hour unitgraphs (with Q/Qp as ordinate and T/T_L as abscissa) were developed and were dimensionalised by multiplying the ordinates by the unitgraph peak Qp and the abscissae by T_L. Qp is estimated from $K_U A/T_L$ where A is basin area and K_U is a constant for each dimensionless unit hydrograph type.

The hydrograph of runoff resulting from a storm of given severity is calculated from the dimensionalised unitgraph and the net storm rainfall depth (taking account, if necessary, of the most probable of storm shape) by conventional unit hydrograph techniques.

The South-African method is built on a considerable volume of data and produces hydrographs for design purposes which are consistent with results obtained using statistical methods. The results of this study should not be transposed to arid areas elsewhere because the configuration of widespread storm rainfall producing runoff events is not well understood in much of the arid zone. Simplified hyetograph shapes assumed to vary within a basin only as to total rainfall depth, and convoluted with unitgraphs will produce runoff volume and peak estimates to which a return period is difficult or impossible to assign.

12.6.3 Empirical Formulas

12.6.3.1. Creagar Envelope Curves

Creagar (1945) derived his flood formula from an envelope curve of worldwide (mainly United States) historic flood peak maxima, as follows:

$$Qp = 46CA^{(0.894A^{-0.048})}$$

Where C is a coefficient whose value depends upon catchment characteristics, A is the catchment area in square miles and Qp is in ft^3/s. No particular significance attaches to the formula other than that it produces, for a particular value of C, an envelope which fits the limit of all floods experienced in a given area or in the world. C = 100 is commonly used for spillway designs but this coefficient cannot be assigned a particular return period. Applying this formula to Wadi Najran in Saudi Arabia, the station used for previous examples, produces a value of 14,350 m^3/s.

12.6.3.2 Rational Formula

This formula is written $Q = CIA/3.6$. Where C is a coefficient ranging from .05 to .95 depending upon catchment type and wetness, I is rainfall intensity mm/hour for a duration equal to the time of concentration T_c of the basin and A is the area in Km^2.

This formula is useful for small areas, not greater than storm size, for which point rainfall may be taken to approximate basin rainfall. It should not be used for basins larger than, say, 5-10 km^2. It should be made clear that the estimated intensity I is the mean for the duration considered. In practice, the storm intensity-duration profile is highly complex. Storms having the same depth and duration corresponding with a given return period will produce a wide range of Qp values each having a different return period.

13. CHARACTERISTICS OF SEDIMENTATION IN ARID ZONES

13.1 GENERAL CONSIDERATIONS

Sedimentation in rivers, natural and man-made lakes, the sea, on flood plains and in internal drainage basins is the product of erosion caused by precipitation and surface runoff, by tide, wave and seiche action and by wind.

Wind (aeolian) erosion and deposition are of concern in arid zone hydrology and agriculture:

- when wind-transported material is trapped in lakes or reservoirs
- as surface deposits providing readily-available material for rainfall runoff erosion
- by altering surface topography with consequent effects on infiltration and recharge, runoff, depth to water table, burial of springs and suppression of vegetation
- when mobile sand dunes threaten to obliterate agricultural development land.

Where aeolian erosion and deposition are concerned we can only draw attention to the existence of these phenomena, which are generally not susceptible of quantitative analysis.

In this Section we are concerned only with fluvial sedimentation, that which results from erosional processes caused by precipitation and runoff.

The units of sediment concentration and discharge used in this Paper are:

C - the concentration of sediments expressed as weight of sediment per unit volume of water-sediment mixture, kg/m^3.

Qs - the weight of sediment per unit of time, kg/s.

Other units are sometimes used in the literature, e.g. C as kg/tonne, percent, per thousand, ppm (parts per million) or N/m^3 ($Newtons/m^3$). The relationships between these units are:

$$1\ kg/m^3 = 1g/l = 0.1\% = 1‰ = 1000\ ppm = 9.81\ N/m^3.$$

In this Section, discharge (Q) refers to the volume discharge of the water-sediment mixture, sediment discharge (Qs) refers to the discharge weight of the sediments themselves and C is the concentration (= Qs/Q).

13.2 THE EFFECTS OF CLIMATE ON SEDIMENTATION

Detailed discussion of arid zone climatic regimes and the magnitude and variability of climatic elements is given in preceding Sections. The following discussion relates climate to erosion and sedimentation processes.

In winter rainfall areas, rainfall is generally the result of frontal activity associated with cyclonic depressions and intensities are generally low with rainfall covering large areas. However with increasing aridityleaward of mountain regions, the climate approximates (in all except ambient temperatures) the characteristics of summer rainfall areas, with conditions conducive to arid zone sedimentation.

In summer rainfall areas rainfall is largely the result of air-mass convergence or convective-type storms and is of generally high intensity. It is the floods resulting from these which produce the highest sedimentation yields.

The characteristic climate extremes of very low and very variable rainfall, high wind velocities, very high diurnal temperature and temperature range, have resulted in the elimination of all except highly adapted xerophytic vegetation. The tendency for wet and dry years to occur in cycles (see Section II) results in progressive elimination of all other vegetation unless a nearby source of repopulation of this exists. Where man and his animals are superimposed on this sensitive ecosystem the opportunity for establishment of vegetal cover is reduced.

That the ecosystem has a certain robustness however is evidenced by the fact that it still exists for man and his animals to damage after some ten thousand years. Man has destroyed trees for firewood, construction material and animal fodder and his animals have overgrazed and overbrowsed. Man has also played a notably beneficial role, by constructing terraces and bunds for soil and water conservation and for flood or local run-off irrigation, for several thousand years.

The relevance of this to sedimentation is that now, and for a long time past, the combination of climate and man's action, together with thousands of years of strong, dry winds and occasional rain has already removed much of what soil there was. Man has however, in many places, constructed terraces to trap the soil downstream to conserve it. Soil is used in the sense, defined by Pemberton - "unconsolidated mineral and organic material underlying the surface that has been sufficiently modified and acted upon by physical, chemical and biological agents that it will support plant growth". In the geologically recent past the soil removed by wind and water in arid zones included a high proportion of clay and silt fractions. The first is now located intermittently and is rarely abundant, so that arid zone erodible material is now of coarser particle size and is generally non-cohesive, except in man made and natural alluvial terraces and seasonal lake beds.

Arid lands tend therefore to be deficient in easily transportable fine sediment while at the same time the almost total lack of natural protection against erosion makes whatever material is easily transportable very readily available. The main sources of erodible material are now, in decreasing order of importance: wide fluvial plains (particularly those of large exogenous rivers such as the Nile, Indus, Tigris and Euphrates), more or less ancient terrace field systems, and lakes or valley sediments subject to rejuvenation drainage (ie resuming an actively eroding regime). In hill country that has not been terraced, most easily erodible fine material has virtually disappeared apart from remnants of Quaternary alluvial terraces on the flanks of valleys where flood flows seldom reach, so that it is mainly what is slowly being manufactured by normal weathering processes that is available. Terracing of hill sides is normally limited to semi-arid areas contiguous with downstream arid areas. It can be said as a generalisation that effective and wellmaintained terracing virtually eliminates both the erosion of the trapped fine sediments and the high discharge rates capable of transporting them. The role played by terracing has already been described in Section 3.1. Neglect (and subsequent inevitable collapse) of terracing where it exists would cause irreparable damage both to the agricultural productivity of the terraced hillsides themselves and to downstream areas which would receive the sediments accumulated during many hundreds of years.

13.3 THE SEDIMENTATION PROCESS IN ARID LANDS

The amount of sediment that can be transported by a flow of water is a function of the discharge, channel geometry, mean velocity, composition of bed material, granulometry of the material transported, water temperature and other variables at any given moment and cross-section of flow. The chief complications in arid zones are:

(i) the highly transitory hydrologic and hydraulic regime;

(ii) the great importance of extreme events due to their rare occurence, relatively short duration and the very high energies associated with them;

(iii) the insignificance of base flow, where it occurs, in relation to flood flows;

(iv) the asymmetry of the probability function of annual sedimentation, whose mode tends towards low values but whose variance is such that, during one year in 20 on average, the sediment load may be 20 times the median annual value.

In any arid area, but particularly in those in which high-intensity convective rainfall is predominant, variability of point rainfall processes is susceptible of straight-forward probability analysis as discussed in Section II. Uncertainty increases enormously when estimating areal rainfall and still further when taking rainfall intensity into account. If it were possible to obtain reliable areal and intensity data for rainfall the only bar to a straight-forward deterministic relationship between rainfall and runoff would be the practical impossibility of knowing in sufficient detail the surface characteristics of the catchment - slopes, material, depth to impermeable rock or water, vegetational cover etc. Since run-off depends on many variables other than rainfall its variability is considerably greater than that of rainfall and its probability distribution at any point is only reliably predictable on the basis of long series of flow observatons. Sedimentation depends on all the variables that affect run-off, including the impact of raindrops as the first link in the erosional chain and a complex variable termed erodibility. Not enough is known about these variables in arid zone conditions and since good sediment transportation data are few, and at best extremely fragmentary, the uncertainties are even greater.

13.4 FORMS OF SEDIMENTS

In sedimentation studies various forms of sediment are recognised and defined according to their behaviour. The definitions based on Einstein's work given here are preferred but it must be emphasised that some of them are of minor importance in arid zone sedimentation studies.

Contact load is defined as material rolled or slid along a stream bed in continuous contact with it; saltation load is material bouncing along the bed or moved by bouncing particles: together these make up bed load. Suspended load is material in suspension. Einstein also defines bed material load in suspension as composed of material of particle sizes corresponding with those found in appreciable quantities in the shifting portion of a stream bed. Wash load is that part in suspension consisting of particles smaller than those found in the shifting portion of a stream bed (see Section 14.2).

In arid zone drainage channels, with beds of unconsolidated and often coarse sediments, the bed material load may play a major role behaving in the same manner as a very viscous fluid. Sediment in transport, either as bed load or bed material load in suspension, may occupy a large part of the cross-section of flow at and just before peak discharge. However the greater part of sediment transport is of short duration, reducing rapidly with reduction in discharge. The amount of material moved, although considerable, may travel a comparatively short distance or may result in a lowering of a considerable length of the channel bed (see Section 13.8).

13.5 POTENTIAL AND ACTUAL SEDIMENT TRANSPORT

The instantaneous rate of sediment transport depends on the hydraulics of the flow, the erodibility of the channel and on the quantity of sediment available for transport from the catchment contributing to the flow. It must be remembered that the fluviatile erosion, which is the source of sedimentation with which we are mainly concerned, is initiated by rainfall. Shallow, low-velocity sheet (or overland) runoff resulting from rainfall of whatever intensity can only remove the finer erodible materials. A denuded hard-rock catchment may already have suffered such devegetation and consequent erosion that there remains insufficient erodible material to satisfy the capacity of even moderate run-off to transport solids. The higher rainfall areas of a catchment may be so protected by dense vegetation and/or terracing or flood control structures that the effect is the same. Conversely a denuded catchment with large expanses of Quaternary deposits or erodible marls may be able to satisfy the sediment transport capacity of run-off under all run-off conditions.

It is clearly essential that the nature of the surface of the catchment concerned should be known in sufficient detail, from aerial photography and ground reconnaissance, for the identification and estimation of areas with varying degrees of erodibility.

13.6 RELATIONSHIP BETWEEN DISCHARGE AND SEDIMENT TRANSPORT

There is a widespread tendency to rely on empirical relationships which express average annual sediment weight as an exponential function of the annual flow volume. Apart from questions of availability of sediment this approach needs caution in conditions where virtually all discharge derives from sporadic spatial and temporal combinations of discrete storm events. The only satisfactory method is the establishment of relationships linking instantaneous values of sediment transport rate or amount with hydraulic parameters derived experimentally or analytically, notwithstanding the inevitable simplifying assumptions, approximations and imprecisions involved in the process. These relationships may then be applied to an estimated instantaneous hydrograph or flow-duration curve to provide an estimate of total sediment transport.

13.7 THE SEDIMENTATION OF STORAGE STRUCTURES

Under conditions of rare, highly variable and very strictly seasonal runoff events the whole of a flood may reach a storage structure when the latter is empty or nearly so. The theoretically normal mechanisms of reservoir sedimentation: of bottomset fine materials over the whole reservoir area; of foreset coarse materials deposited on relatively steep slopes at the upstream end; and of topset material ultimately deposited over the prograding foreset beds, (Figure 13.1) do not always exist. For this normal mechanism to occur requires the presence of some still water. Consequently the sediment profile of an arid zone reservoir after some years may be very different from that normally found in temperate areas. There are relatively few well documented examples of reservoirs in arid zones deriving their water from arid catchments. However the results of studies in Tanzania (Anders Rapp et al,1972) have shown that a different depositional regime prevails in semi-arid conditions where reservoirs are empty for long periods (Figure 13.2). Study of the characteristics of arid zone reservoir sedimentation is particularly important for the design and operation of "spate breaker" structures intended to detain endogenous flood waters for controlled flood irrigation or enhancement of groundwater recharge potential.

Under conditions of floods generated within the arid zone and carrying almost exclusively coarse sediments there is a different depositional process. After the first few floods have deposited a coarse-grained and irregular bottom-set, all succeeding floods arrive on a progressively increasing area of highly permeable coarse materials: this area increases by a delta formation process in an upstream direction and laterally within the reservoir area. The effect on succeeding floods is two-fold:

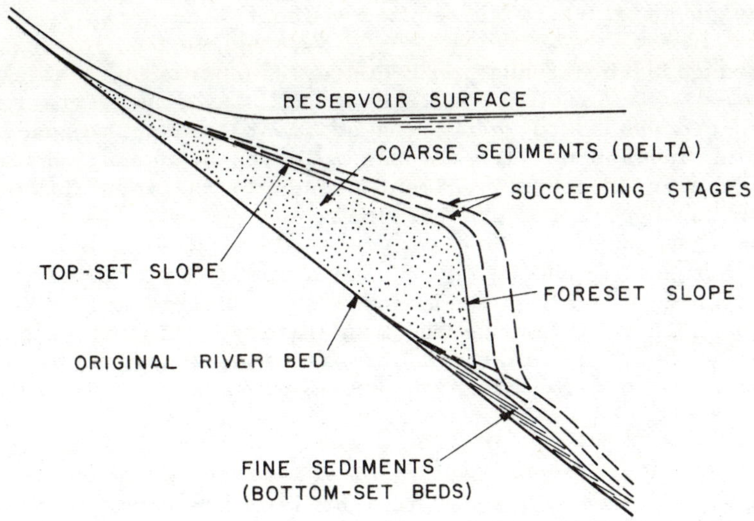

Fig. 13.1 Reservoir delta form (after ASCE Sedimentation Manual, Fig. 5.4.4)

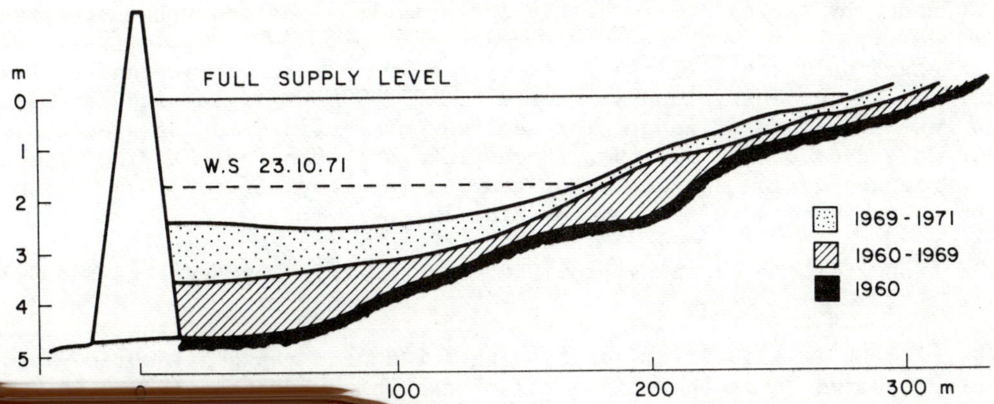

Fig. 13.2 Long profiles of Kisongo reservoir, 1960-1971. The lack of deltaic growth is noteworthy.
Vertical exaggeration: x 20

These figures illustrate the different sediment depositional regimes of reservoirs on perennial and ephemeral water-courses.

Figure 13.1 shows the stratification and mode of accumulation of sediment in reservoirs ordinarily operated for long periods of time near full supply level. Here the coarse sediments are deposited on reduction of inflow velocity due to the static water body.

Figure 13.2 shows stratification in a reservoir (in the semi-arid zone of Tanzania) which is empty or nearly so for long periods of time. Here the highly sporadic inflow spates and their sediment load are disturbed more evenly over the reservoir floor in the relative absence of velocity retardation by impounded waters.

(i) they tend to increase in width, with reduced velocity and depth

(ii) infiltration increases with attenuation of water discharge.

The effect of each process is to reduce the sediment transport capacity of the flowing water, the reduction increasing with the accumulation of deposited sediments. The result is the sudden deposit of coarse sediment masses in the form of bars whose height is dependent on the depth of water and bed load.

The rate of reservoir sedimentation depends on a number of climatic, geographic and human factors in addition to the size, shape and operational method of the reservoir. These factors make it impossible to transpose the mean rate of sedimentation from one basin to another, especially when the variability of arid zone runoff is such as to require a very long period of record to obtain an annual or seasonal mean runoff rate with any precision.

However, it is useful to study the results of reservoir sedimentation studies carried out in arid regions in order to gain an insight into the possible scale of the sedimentation problem. Table 13.1 shows, for illustrative purposes, the sedimentation which has taken place in a number of reservoirs located in arid and semi-arid parts of the world.

The data for the United States relate to reservoirs in the arid and semi-arid south-west ranging in initial storage capacity from 6 to 1900Mm^3. Runoff data are not available but, although sediment inflow for individual years is not known, the accumulation of sediment is shown by Table 13.1 to be highly variable as between the survey periods.

Beverage and Culbertson (1964) have reported "hyperconcentrations" of suspended sediments (defined as concentrations in excess of 400,000 ppm) from tributaries of the Rio Grande in New Mexico and Arizona. The same authors defined mudflows as those with concentrations in excess of 800,000 ppm. Such flows are comparable to the fine sand and silt flows experienced during current meter gauging in alluvial wadi beds of the Middle East, when the channel bed material becomes fluidised.

A selection from the examples from USGS records cited by Beverage and Culbertson is given in Table 13.2.

There are many examples from survey data for a wide range of debris - sediment traps and storage resrvoirs in the arid parts of the United States. These show that complete sedimentation of smaller capacity reservoirs may occur within a matter of hours. In the Report of the Working Group of the Technical Committee on Surface Water (Australian Water Resources Council) 1969, a number of examples of significant sedimentation of major storage dams is given including:

- Loss of 58% of the original 112Mm^3 storage capacity of MacMillan Reservoir in New Mexico over a period of 46 years.

- Accumulation of 1730Mm^3 in Lake Mead, USA, (4.6 percent of its capacity) over a period of 14 years.

- Accumulation of 52Mm^3 in Fodda Dam reservoir, Algeria, (22 percent of its initial capacity) after 18 years.

These examples illustrate the importance of assessing the reservoir sedimentation rate when contemplating storage solutions to water resource development problems. The particular difficulty is the arid zone characteristic of high variability of emphemeral channel flow in time and space. The results of a stochastic generation study carried out for Wadi Najran in Saudi Arabia have been used in this publication to illustrate many of the problems of water resource assessment and development. Figure 13.3 shows the estimated probability distribution of annual sediment discharge. Based on a large number of samples and extrapolation beyond the range of data by calculation of Einstein's Bed Load Function. It is important to realise that, the risk of project failure cannot be estimated by calculating the useful life of the reservoir on the basis of the mean annual sedimentation rate.

13.8 THE SEDIMENT PROCESS IN ALLUVIAL CHANNELS

Much the same depositional process occurs where flood flows traverse coarse, unsaturated sediments with sediment transport capacity reduced progressively by bed losses or by overbank spill of the upper layers of water containing less sediment. Bars formed in active reaches of alluvial channel beds are removed by subsequent floods with adequate sediment transport capacity.

Observations of temporal changes in bed level in arid zone alluvial channels show that the volume of bed material moved downstream during flood periods may be very large indeed. Investigations have shown that scour and deposition may occur over a long tract of alluvium. For example, the average bed level at observation boreholes covering a 50 km long tract of channel alluvium about 200m wide in W. Bishah, Saudi Arabia was lower by about half a metre following floods with a peak discharge corresponding with a return period of about 5 years, in 1968. There is no evidence of the scoured material having been redeposited elsewhere laterally or longitudinally within the same tract and the conclusion may be drawn that some 50 km x 200 m x 0.5 m = 5 Mm^3 of bed material was removed from the tract and deposited downstream during this period. Consideration has been given to the construction of a dam within this reach for irrigation and recharge control. If such a dam had been in existence its useful life would undoubtedly have been dramatically reduced during this period.

Measurement of suspended sediment transport in the Hablehrud River in Iran (UNDP/FAO, 1971) have shown that very high concentrations and sediment runoff may occur during flood flows. At Bonekuh gauging station, where the drainage area is 4095 km^2, the average runoff for a 22-year period was estimated as 250 Mm^3 (equivalent to about 60 mm). In the one hydrological year for which rainfall data are available the mean basin precipitation was 407 mm and the runoff 67 mm. As part of an intensive hydrological study, measurements of water discharge Q, sediment discharge Qs and salt concentrations were made at one-hour and sometimes half-hour intervals during two separate flood periods. For the second flood the results for a 24 hour period were:

Total runoff	=	2,32 Mm^3
Average discharge $\bar{Q}$	=	26.9 m^3/s
Peak discharge Qp	=	120 m^3/s
Total suspended sediment discharge	=	1.04 M. tons.
Average suspended sediment discharge Qs	=	12000 kg/s
Maximum suspended sediment discharge Qsp	=	84000 kg/s
Average Concentration C	=	447 kg/m^3
Maximum Concentration C	=	870 kg/m^3

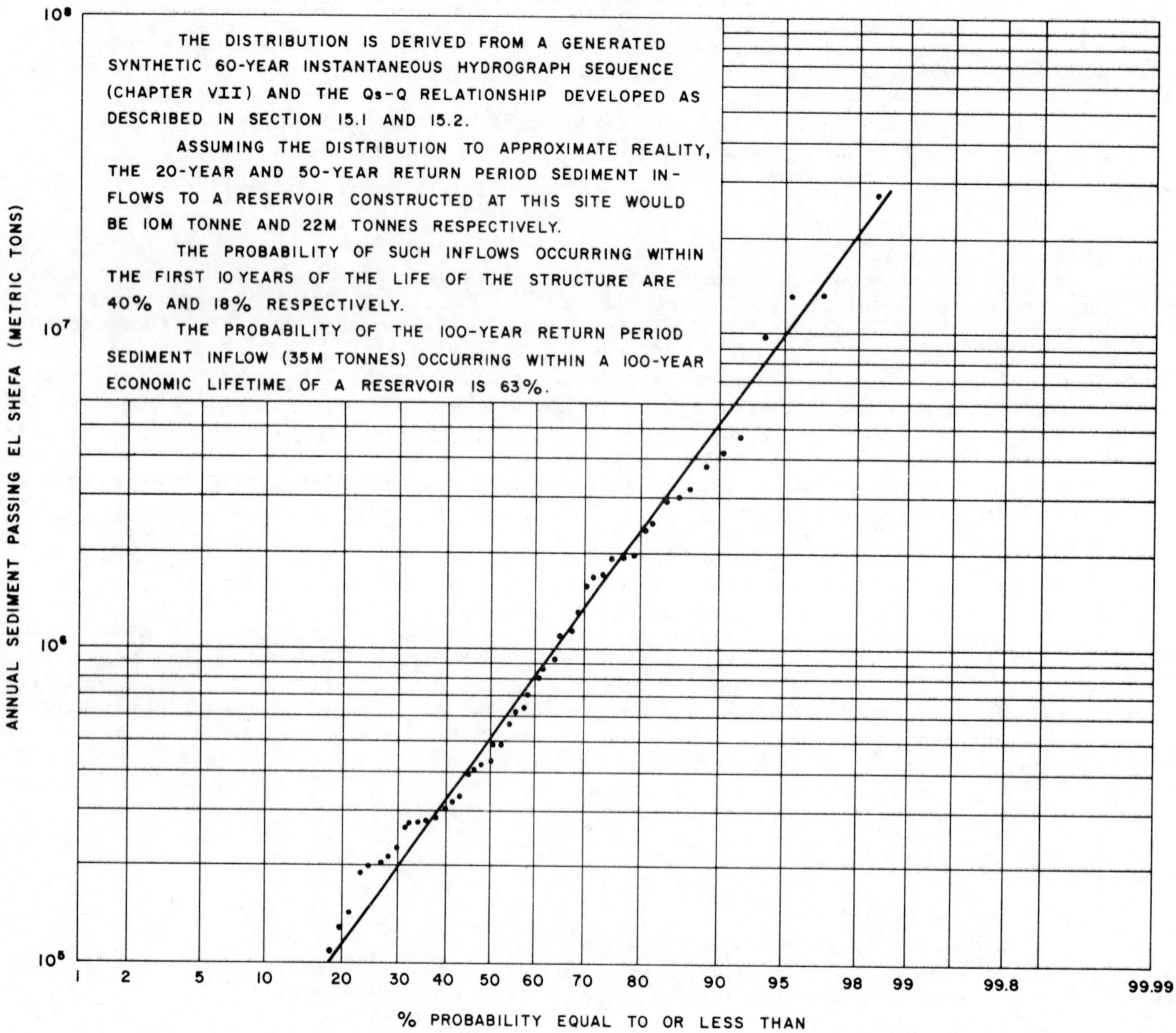

Fig. 13.3 Distribution of sediment runoff at El Shefa - 60 year sequence

TABLE 13.1

SEDIMENTATION OF RESERVOIRS IN SEMI-ARID ZONES

BASIN	NAME	STATE & COUNTRY	NET DRAINAGE AREA km^2	SURVEY DATE	PERIOD YEARS	CAPACITY Mm^3	SEDIMENT ACCUMULATION Mm^3	ANNUAL UNIT RATE m^3/km^2	ANNUAL UNIT RATE t/km^2	C/I(1)	ANNUAL STORAGE Loss(2) %
COW BAYOU 4	BRAZOS	TEX. USA	12.6	7.56	-	2.36	-		-	1.24	
				9.69	13.2	2.26	0.10	601	579	1.19	0.3
				7.75	5.8	2.23	0.03	410	506	1.18	0.2
JEMEZ CANYON	JEMEZ R.	N.M. USA	2570	10.53	-	144.64	-	-	-	3.01	
				8.59	5.9	140.52	4.12	272	762	2.93	0.5
				12.65	6.3	139.21	1.31	81	110	2.90	0.1
				1.75	9.1	130.93	8.28	354	480	2.73	0.6
UPPER HONDO	RIO SALADO CREEK	N.M. USA	121.6	5.59	-	6.136	-	-	-		
				10.59	.42	6.104	0.032	627	432	x	1.2
				5.61	1.58	6.042	0.062	324	226	x	0.6
				3.62	.83	5.945	0.097	961	673	x	1.9
				2.63	.92	5.919	0.026	232	166	x	0.5
				1.65	1.92	5.861	0.058	248	173	0.97	0.5
				5.72	7.33	5.334	0.527	591	x	0.89	1.2
ROOSEVELT	SALT R.R TONTO CREEK	ARIZ. USA	14400	5.09	-	1878.4	-			1.886	
				12.14	5.7	1845.4	33.0	402	509	1.853	0.3
				10.16	1.8	1801.8	43.6	1682	2084	1.809	1.3
				9.25	8.9	1759.5	42.3	330	417	1.767	0.3
				1.35	9.3	1749.8	9.7	72	90	1.757	0.1
				1.39	4.0	1725.7	24.1	418	529	1.733	0.3
				1.46	7.0	1704.9	20.8	206	260	1.712	0.2

TABLE 13.1 (Continued)

NAME	BASIN	STATE & COUNTRY	NET DRAINAGE AREA km^2	SURVEY DATE	PERIOD YEARS	CAPACITY Mm^3	SEDIMENT ACCUMULATION Mm^3	ANNUAL UNIT RATE m^3/km^2	ANNUAL UNIT RATE t/km^2	C/I(1)	ANNUAL STORAGE Loss(2) %
IKOWA	MAJENJEULA/ KINYASUNGWE	CENTRAL PROV. TANZANIA $\bar{P}a$=570 mm	640	57	-	3.807	-	-			
				60	3	3.111	0.696	362	x	x	6.13
				63	3	2.740	0.371	193	x	x	3.23
				69	6	2.315	0.425	111	x	x	1.85
IMAGI	KINYASUNGWE	"	1.5	26	-	0.174	-	-			
				71	45	0.136	0.038	601	x	x	0.52
M.SALATU	"	"	8.67	44	-	0.388	-	-			
				71	27	0.300	0.095	406	x	x	0.90*
MATUMBULU	"	"	18.12	60	-	0.383	-	-			
				71	11	0.264	0.119	729	x	x	3.4
KISONGO	L.MANYARA DRAINAGE	NORTHERN PROV. TANZANIA $\bar{P}a$ 870 mm	9.3	60	-	0.121	-	-			
				69	9	0.084	0.037	442	x	x	3.34
				71	2	0.072	0.012	645	x	x	4.75
UMBERUMBERKA	DARLING	NSW. AUSTRALIA	407	15	-	13.57	-	-			
				61	46	8.64	4.93	263	x	x	0.79

* 8750 m^3 excavated from reservoir in 1953

Note (1) C/I is ratio of reservoir capacity to average annual inflow.
(2) Annual storage loss as a percentage of the original capacity.

TABLE 13.2
SUSPENDED SEDIMENT CONCENTRATIONS > 400,000 ppm
(From Beverley and Culbertson, 1964)

Station	Date	Time	Water Temp. in °F	Approx Discharge in cfs (f^3/s)	Suspended Sediment Concentration in ppm.	Concentration of Sands (greater than 0.062mm) in ppm.	Remarks
Rio Puerco below Cabezon, N.Mex	Aug. 6, 1951	0115	-	340	414,000	-	-
Rio Puerco near Bernario, N.Mex	Jul. 24, 1949	0600	70	1,980	418,000	-	-
		0700	70	2,380	412,000	-	-
Rio Salado near San Acacia N.Mex.	Aug. 13, 1953	0200	68	5,680	405,000	230,000	-
Paria River at Lees Ferry, Ariz.	Jul. 9, 1950	1300	-	218	480,000	-	Peak of 840 cfs at 0730h
	Jul. 10, 1950	1500	-	455	515,000	-	Peak of 1,450 cfs at 1200h
	Aug. 27, 1952	0830	-	632	646,000	-	Peak of 1,450 cfs at 0300h
		1100	-	316	533,000	-	Peak of 1,450 cfs at 0300h
		1320	-	280	418,000	-	On crest of smaller peak
	Jul. 14, 1953	1730	-	127	599,000	-	Peak of 259 cfs at 1200h
		1830	-	107	506,000	-	Peaks of 259 cfs at 1200h
		2000	-	94	441,000	-	Peak of 259 cfs at 1200h
	Jul. 16, 1953	1000	-	720	587,000	-	Peak of 920 cfs at 0830h
		1100	-	610	546,000	-	Peak of 920 cfs at 0830h
	Jul. 18, 1953	1745	-	242	489,000	-	Peak of 1,440 cfs at 1000h
	Aug. 1, 1953	0850	-	1,060	530,000	-	Peak flow
		1130	-	560	455,00	-	-
	Jul. 25, 1954	0900	-	3,310	514,000	-	Peak flow
	Oct. 8, 1954	1630	-	435	427,000	205,000	-
	Jul. 25, 1955	1230	-	288	529,000	-	Peak of 491 cfs at 0830h
	Aug 26, 1955	1110	75	408	468,000	300,000	Peak of 1,780 cfs at 0530h
	Jul. 29, 1956	1930	-	1.1	400,000	-	Peak of 30 cfs at 1330h
Little Colorado River nr. Cameron, Arizona.	Aug. 25, 1957	0900	70	5,700	620,000	-	Peak flow from Moenkopi Wash.
	Aug. 22, 1958	1530	74	5,000	438,000	-	Peak flow from Moenkopi Wash
		1600	75	4,400	415,000	-	Peak flow from Moenkopi Wash

14. SEDIMENT TRANSPORT MEASUREMENT

14.1 GENERAL

Sedimentation is a major factor affecting the design, efficiency and useful life of a range of structures such as:

- detention storage dams required to serve as spate-breakers for semi-controlled irrigation and/or recharge projects,

- dams creating reservoirs whose purpose is to regulate the supply of water to irrigable areas.

- dams creating reservoirs whose function is essentially one of flood protection,

- diversion barrages whose purpose is to divert flood waters into flood irrigation basins or to control the distribution of base flows,

- terrace formation structures whose purpose is to store sediment for soil and water conservation etc.

In considering methods of sediment measurement it is important to have these engineering solutions to agricultural development problems in mind, in order that the hydrologist's measurement programme is capable of producing the required results. It is also valid to consider whether, in view of the low degree of correlation between water and sediment discharge, water discharge and concentration and the relatively acceptable fit of formulas such as the Einstein Bed Load Function (described in detail in Section 15) to such measurements as have been made, it is worth-while embarking on any sort of sediment measurement programme other than that of obtaining the basic data required for such formulas.

However, despite the possible usefulness of sediment load formulas, a runoff station at a proposed dam or diversion site should include equipment for water and sediment discharge measurement capable of producing acceptable results over the discharge range. Such stations must be staffed with well-trained observers. The measurement of water discharge is difficult enough and is subject to large errors of interpretation. The measurement of sediment discharge is even more difficult and the interpretation of results subject to even greater uncertainty. Such interpretation is required to produce conclusions (or ranges of uncertainty) as to the useful life, and thus the economic viability, of structures.

14.2 SEDIMENT TRANSPORTATION MECHANICS

Rigid channels in rock or cohesive soils have relatively unchanging geometric and roughness properties which can be quantified without undue difficulty.

However, in the case of flows contained in channels composed of non-cohesive granular material, the interaction between the flow and boundary moulds the geometry of the channel and determines its hydraulic roughness. In these conditions large and small-scale geometric properties of channels and the depth, velocity and sediment transport are interdependent.

In mobile bed conditions this inter dependence is a major obstacle in development of empirical relationships between discharge depth, velocity, bed sediment size and sediment load, since there is no single independent variable with which sediment discharge is uniquely related (see Section 15).

Such relationships are, however, essential for estimation of sediment transport at runoff gauging stations since there is little possibility of reliable estimation by continuous sampling. There is a large body of literature describing relationships, derived experimentally in laboratory flumes and from measurements in alluvial channels, between sediment discharge and hydraulic and channel parameters. The Committee on Sedimentation of the Hydraulic Division of the American Society of Civil Engineers has prepared a Sedimentation Manual (1975) which should be regarded as a standard reference on the subject.

In the measurement and estimation of sediment discharge it is necessary to adopt a standard terminology. The definitions recommended and used in this publication, are as follows:

- Bedload - the material which moves as contact load or saltation load (Section 13.4) in almost continuous contact with the stream bed, being rolled or pushed along the bottom by the force of water.
- Suspended load - sediment that is supported by the upward component of turbulent currents and that stays in suspension for appreciable lengths of time.
- Bed material - the sediment mixture of which the channel bed in composed.
- Wash load - that part of the suspended load which consists of grain sizes finer than those found in the bed.
- Bed material load - that part of the total sediment load which consists of grain sizes represented in the bed and equals the sediment transport capability of the flow.
- Total sediment load - bed load plus suspended load or bed material load plus wash load.

Table 14.1 illustrates these terms and the manner in which they overlap.

TABLE 14.1

(After Cooper and Peterson, 1970)

Sediment Load Terminology

<table>
<tr><td colspan="2" rowspan="2"></td><td colspan="2">CLASSIFICATION SYSTEM</td></tr>
<tr><td>BASED ON TRANSPORT MECHANISM</td><td>BASED ON PARTICLE SIZE</td></tr>
<tr><td rowspan="3">TOTAL SEDIMENT LOAD</td><td>WASH LOAD</td><td rowspan="2">SUSPENDED LOAD</td><td>WASH LOAD</td></tr>
<tr><td>SUSPENDED BED MATERIAL LOAD</td><td rowspan="2">BED MATERIAL LOAD</td></tr>
<tr><td>BED LOAD</td><td>BED LOAD</td></tr>
</table>

Turbulence in uniform channels is generated in the water contact with bed and banks in the form of large eddies which continually shear off from the boundary, move upwards and downstream degenerating into smaller and smaller eddies and finally decaying as microscopic vibrations and dissipate as heat. The overall result of the turbulent currents is a complete mixing or transfer of momentum (or any other quality) carried in the flow (Oltman, 1962).

In the case of alluvial streams some sediment particles are carried upwards by the turbulent currents against gravity acting to return the particle to the bed. The net effect of the turbulent vertical velocity acting on the bed is to produce a gradient in the quantity of particles in suspension, decreasing with vertical distance upwards from the bed, at which the sand particle concentration is a maximum.

Gravity makes sediment particles settle in water. The rate of velocity of settlement of a single particle in still water depends on its shape, specific gravity and size and is called the fall velocity w.

For quartz sphere diameters less than about 3×10^{-4}m w varies approximately as the square of the diameter and for larger particles, approximately as the square root of the diameter. In the transitional zone, 2 to 5×10^{-4}m, w values range between the square and square root variation. Figure 14.1 shows an example of how the fall velocity controls the vertical distribution of various sizes of suspended sediment at a single vertical. Figure A.4.8 shows the relationship between the diameter of quartz spheres and fall velocity in water.

The true fall velocity of a sediment particle depends upon temperature, sediment concentration, turbulence, particle shape etc. and may differ considerably from the values represented in Figure A.4.8 (Annex A.4) for the same sieve size.

Figure 14.2 shows a typical stream cross-section and the generalised distribution of velocity, sediment concentration and sediment discharge. This illustrates the problem faced by the hydrologist in arid zones where the concentration gradient, and thus the measurement problem, is greatest.

Estimation of sediment discharge is normally based on the relationships between water discharge Q and sediment concentration C established on the basis of sediment sampling throughout the discharge range. In some channels this relationship may be relatively clearly defined and this is more likely to be so in alluvial channels in which the sediment in suspension is a maximum for the hydraulic and physical parameters of the flow and channel because the supply is unlimited.

In stable channels, for example those eroded in rock, the concentration for a particular discharge may depend more on such factors as the sediment supply, the hydrograph shape, the intensity of rainfall causing the flow and the Q vs C relationship may be obscure or non-existent. For many arid zone gauging stations the supply of sediment is virtually unlimited and the sediment rating curve may be reasonably well-defined although the position of the Q vs C curve within each flood period and seasonally may be different as discussed in Section 15.1.5 and illustrated on Figure 15.3.

14.3 FIELD OPERATIONS

14.3.1 Commonly Used Sampling Techniques

Because of the difficulty in directly measuring quantities of load in suspension, common practice is to measure water discharge and concurrent concentrations of suspended load and to compute the suspended load from the relationships established, for desired intervals of time. Therefore, sediment measuring stations must always be stream gauging stations at which continuous records of water stage and virtually simultaneous measurements of water and sediment discharge are collected.

Field operations for an adequate programme of sediment sampling require:

- the selection of appropriate equipment for the gauging station,
- the scheduling of sample collections by hydrologist and local observer and

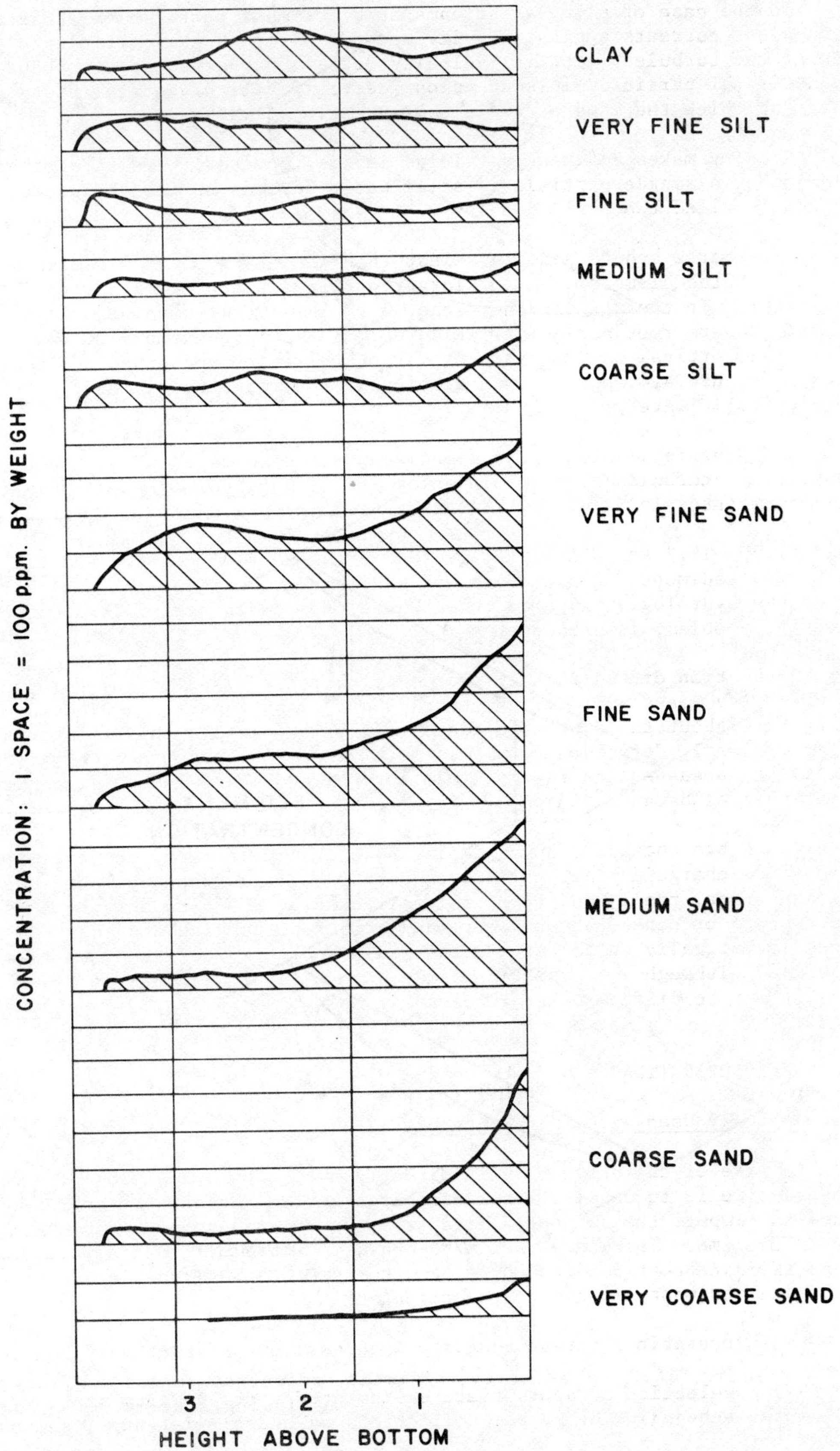

Fig. 14.1 Distribution of sediment in the Missouri River at Kansas City, USA

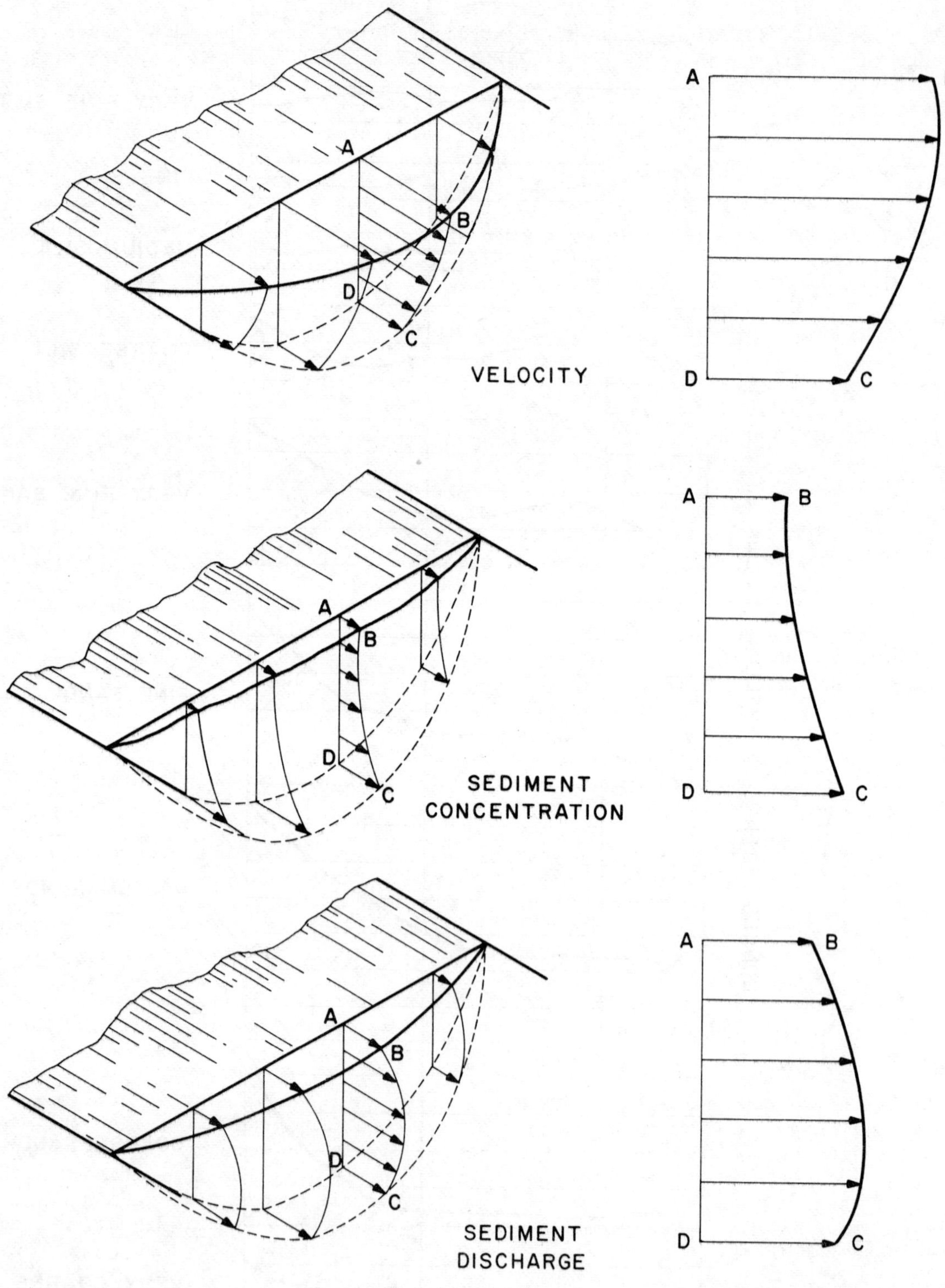

Fig. 14.2 Velocity, sediment concentration and sediment discharge in flowing streams

- the transport of sample to a central laboratory for analysis.

Because decisions about field procedures hinge on the method used to compute a record of sediment transport it is appropriate at this point to discuss methods of computation. The method commonly used by hydrological agencies is to establish discharge/concentration (Q/C) relationships (sediment rating curves) for the station and to apply these relationships to discharge rates at points on the hydrograph selected according to the rate of increase or decrease in discharge.

The sediment load consists of the suspended load and the bed load. The suspended load may be estimated from the sediment concentration of samples of the water-sediment mixture taken from the stream plus an estimate of the highconcentration in the unsampled zone within a few centimetres of the stream bed.

Two methods are used to collect samples: the "equal-transit-rate" (ETR) and the "equal-discharge increments" (EDI) methods. The scheme of sampling for each of the two methods is shown by Figure 14.3.

In the ETR method the observer selects verticals equally spaced across the section and, using an equal rate of lowering and raising the sampler, collects one sample at each vertical. The lowering and raising rates need not be the same but each should be constant and the total transit time should about right for the depth and stream velocity to avoid overfilling the sampling bottle. The transit rate should be the same for each vertical. The sampler is designed to admit water-sediment mixture at a rate nearly equal to the velocity of the stream at the sampler intake. Thus, the sample from each vertical in the ETR method is automatically weighted by discharge and the composite of all the samples is a nearly-correct sample of the mean stream concentration $\bar{C}$. The number of verticals into which the observer divides the whole section depends on the degree of variation of concentration across it and the rate of change of stage.

The number of verticals must be decided by the hydrologist on the basis of the mean stage-discharge relationship and past experience of the probable range of variability in concentration across the section. The ETR method requires only an estimate of the maximum mean vertical velocity to compute the constant transit rate.

In the EDI method the observer makes a measurement of water discharge and from it determines spacing of verticals such that the verticals divide the water discharge into approximately equal segments. He then lowers and raises the sampler from top to bottom and back at the centroid vertical of each equal discharge zone. The quantity of sediment in each centroid sample should be proportional to the water discharge in the segmental section - if the location of verticals has been correctly determined. Equal volumes of water-sediment sample should be taken by ensuring that the transit rate and total transit time are properly selected for the centroid depth and velocity. The composite sample represents the mean concentration in the stream section.

The method recommended for use by hydrological agencies in arid zones is the ETR method with sample spacing depending on flow conditions. The transit rate may be estimated using the alignment chart given as Figure 14.4 and modified as necessary during sampling on the basis of experience gained. In practice the EDI method can only be applied where the rate of change of discharge permits rapid calculation of discharge distribution across the section or where this distribution has been determined from previous measurements.

There are two basic sediment sampler types - those with a valve for control of the intake and those with an open, uncontrolled intake. The former, called a point-integrating sampler, is for collecting a sample during a finite time interval, at a point in the vertical or for taking a depth-integrated sample over part of a stream vertical and is used mainly on deep streams or for special field investigations. The latter,

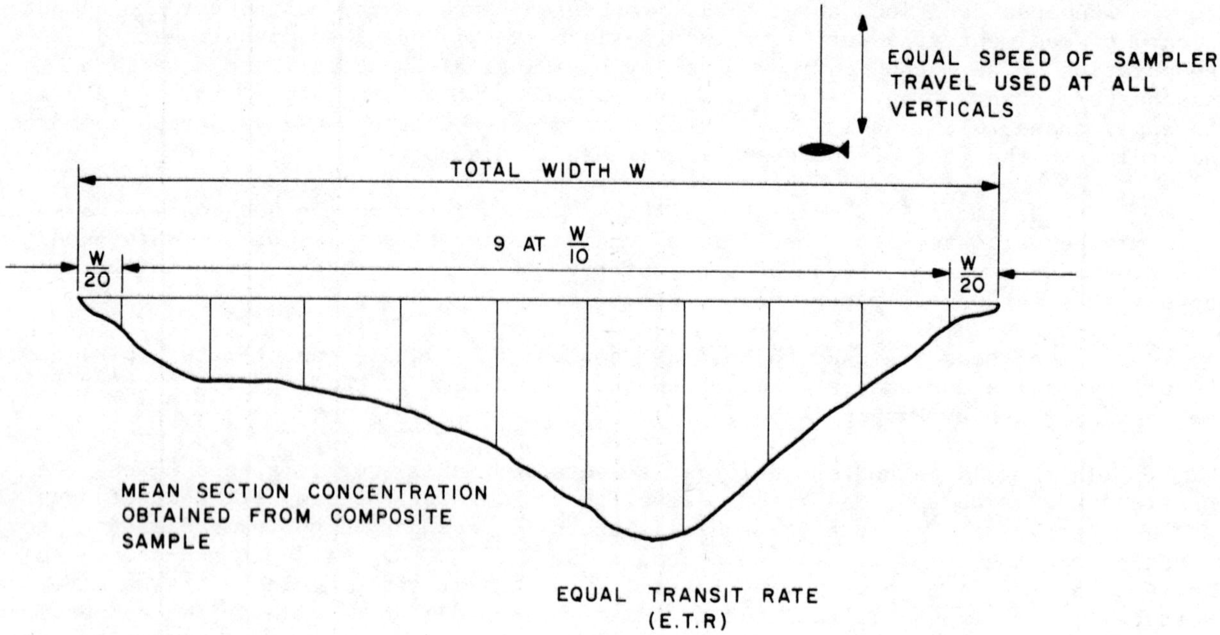

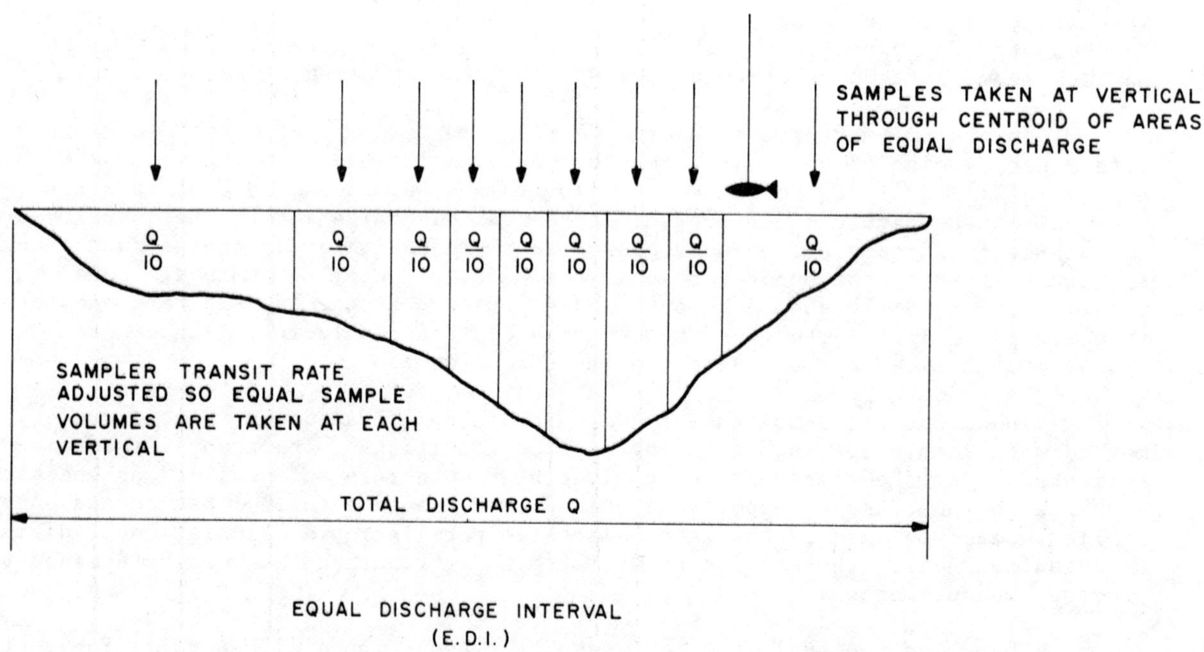

Fig. 14.3 Two methods of measuring mean sediment concentration

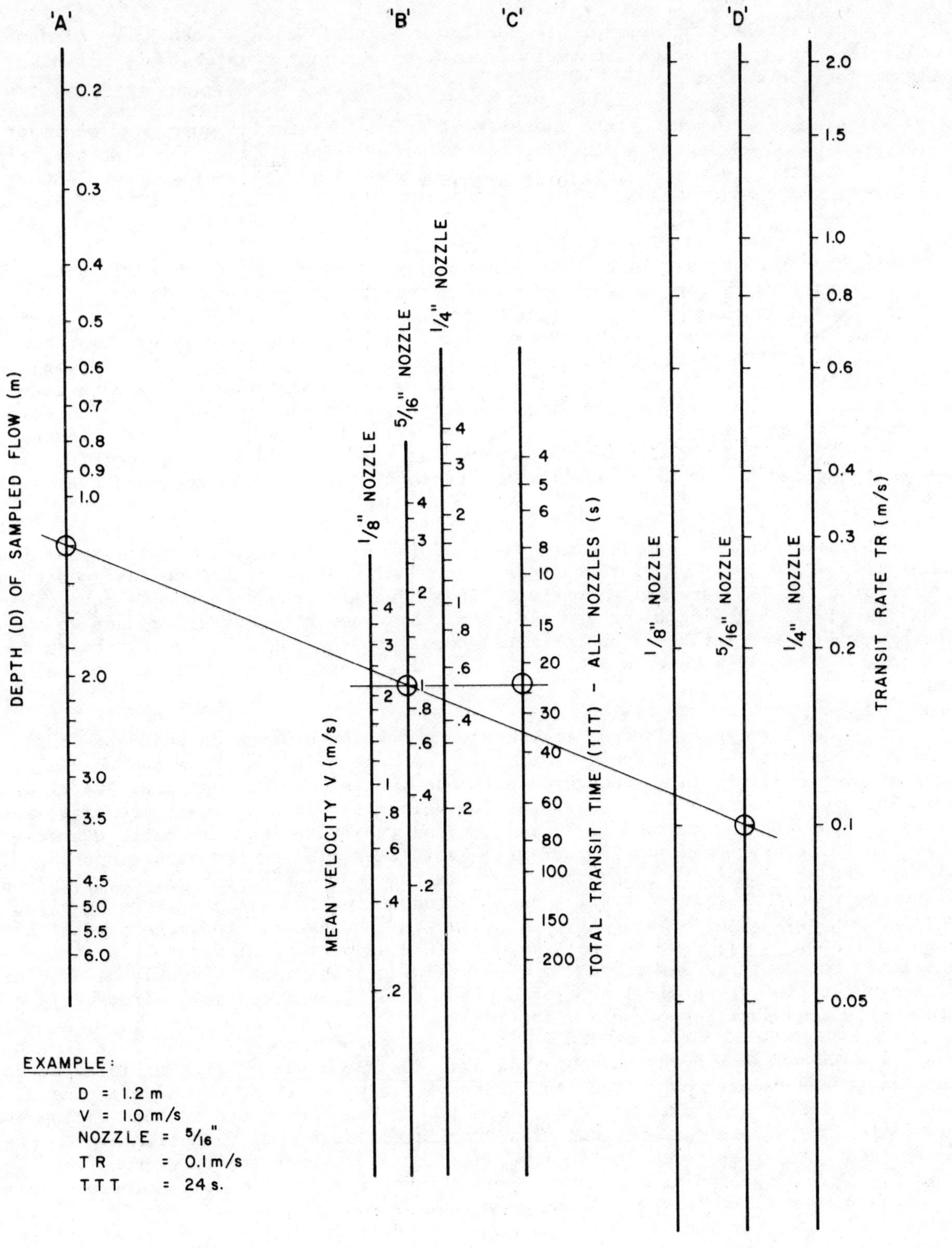

Fig. 14.4 Suspended sediment sampling equal transit rate method alignment chart for estimating transit time and rate

called a depth-integrating sampler, is much more commonly used. Both kinds of samplers are made for cable suspension and various depth-integrating sampler types are also available for rod mounting.

Samplers in common use are the US P-61 Point-Integrating suspended sediment sampler, the US DH-48 Depth-Integrating suspended sediment wading-type hand sampler and the US D-49 Depth-Integrating suspended sediment (cable suspension) sampler (ASCE, 1977). Rod-mounted hand samplers may also be locally made variations of the DH-48 or Nilsson (1969) type.

Another type in use is the Neyrpic sampler which may be operated in either the point or depth integrating modes. For point integration, suspension is by air supply cable through which compressed air is fed from a cylinder. The inside of the instrument, the sampling bottle and nozzle are pressurised by a continuous supply of compressed air which prevents inflow of water until the instrument is positioned at the sampling point. The air supply is then shut off to allow inflow of water-sediment in order to take a one-litre sample. Duration of sampling, t-seconds, must not exceed twice the reciprocal of the point stream velocity V-m/s (i.e. $t \ngtr 2/V$). For the depth integration the air supply is not used and the sampler is operated as for the US D-49 with the time constraint applied to the mean velocity and depth at the vertical to avoid over-filling the sample bottle.

US DH-48 and D-49 samplers are designed to permit changes in the size of intake nozzles 1/4", 3/16", or 1/8" internal diameter. All USGS samplers are designed to use standard pint (0.473 litre) milk bottles. They have air exhaust ports to carry away the air displaced by inflowing water and sediment. All sampler intakes are made so the velocity through the intake is approximately equal to the stream velocity at the end of the intake. The D-49 sampler weighs 62 lbs. (≈ 28 kg) and the Neyrpic sampler weighs 93 kg.

The sampler type selected for use at any station depends on the depth and velocity of flow at the sampling section. In the case of US samplers the nozzle diameter should theoretically be large enough to pass the largest sediment particle carried in suspension at each station. In practice flow velocities in arid zones are often such that small boulders are carried in suspension and the criterion for nozzle diameter is that which governs the sampling duration to avoid over-filling the sample bottle. The D-49 type can be used up to 5m depths - for greater depths the pint milk bottles would fill completely, and begin carrying a circulation current, during the time required for lowering and raising the sampler from top to bottom and back at the stream vertical. During this time density differences could cause displacement of water by sediment particles. For depths greater than 5m a point-integrating sampler should be used either at fixed points in the vertical or, for depth 5 to 10m, by integrating from surface to bottom only, then shutting off the water intake.

Sampling frequency during the progress of a flood varies with rapidity of change in discharge and related concentration. For the flashy sediment characteristics of arid zones, the observer should take frequent samples during each flood on both rising and falling stages. During rare periods of approximately steady flow, daily samples are normally adequate. Very often an observer has to resort to relatively crude techniques due to the need to take discharge measurements interspersed with collection of sediment samples. In such cases a single depth-integrated sample should be taken by cableway or from a bridge at a vertical in the high-velocity, deep central portion of the stream channel. This one sample is much more valuable than a large number of more conveniently collected hand samples taken at a shallow, slow-velocity segment of the stream cross-section near the water's edge. Since both water and sediment discharge measurements cannot be taken concurrently the water discharge appropriate to a particular sediment sample has to be inferred from the rating curve or by interpolation between consecutive discharge measurements. Major errors can result from assignment of sample concentrations to discharges measured during rapidly rising and falling stages.

If possible, samples should always be taken at the same verticals used for discharge measurement which should themselves be at fixed distances from the station "initial point". In this way relationships between concentration, velocity and depth can be established with respect to each vertical.

Equipment currently in use is not capable of collecting samples near the stream bed. In the arid zone, where much of the total sediment load is transported near the mobile interface of water and sand bed, this can be a serious shortcoming. Total sediment discharge may be estimated from observations of suspended sediment concentration, particle size distribution of suspended sediment, velocity distribution, stream depth, water temperature and bed sediment size for sand-bed channels (Colby et al, 1955, 1957, 1961). An example of the Colby 1957 procedure is given as Example A.4.6. It is total sediment discharge Qs which should be regressed on water discharge Q to produce the sediment rating curve.

14.3.2 Other Sampling Techniques

Other sampling equipment and techniques include:

a. The US Single Stage Suspended Sediment Sampler which is fixed in position at the sampling station and samples automatically at a constant stage corresponding with the level of the inlet tube.

The ASCE Sediment Engineering Manual states (Chapter III.3) that single stage samplers obtain a sample on the rising stage and the vertical intake design (to reduce the risk of plugging) does not take an accurate sample when sand particles are present. Flow into the sample bottle is governed by the syphoning principle and ceases when the water level in the bottle is at the bottom of the air exhaust pipe. The rate of flow into the bottle is independent of the rate of flow of the stream and the sample is therefore not truly representative of the sediment concentration in the flow at the intake point.

Several such sample bottles may be fixed to a recorder house or bridge set to obtain samples covering a range of levels. The difficulty lies in assessing the factor to be applied to the sample concentration in order to represent the mean section. The discharge to which the sample concentration relates is approximately that corresponding with stage at the intake pipe weir. Data based on samples taken from these devices should be used with caution. They produce qualitative information somewhat similar to that of a gauging station stilling well found silted after a single flow peak. The volume of sediment in the sample can be measured but the result is open to a wide range of interpretation. The use of single-stage samplers in arid zones is not recommended.

An automatic sampler operating on similar principles has been designed by Schick et al (1967) and used by him in the Negev desert (1971) and by Rapp et al (op.cit.) in Tanzania. The Hayim 7 automatic sampler consists of a stream-lined, "bridge pier" shaped metal housing containing a rack of 12 one-litre sample bottles. Samples are collected via tubes of 7mm plastic whose intakes are at varying stages above the channel bed.

b. The pumping sampler is a device which produces a continuous record of sediment concentration by periodic sampling at a fixed point. Although it does not produce a spatially integrated concentration value, it does provide a measure of temporal changes in concentration. The measuring system is calibrated by periodic sampling over the entire cross section to produce a spatial correction coefficient. In this way a continuous sediment transport record is produced. Samples are automatically collected in bottles at predetermined frequencies. The commencement of the sampling process is initiated by activation of a float-switch at a predetermined water level.

c. Nuclear sensing devices have been developed which determine the density of the water-sediment mixture. At the present time these devices are not sufficiently well developed for them to be recommended in this publication.

Despite some progress in development of new instruments and techniques for sediment sampling, it is considered that, in arid zone conditions, application of well-established methods using well-tried instruments and well-trained field staff will produce the best results in the immediate future.

14.4 SEDIMENT SAMPLING PROCEDURE

14.4.1 Equipment

a. D.H. 48 or similar - Hand Sampler.
For wading measurements by depth integration.

b. D.49 or similar - Cable Suspension Sampler.
For bridge and cableway measurement by depth integration.

c. Nozzles for (a) and (b) - 1/8", 3/16" and 1/4" diameter.

d. Sampling bottles - 1 U.S. pint (= 473 ml).

e. Plastic sample containers and crate.

f. Lables - scotch tape.

g. Ball-point pen.

h. Current meter equipment, including winch for bridge and cableway measurement.

14.4.2 Field Procedure (assuming relatively slow rate of change in discharge)

a. Make a current meter measurement using standard procedures.

b. Estimate the mean velocity at each vertical.

c. Using the alignment chart (Figure 14.4) or by experience select depth integrating transit rate and nozzle diameter for the velocity and depth corresponding with maximum discharge per unit width.

Transit rates in excess of 0.5 m/s should be avoided if possible and the selection of the nozzle should be made with this in mind. Suppose, for example, the depth is 4.0m and the mean velocity is 1.5 m/s. Figure 14.4 shows that the appropriate transit rates are as follows:

with 1/8" dia. nozzle - 0.23 m/s
with 3/16" dia. nozzle - 0.50 m/s
with 1/4" dia. nozzle - 0.94 m/s

Of these the 1/8" or 3/16" diameter nozzle would be selected depending on bed-material particle size distribution. For the drainage channel used to illustrate the Einstein Bed Load Function method of sediment transport estimation (Figure A.4.6) 3/16" (4.8mm) corresponds with the 75% (by weight) size.

d. Carry out sampling using the Equal Transit Rate (ETR) method at equi-distant verticals.

The number of verticals will depend upon the discharge rate and distribution across the section. Normally, samples are required at a minimum of five verticals for the range of discharge. When the rate of discharge change is high the number of sampling verticals must be reduced to allow for increased frequency of sampling. The sampling procedure for both new stations and for routine must be determined and revised as necessary for each station according to flow regime characteristics found from experience.

Lower and raise the sampler from the water surface to the bed and back at the constant transit rate. The total transit time should be checked by stop-watch at each vertical. When using a winch the vertical rate of movement of the sampler at one revolution per second should be calculated and the rate of turning the handle estimated for the required rate.

Before commencing at each vertical, ensure that the sampling bottle gasket is forming a proper seal by placing a finger over the exhaust port and blowing into the nozzle. Ensure the nozzle and exhaust port are clean. The section on which the selection of transit rate has been based (i.e. that with the greatest velocity-depth product) should be sampled first. After the sample has been collected check the level of the water in the sample bottle. If the bottle is more than 3/4 full, decant and store the sample in one of the plastic containers provided and repeat the measurement at a moderately increased transit rate. If the bottle is still overfull either the transit rate or the nozzle diameter have been wrongly selected and one or the other should be changed.

For new stations each sample should be placed in a separate plastic container. A decision should be made in the laboratory whether or not to composite the samples. Depending upon the results of sediment sample analysis over a wide range of discharge, a decision should be made as to whether future samples may be composited in the field.

e. Take the river water temperature.

f. Record sampling data on sampling bottle label. Blank scotch tape labels should be affixed to empty plastic sample bottles in the office before returning them to the field. The information to be written on each bottle in the field is:

Station No. and Name.

Date:
Time:
G.H.:
Dist. from Initial point:
Water Temperature;
Initials of Observer:

Sample bottles should be sealed, replaced in crate and kept in the shade until they reach the laboratory.

14.5 LABORATORY ANALYSIS METHODS

Analysis of sediment samples is required to determine the concentration and particle size distribution of the sample. The concentration is calculated as the weight of sediment divided by the volume of sediment-water mixture and expressed in milligrams per litre (mg/l) or kilograms per cubic metre (Kg/m^3). ($1\ kg/m^3 = 1000\ mg/l$).

The main methods of determining the sample concentration are by filtration and by evaporation.

14.5.1 Evaporation

a. Determine the total weight of sample plus bottle to the nearest 0.5 gm and deduct the tare weight of the bottle.

b. Allow the sample to stand for a period of time for settlement of the sediment from suspension.

c. Decant the sediment-free liquid.

d. Wash the remaining sediment from the sample bottle onto an evaporating dish of known (tare) weight.

e. Oven-dry the sample in the evaporating dish until all visible water is lost, then heat at 110°C for one hour.

f. Cool the evaporating dish in a dessicator cabinet.

g. Weigh the dish and contents to the nearest .0001 gm.

h. Subtract the tare weight to obtain the net sediment weight.

i. Compute the concentration

$$C = A. \frac{\text{net weight of sediment x 1000}}{\text{weight of water-sediment mixture.}}$$

Where A is a factor which corrects for the difference in specific weights of water and water sediment mixture (Table 3.6 in ASCE Sedimentation Engineering Manual.

14.5.2 Filtration

a. Transfer the sample to a Winchester bottle and place inverted on a rack above the filter assembly, comprising Gooch crucible (previously tare-weighted), funnel and flask.

b. Filter the sample through the crucible by gravity or using suction applied by vacuum bottle.

c. Oven-dry the crucible and contents.

d. Following weighing procedure described for the evaporation methods.

14.5.3 Particle Size Distribution

A variety of method are available for this:

14.5.3.1 Sieve-pipette method

Samples are withdrawn at pre-determined depths and times from the water-sediment mixture after sieving to remove all sands and after thorough stirring. The pipetted samples contain all sediment sizes below that corresponding with the sampling depths, times and the particle diameter corresponding with the fall-velocity. Samples are individually dried and weighed to determine the concentration of sediment sizes below each of these particle diameters. Results are converted to percentages of the total sediment weight and plotted on particle size analysis diagrams.

14.5.3.2 Visual-Accumulation Tube Method

The particle size distribution of the sand fraction is determined by introducing the sand at the top of a vertical glass cylinder full of clean water and following the top of the accumulating deposits at the bottom of the tube with a manually operated pointer. The pointer is linked to a chart-drive mechanism which produces a trace on a calibrated chart from which percentages of sand particles finer than specified values can be calculated. Analysis of the silt and clay fraction may be accomplished by the pipette, bottom-withdrawal tube or hydrometer method.

14.5.3.3 Bottom-withdrawal Tube Method

The sand fraction is removed by sieving and analysed by means of the Visual Accumulation tube method. The silt and clay fraction is dispersed in a vertical glass tube by thorough mixing and samples are withdrawn from the bottom at predetermined time intervals. The percentage of sediment in each sample is plotted against time to form an Oden curve the particle size distribution determined from the Y-axis intercepts of tangents to the curve.

14.5.3.4 Sieve and Hydrometer Method

For approximate size distribution the sand fraction may be determined by sieving and the silt and clay fraction by hydrometer methods. Fluid densities measured by sensitive hydrometers at predetermined time intervals after thorough dispersal of the sediment in a beaker of distilled water are interpreted as representing the proportion of silt to clay.

Full description of laboratory methods is given in Chapter IIIE of the ASCE Sedimentation Engineering Manual (op.cit.).

Arid zone fluvial sediments are characterised by a high proportion of sands, for which the Visual-Accumulation and Sieve plus Hydrometer Method are generally adequate.

14.6 VOLUMETRIC METHOD OF SEDIMENT SAMPLE ANALYSIS

In order to reduce the labour involved in the determination of concentration by the filtration and evaporation methods of total sediment analysis, the results of either of these methods may be used to calibrate the results obtained by simple volumetric measurement. The result produces a relationship which can be applied to the bulk of the samples to which only the latter, simpler method is applied. The procedure is:

a. The bulk-volume concentrations of all samples are estimated by pouring the water-sediment mixture into clear plastic cones with an apex angle of 60° and measuring (by scaling) the volume of the sample and of the water-sediment mixture below the level to which settlement has taken place after a selected period, usually 24 hours.

b. The sediment weights of a proportion of the samples covering a wide range of concentration are then determined by laboratory analysis as outlined in Section 14.5.

c. The weight concentrations (C) are then correlated with the bulk volume concentrations (Cv). The regression constants are derived and applied to the remainder of the sample Cv values.

d. The samples, whose weight concentration is determined in (b), are also analysed for particle size distribution.

A typical C/C_v relationship is illustrated by Figure 14.5.

The method produces quick results of adequate accuracy for large numbers of samples taken during prolonged flow periods. Considerable staff time and costs, of transportation of the samples to the laboratory and of the analysis themselves, are saved.

14.7 COMPUTATION OF SUSPENDED SEDIMENT LOAD

Examples of relationships between water discharge and sampled suspended sediment concentration and sediment discharge are shown on Figures 14.6 and 15.1. These graphs illustrate the wide variability about the mean relationships at a station where the channel is stable but upstream of which the wadi bed is composed substantially of sand.

The use of this relationship (Figure 14.6) for computing suspended sediment load in combination with the discharge hydrograph (Figure 14.7) is demonstrated by Example 14.1.

To save computation time, the calculation may conveniently be performed together with the computation of daily water discharge volume using the same hydrograph coordinates of time and discharge rates.

In the example computation the sediment discharge is based on the sampled concentration without the Colby correction for sediment concentration in the unsampled zone (Example A.4.6).

14.8 BED LOAD

The sediment transport calculation described above related to suspended sediment only and does not include bed-load. Generally, the amount of bed-load transported by an alluvial channel in arid zones is of the order of 5-25% of the suspended load. It cannot be measured satisfactorily and may therefore be accounted for only by using empirical formulas based on field and laboratory research data.

These methods are discussed in Section 15 and are described in detail in the ASCE Sedimentation Engineering Manual (Chapter II H).

In the absence of adequate data for these calculations the bed-load may be estimated from Table 14.2 (taken from the ASCE Sedimentation Engineering Manual, Table 3.2, "Maddock's Classification for Determining Bed Load").

Yang (1973) has developed a dimensionless unit stream power equation for computation and prediction of total sediment concentration. Comparison between the measured results from laboratory flumes and natural rivers showed that, for 1259 data sets, the average computed result was only 3% higher than the measured result.

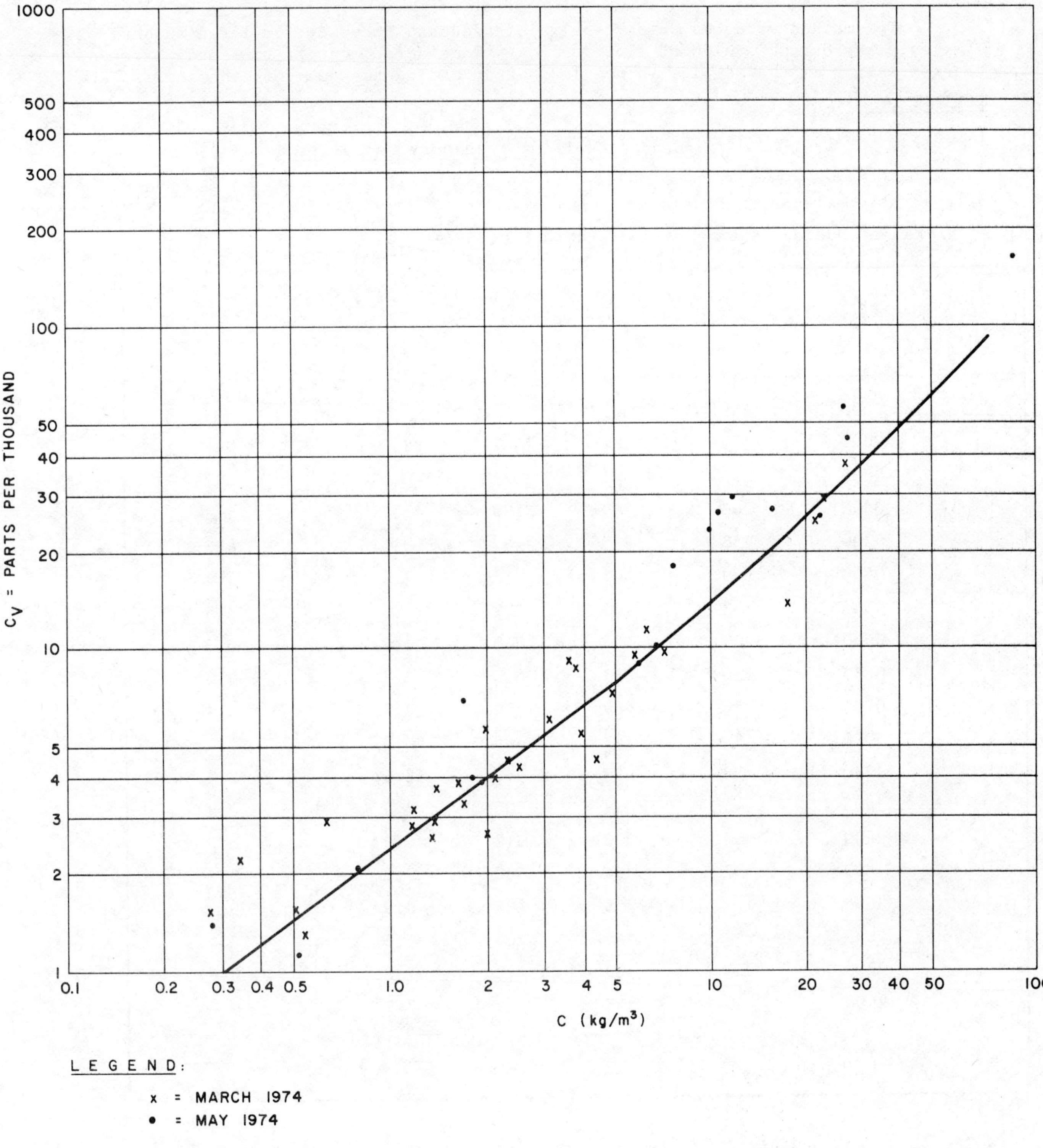

Fig. 14.5 Sediment sample analysis: bulk volume concentration vs. weight concentration

EXAMPLE 14.1

Suspended Sediment Transport Calculation

W. Najran at Mudiq: Rating period 2-26th May, 1974

Date	Time	Q m^3/s	Qs kg/s (1)	$\bar{Q}s$ kg/s	time s x 10^3	Qs t x 10^3
9.5.74	0000	159	603			
	0005	250	1821	1362	0.3	0.4
	0015	270	2051	1936	0.6	1.2
	0100	240	1709	1880	2.7	5.1
	0200	150	825	1267	3.6	4.6
	0400	79	305	565	7.2	4.1
	0545	52	160	232	6.3	1.5
	0600	103	461	310	0.9	0.3
	0630	126	630	546	1.8	1.0
	0700	118	569	600	1.8	1.1
	1000	68	242	406	10.8	4.4
	1330	38	98	170	12.6	2.1
	1405	43	119	108	2.1	0.2
	1600	40	106	112	6.9	0.8
	1800	37	94	100	7.2	0.7
	1805	74	276	185	0.3	0.1
	2000	48	141	208	6.9	1.4
	2200	34	83	112	7.2	0.8
	2400	30	68	76	7.2	0.6
					86.4	30.4

NOTE: (1) $Q_s = 0.3495Q^{1.55}$ (See Figure 14.6)

(2) Total suspended sediment runoff = 30 400 tonnes.

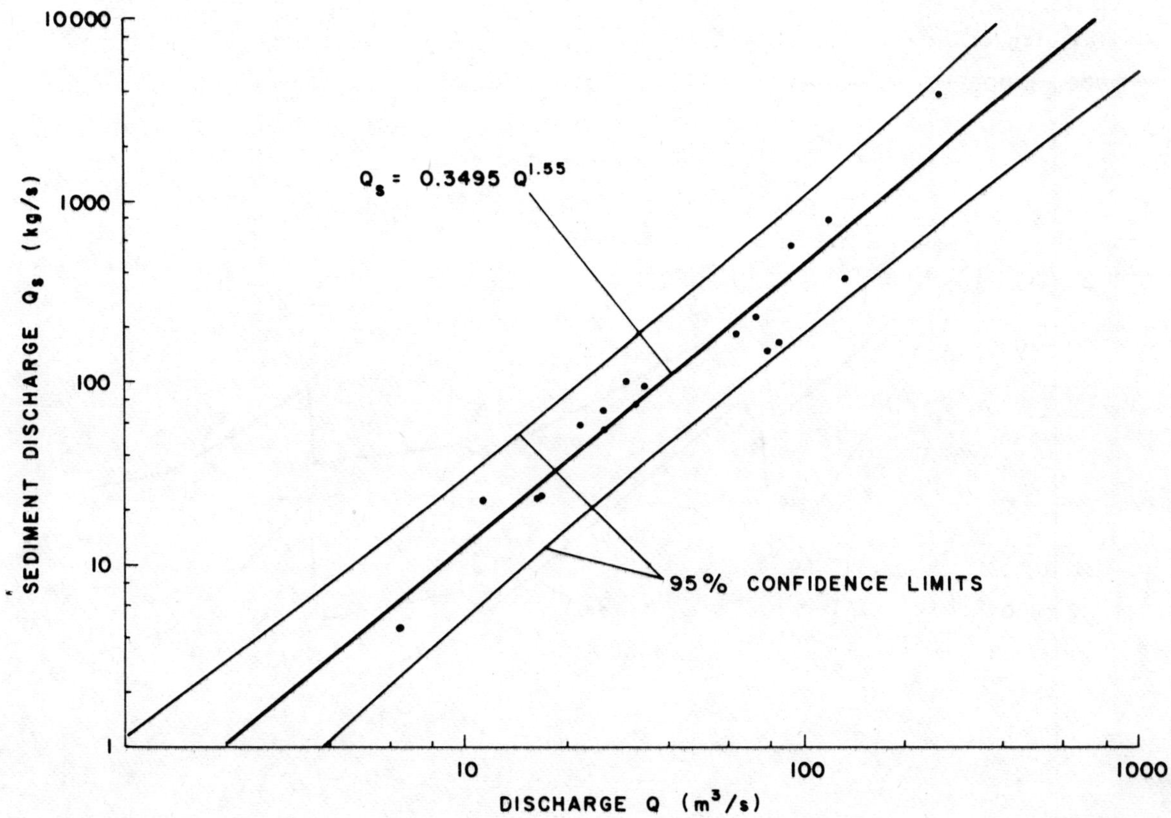

Fig. 14.6 Sediment discharge Q_s vs. water-sediment discharge Q
Wadi Najran at Mudiq - 6-25 May 1974

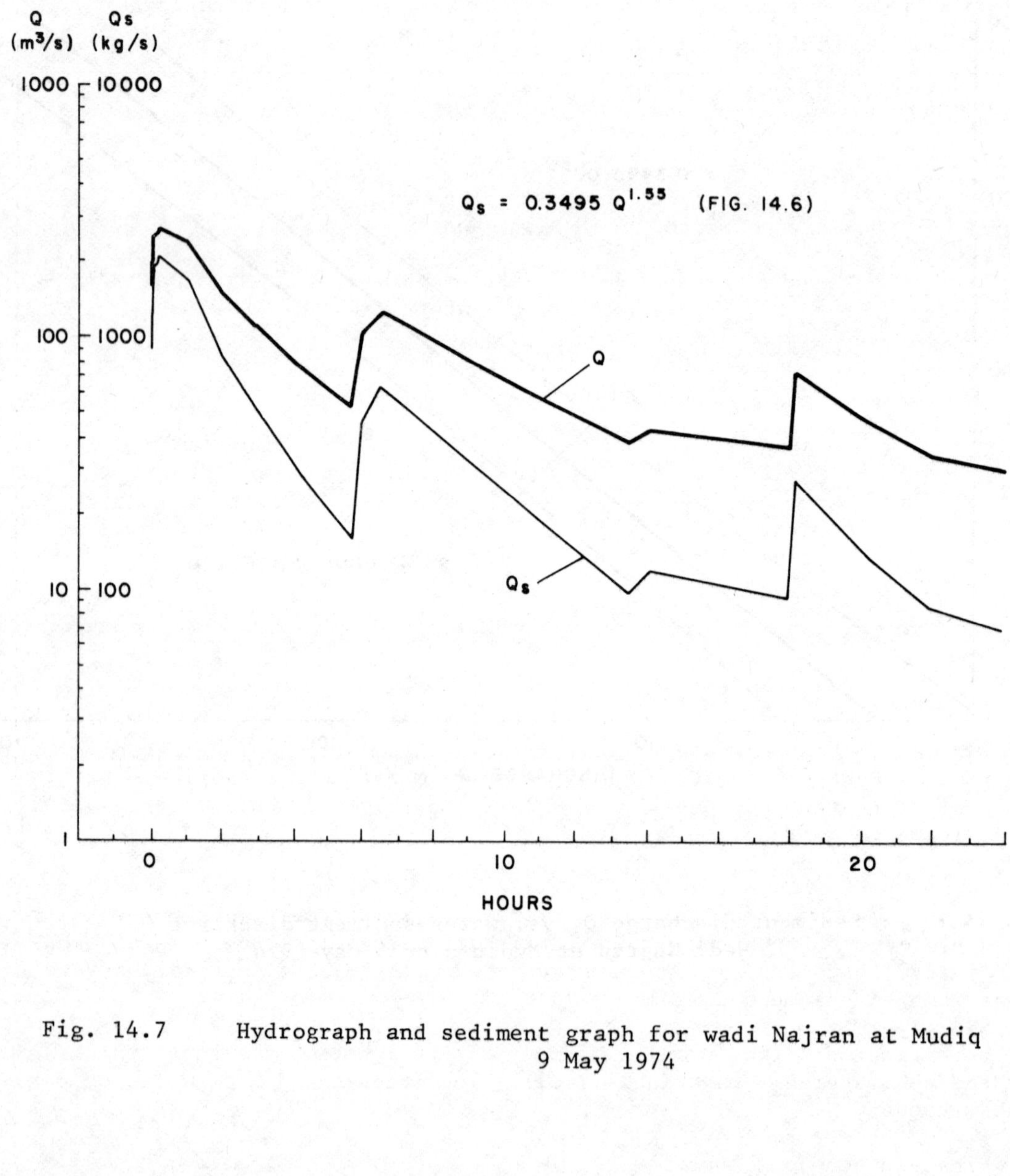

Fig. 14.7 Hydrograph and sediment graph for wadi Najran at Mudiq 9 May 1974

TABLE 14.2

Maddock's Classification for Determining Bed Load

Concentration of suspended load, in parts per million (1)	Type of material forming channel of stream (2)	Texture of suspended material (3)	Bed load discharge, in terms of suspended sediment discharge, as a percentage (4)
Less than 1,000	Sand	Similar to bed material	25-150
Less than 1,000	Gravel, rock, or consolidated clay	Small amount of sand	5-12
1,000-7,500	Sand	Similar to bed material	10-35
1,000-7,500	Gravel, rock or consolidated clay	25% sand or less	5-12
Over 7,500	Sand	Similar to bed material	5-15
Over 7,500	Gravel, rock, or consolidated clay	25% sand or less	2-8

14.9 SEDIMENT TRANSPORT FORMULAS

The method of suspended sediment transport calculation described in Section 14.7 assumes that adequate suspended sediment samples are taken as routine, that they cover the range of discharge and that there is an acceptably good relationship between sediment load and discharge during flood periods.

These conditions are seldom fulfilled and it is often necessary to resort to estimation of total sediment transport by means of empirical and theoretical relationships developed by research engineers and scientists. Such relationships are founded on laboratory flume data and field measurements and progressive improvements are made as more measurements become available.

For arid zones the formulas of most direct interest are those which attempt to predict the bed material load (Table 14.1). The following are examples of potentially suitable methods:

a. Einstein bed-load function
b. Einstein - Brown formula.
c. Engelund - Hanson formula.
d. Inglis - Lacey formula.
e. Meyer - Peter formula.

These methods are discussed and example computations given in Section 15.

15. SEDIMENTATION DATA PROCESSING AND ANALYSIS

15.1 SEDIMENT RATING CURVES

15.1.1 Introduction

In any study of sedimentation the hydrologist is faced with the task of estimating the instantaneous sediment transport rate at a given cross-section of a water-course and more generally, the sediment volume over a period of time. Estimates may be based upon the results of sediment concentration sampling and discharge measurements, as described in Section 14.

The procedure currently used to analyse the sample data consists of developing a regréssion of sediment concentration C or load Qs on other measurable variables and channel parameters. A regression of this type is commonly termed a sediment-rating curve which, when applied to a long-term flow-duration curve, produces a sediment-duration curve, the area of which represents an estimate of a long-term average sediment yield.

The hydraulic and physical variables that determine the sediment concentration at a given discharge are:

i) The magnitude and distribution of velocity V within the cross-section of flow.

ii) The diameter d_i of the bed material particles.

iii) The viscosity ν of the water, which determines the fall velocity

$$w = \frac{d^2 g \Delta \gamma}{\nu \gamma} \quad \text{(Stokes)}$$

Where g is gravity acceleration, $\Delta\gamma$ is the difference of specific weights of particles γs and water, γ is the specific weight of water).

Therefore the equation for concentration C may be written as:

$$C = f(V, d, \frac{d^2 g \Delta \gamma}{\nu \gamma})$$

assuming that the availability of erodible sediment material is unlimited. If this latter condition is not met a correction factor should be introduced into the relationship.

At present, functional relationships involving these variables and parameters and capable of being used to predict sedimentation rate have not been firmly established. Only global relationships of the type $C = f(Q)$, where Q is the water-sediment discharge; or $C = f(V,d,t)$, where V is velocity, d is depth and t is temperature, are used at present. The main reasons why only coarse relationships of this type can be established are:

i) Lack of data on velocity and sediment concentration distributions in the cross-section, and their inter-relationship. No measuring instruments yet exist to measure simultaneously the instantaneous values of V and C.

ii) The availability of transportable material is a characteristic of the watershed and the river channel, but not of the hydraulic conditions.

At present, the best way of making use of concentration sample data is to correlate the simultaneous values of concentration and discharge. Calculation of sediment discharge $Qs = CQ$ then produces an estimate of sediment yield integrated over the time period of the sediment discharge. The expression $Qs = f(Q)$ is the sediment rating curve and expresses the mean instantaneous value of sediment discharge in a given cross-section for any discharge. Measured Qs values depart considerably from the mean relationship as shown by Figure 15.1.

It should be noted however that the continuity principle which applies to liquid flows (i.e. the rate of storage change ΔS = difference in rates of inflow and outflow $(Q_I - Q_o)$ is not applicable to sediment discharge. This is because aggradation (or degradation) may occur between the two sections considered, e.g. when an enlargement (or a constriction) occurs resulting in a velocity decrease (or increase) with deposition (or scour) of sediment material. These considerations must be taken into account whenever possible in choosing gauging sites and especially when considering transposition of data or relationship to another site.

15.1.2 Regression of Sediment on Flow Discharges

Research into sediment-discharge relationships, by plotting Q vs C or Q vs Qs on different types of graph paper demonstrates in all known cases that the best grouping and linear trend of the data points is produced by a bi-logarithmic plot. The scatter is usually large (of the order of one power of 10, i.e. one log cycle), and is approximately evenly distributed over the range covered if there is a sufficient number of points (Figure 15.1.).

The equation that defines a linear regression on bi-logarithmic paper is of the form:

i) $Qs = aQ^b$ or $\log Qs = b \log Q + \log a$

ii) $C = aQ^{b-1}$, as $C = \frac{Qs}{Q}$, or $\log C = (b-1) \log Q + \log a$

(note: logarithms are to base 10)

The first relationship always shows the best correlation.

Coefficient a has no particular range of values and is proportional to the width of the channel: the sediment discharge per unit width W of channel is then given by $q_s = \frac{aQ^b}{W}$

On the other hand coefficient b has a physical lower limit: b equalling unity implies that the concentration C is constant and independent of Q and if b exceeds 1, C increases with Q. For a given hydraulic bed, the turbulence which creates the forces of suspension of the particles also increases with Q. This implies that b must be greater than 1. In practice b is generally between 1.4 and 2.8.

Since C cannot exceed unity, care must be taken not to extrapolate the Q-C relationship beyond the range of data by extension of the regression line only.

15.1.3 Adjustment of the Theoretical Regression Line

Calculation of the regression constants by logarithmic transformation of the sample pairs Qs_i and Q_i produces the equation $Qs = aQ^b$ as the best least-squares fit of the logarithmic values, not of their actual values. The scatter represented by the confidence intervals is asymmetric about the regression line. The interval between the upper confidence limit and the regression line is greater in terms of absolute Qs than that for the lower limit, for a given value of Q. It is proposed not to use the calculated regression $Qs = aQ^b$ which passes through the point $(\bar{x}, \bar{y})$ given by the means

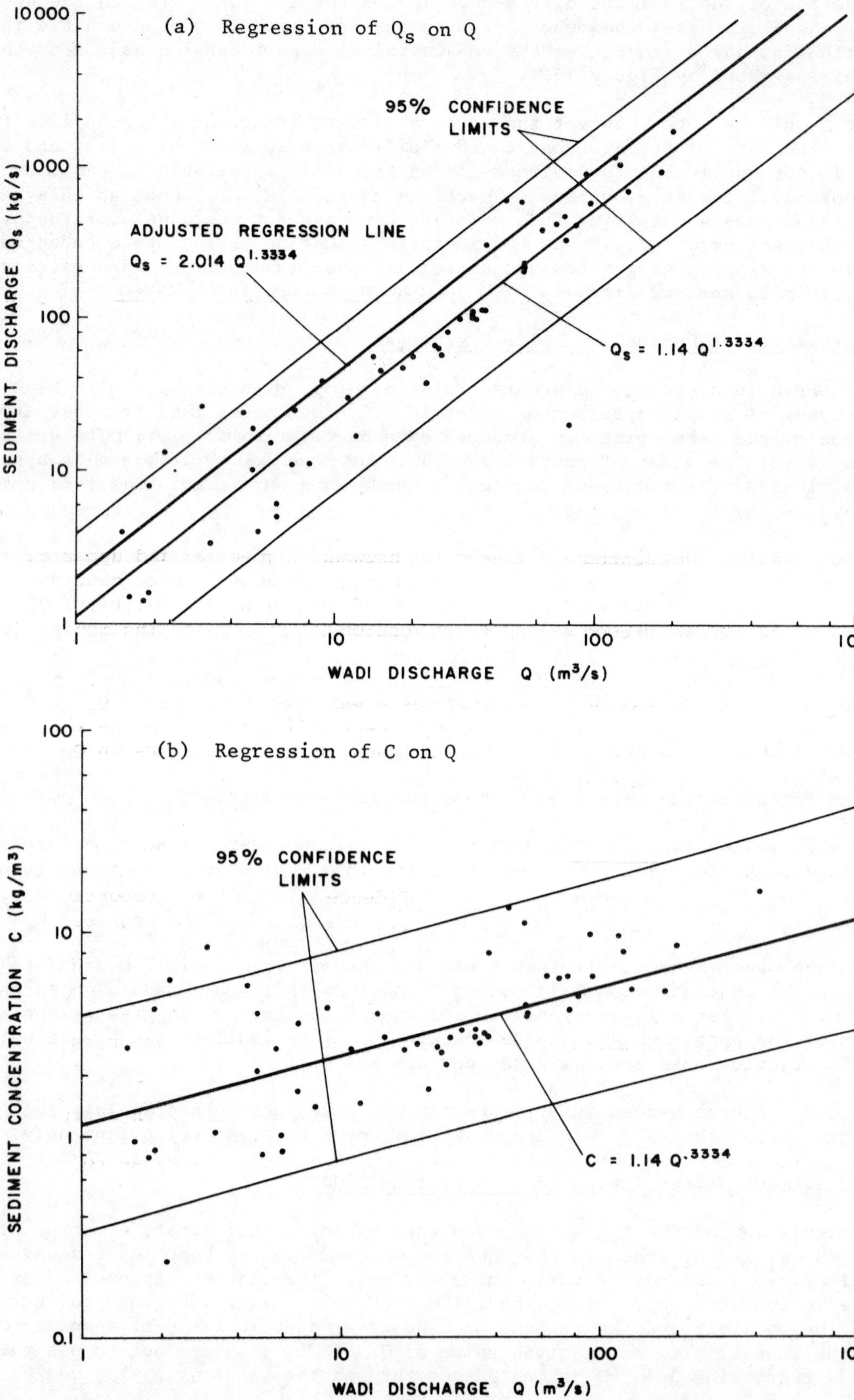

Fig. 15.1 Sediment discharge - Wadi Najran at Mudiq, Saudi Arabia

FIGURE 15.1

Sediment Discharge - Wadi Najran at Mudiq, Saudi Arabia

a. Regression of Q_s on Q

The figure shows a considerable degree of scatter of the data points about the regression line. The 90% confidence interval between the confidence limits of Q_s about the value represented by the regression line is approximately an order of magnitude.

The Q_s error represented by the "area" between regression and upper confidence limit lines is much greater than between regression and lower confidence limit lines. This is because the distribution of sample (and population) Q_s for given Q is highly skewed with the regression line representing median Q_s values.

The "adjusted regression line" produces a weighted estimate of Q_s for given Q, having been drawn through the $\bar{Q}_s$-$\bar{Q}$ intercept but parallel to the regression of logarithmic Q_s and Q values. By this means, the large errors in Q_s predicted from Q tend to be equalised.

b. Regression of C on Q

The scatter of data points and the confidence interval is proportionately the same for C as for Q_s, with respect to Q. The slope of the regression line differs from that of Q_s on Q by the power of unity, since $Q_s = CQ$.

The sample sediment concentration C multiplied by measured Q produces the data points shown in Figure 15.1(a) which reproduces the variability of C about Q. This variability is due to the fact that Q is only one of numerous physical factors affecting sediment concentration.

Sediment transport estimates for particular periods, based on regression equations such as those shown on Figures 15.1(a) and (b) are subject to errors of the relationship for the sampling period and uncertainty as to the validity of applying the relationship to any other period.

of the logarithm but a line parallel to it, passing through the point $\bar{Q}s$, $\bar{Q}$, the means of the actual values.

This method produces an equation expressing a relationship for values (Qs_i Q_i) of which the principal characteristics are:

i) Linear relationship parallel to the plot of the data points on double-log paper.

ii) The line of relation passes through the arithmetic means of the samples.

iii) For any discharge Q the equation gives an estimate of the arithmetic mean of the population of Qs at that discharge, as quantified by the particular confidence interval. The important point to emphasise is that, in adopting the line defined directly by the regression theory ($Q_s = aQ^b$), the mean of calculated values of Qs corresponding with the sample Q would be much lower than the mean of the sample Qs.

The equation for the proposed relationship becomes:

$$Qs = a' \, Q^b \text{ with } a' = \frac{\bar{Q}_s}{\bar{Q}^b}$$

where $\bar{Q}$ and $\bar{Q}s$ are the arithmetic means of the samples Q_i and Qs_i.

15.1.4 Extrapolation of the Sediment Rating Curve

In practice, the sediment rating curve has to be extrapolated beyond the limits of the data sample (Q_i ; Qs_i). Either it has not been possible to obtain samples at higher discharges or sediment discharge rates must be estimated for high discharges of low probability. No standard method of accurate extrapolation yet exists and it is up to the hydrologist to judge how to extrapolate the regression line, derived as described above. It cannot be simply extended linearly for the following reasons:

i) The theoretical upper limit of the concentration C is equal to the specific weight of saturated sediments γw, which ranges from 500 to 1500 kg/m^3, since such concentrations are seldom approached in practice, the Q-Qs relationship on bi-logarithmic paper must be non-linear. The relationship $Qs = aQ^b$ (with $b > 1$) must reduce its slope (defined by b), as it cannot cross the line of equal concentration $Qs = \gamma wQ$, i.e. the coefficient b must reduce at high discharges.

ii) Einstein noted, during his experiments in laboratory flumes, that the sediment-carrying capacity of the flow reduces with discharge after the concentration has reached C = 100 kg/m^3 and consequently that the exponent b should reduce progressively beyond this concentration. The presumed explanation is that increasing shocks caused by impact between the sediment particles reduce flow turbulence and thus the energy available for keeping the sediment load in suspension.

iii) It is possible that beyond a certain discharge all available bed material is in motion and consequently that C cannot increase.

Empirical formulas cannot reliably be used to extrapolate the Q-Qs relationship beyond the range of the data.

Use is recommended of the method of Einstein (Section 15.2) which is based on the physical principles governing sediment transport. It can be applied up to a limit of concentration of about 100 kg/m^3, which is usually greater than that given by the method, for maximum Q.

On Figure 15.2 the sediment rating curve defined by the method of Einstein (for which full calculations are given in Example A.4.1) slightly underestimates the sediment discharge Q_s over the measured range. Its slightly greater slope makes the two relationships defined by the adjusted regression equation and Einstein method cross at Q = 8 500 m^3/s; Q_s = 380 000 kg/s, C = 45 kg/m^3 which is near the upper limit of discharge considered in the particular hydrological study. Therefore the straight line relationship derived from regression analysis would be the best sediment rating curve for practical purposes and the results of Einstein's method do not add any further information for this particular case study (other than that concerning the general validity of the Einstein method). Should sediment discharges have to be extrapolated beyond 380 000 kg/s, the use of the Einstein relationship would be preferred to extrapolation of the empirical curve because its slope is slightly greater and accords better with values encountered in practical experience.

15.1.5 Sediment Rating Curves

The considerable scatter in the plot of simultaneous values of C vs Q or Qs vs Q is attributable to a number of factors:

i) Random errors due to the high variability of concentration in time and space over a finite sampling interval. When a cableway is used to sample the concentration, the flow velocity is such that it may not be possible for the sampler to reach the bed where the concentration is at a maximum. The concentration derived from the sample and ascribed to the vertical, may then be an underestimate of the true concentration as shown by Example 14.1.

ii) Random errors resulting from the analysis of the samples, as in any laboratory process.

iii) Errors due to the imprecision of the stage-discharge relationship caused by large variations both during a flood and from one flood to another.

iv) The fact that the concentration C or the sediment rate Qs are dependent on many more physical variables than discharge alone, as has already been pointed out.

Some of these factors are discussed below.

a) Flash Floods. Flash floods, varying in time and space and superimposed on generally low (or zero) base-flows, are a feature of arid zones. Very large variations in discharge occur over very short time intervals during rising stages. Discharge peaks are followed by relatively smooth recession curves on which other peaks are superimposed, as illustrated by Figure 14.7. Sediment discharge rate varies markedly between rising and recession stages. The turbulent front wave that constitutes the sharp rising stage acts as a "broom" in the channel bed removing, part of the new sediments deposited during preceding floods. This fact is confirmed by visual observation of a wave-front rolling down a dry alluvial channel. The extremely turbulent wave-front, of dark brown colour, carries a mixture of sediments and debris (eg. crop residue, branches, whole trees, etc) deposited by the preceding flood. The hydraulic energy characteristics of the wave-front are much different from those corresponding with steady-state flow of similar magnitude and its carrying capacity is greater.

To demonstrate this feature in natural alluvial channels a great number of samples would be necessary in about equal proportions during rising and falling stages. The difficulties of collecting such samples are two-fold:

i) It is not possible to include sampling of the sharp rising stage in a routine programme of measurement due to unpredictable arrival of the flood wave-front. The presence of a permanent, efficient observer is therefore essential.

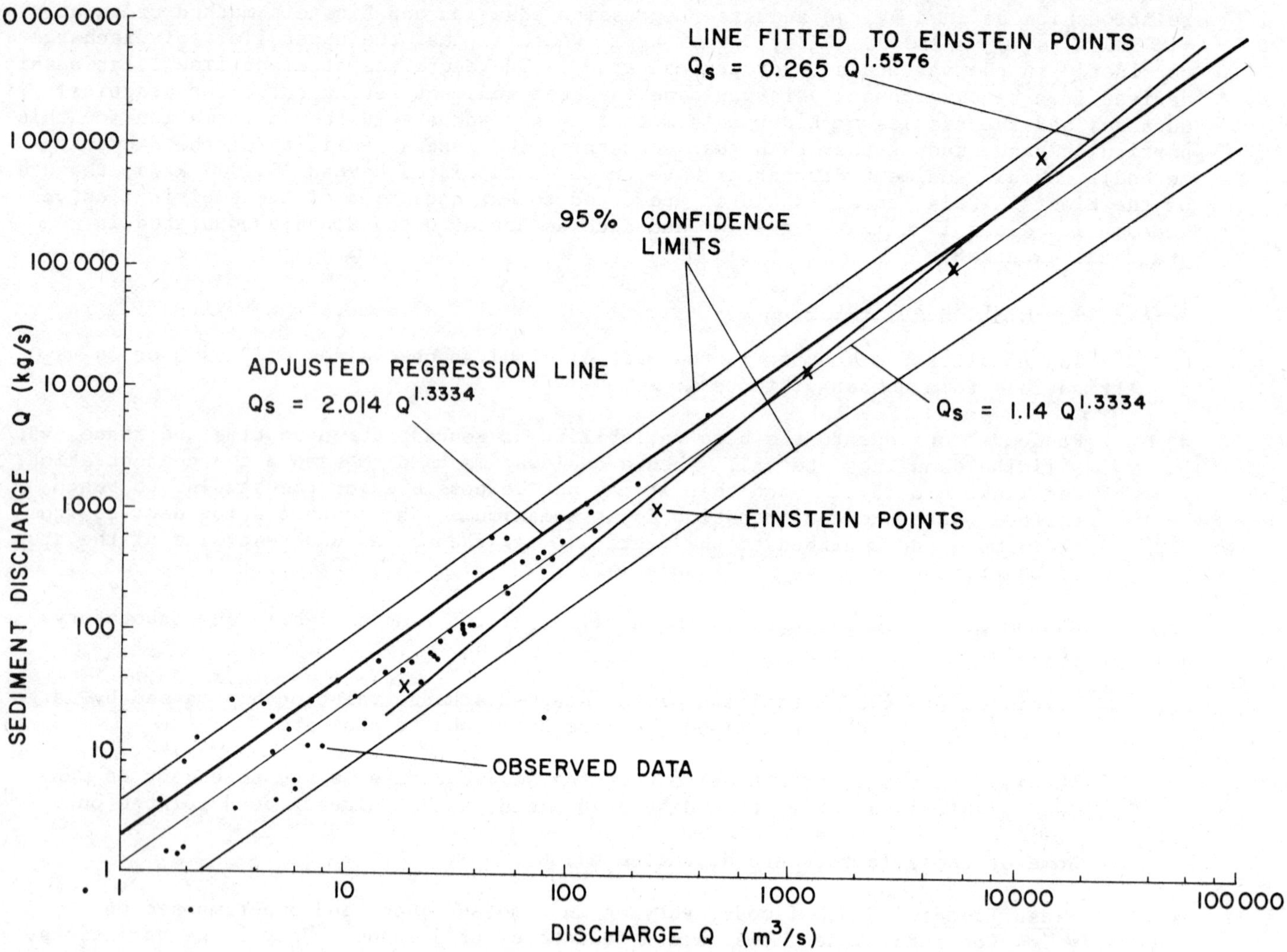

Fig. 15.2 Sediment discharge, Wadi Najran at Mudiq, Saudi Arabia - Extrapolation of regression by Einstein's method

The figure illustrates the results obtained by application of Einstein's method, as shown by Example A.4.1 to data for a Saudi Arabian wadi.

In this case, the intersection of the regression of measured Q_s (= CQ) on Q with the line of the best fit to the Einstein points occurs at about 8500 m^3/s corresponding with Q_s = 350 000 kg/s, C = 41 kg/m^3. The probability of Q = 8500 m^3/s at this site is very low. The Einstein method, in the case illustrated by the figure, services only to support extrapolation of the adjusted Q_s vs. Q regression, which would be preferred for calculating sediment transport.

ii) The relative time lengths of rising and falling stages of the flood are very disproportionate, the recession time being possibly two orders of magnitude longer than the rising stage. The latter may last only a few minutes in violent floods and does not afford time for sampling. However if such a distinction can be made the two sample sets should be plotted separately in order to derive separate $Qs = a Q^b$ relationships (Figure 15.3a) using the regression method presented above.

b) Seasonal Effects. The climate of an arid zone may be dependent on two or more meteorological situations in the course of the year, due to seasonal climatic characteristics (temperature, distribution and intensity of rainfall, etc). The erodibility of the watershed may be dependent on these seasonal variations and different groupings of sample points (C vs Q) or (Qs vs Q) might demonstrate the influence of seasonal parameters. This factor has been demonstrated in rivers of temperate climates (Figure 15.3b).

15.2 ANALYTICAL COMPUTATION OF SEDIMENT LOAD

15.2.1 General Discussion of Methods

Many authors have proposed "bed-load formulas" to compute the sediment discharge Q_s. The term "bed-load formula" is improper because computation is of the bed material load (see Section 13.4 and Table 14.1). Some are not strictly formulas since coefficients are often given in empirical graphical form (as with the Einstein method).

A critical review and comparison of many of these "formulas" has been made by Shulits et. al (1968) and the ASCE Task Committee for the Preparation of the Manual on Sedimentation (Vanoni, 1975) and is not repeated here. These relationships of the sediment discharge Qs, or more often the sediment discharge per unit bed width q_s, are given as functions of:

- The bed composition defined by its granulometry or by a single representative particle diameter.
- The flow parameters such as the tractive force, the discharge, the velocity and/or energy slope.
- The density of water and sediments.
- The hydraulic radius or the flow depth.
- The roughness coefficients (relative to the grain and/or the bed forms).
- The viscosity of water (or its temperature), etc.

These elements are incorporated into the equation in a more or less complex manner. Constants may be derived from laboratory experiments or actual river data.

At present there is no basis on which to recommend one or other of these formulas for a particular field problem. It may be observed for example that a formula derived from a particular condition of flow, geology or climate fits particularly well the data collected in quite different conditions. However, the opposite may also be true and it is also possible that a particular formula may fit low Q and Qs data but would be a poor fit at high values.

The indiscriminate use of any of the formulas is therefore not recommended: it can be dangerous to base feasibility assessments of any engineering projects on the results of applying them. As yet, they should only be used to compare their fit to a set

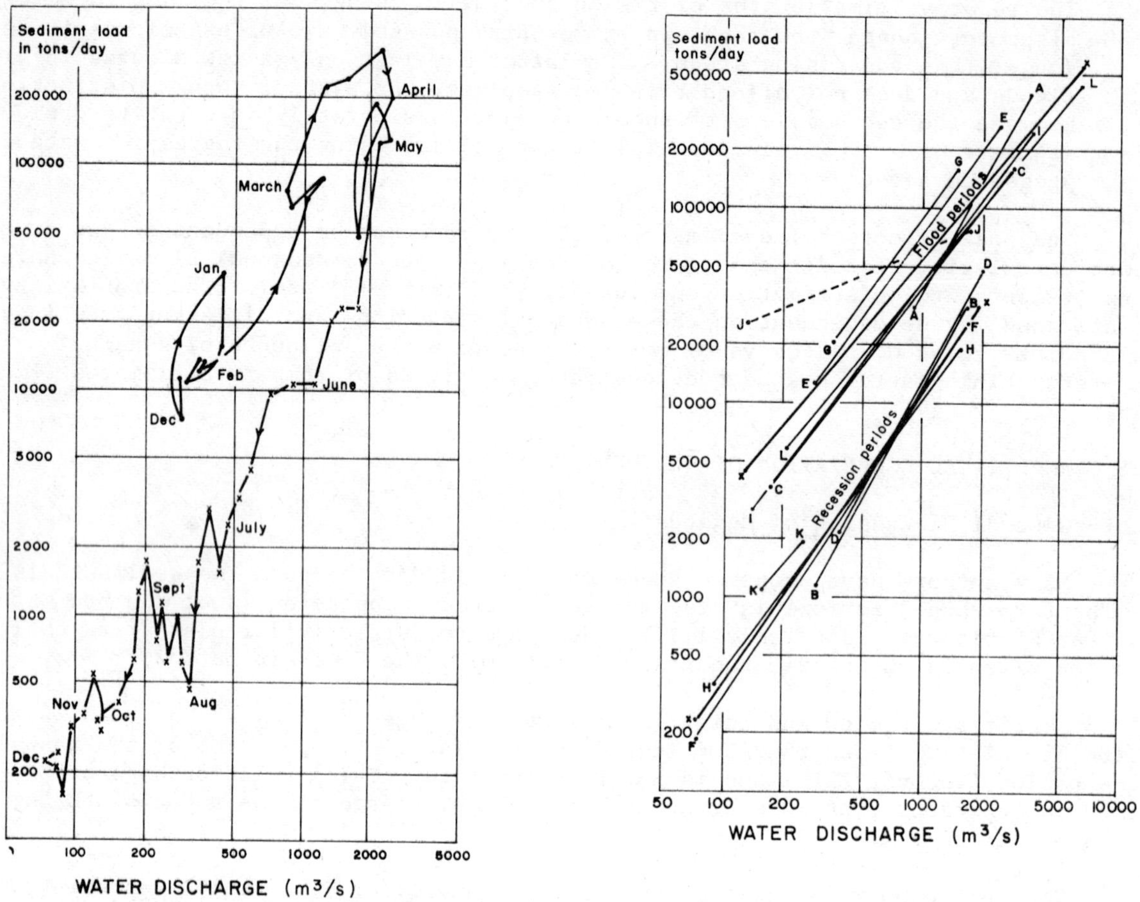

(a) Seasonal effect

Relation between sediment loan and water discharge during the hydrological year 1957/58 at station 1K3 (Stiegler's gorge).

(b) Hysteresis effect

Sediment rating curves for different flood and recession periods at Stiegler's gorge. Coarse lines indicate sediment rating curves computed from all available flood period data and for all available recession period data.

Fig. 15.3 Sediment multiple rating curves - illustrated by data for the Rufiji River Basin at Stiegler's Gorge, Tanzania

(after Rapp, Berry and Temple, 1972)

FIGURE 15.3

Sediment Multiple Rating Curve

The figure shows the change in Q_s vs. Q relationships due to seasonal and hysteresis effects.

a. Seasonal effect

In the case illustrated, sediment concentration is substantially greater during the period December-March, when flash floods occur on the rising limb of the seasonal hydrograph, than in April-June at the time of the seasonal peak. This is due to early flushing out of sediment from the drainage system.

b. Hysteresis effect

The figure shows the variation in Q_s vs. Q relationships as between flood and recession periods. After the flood flow period of highly fluctuating discharge the sediment concentration decreases. The relationship between Q_s and Q, improves during the transition from fluctuating flood flows to a period of recession and is relatively invariable between recession periods. During flood flows the Q_s vs. Q relationship is less stable and variability of relationships between periods is greater.

of actual data (Q-Qs) and possibly to extrapolate an empirical sediment rating curve towards high discharges when the formula has been shown to produce acceptable results in the same geological and hydraulic conditions. When a formula does not fit an actual set of data it should be completely discarded.

The validity of a formula may be assessed by checking that the slope b of the regression line in a logarithmic graph $Qs = aQ^b$ is in the range 2.0 to 3.0 and that this line passes through the group of data points.

The list below gives some well-known "bed-load" formulas:

Ackers and White	Elzerman and Frijlrik	Meyer-Peter and Muller
Blench	Engelund-Hansen	O'Brien
Casey	Haywood	Rottner
Chang	Kalinske	Schocklitsch
Colby	Inglis-Lacey	Straub
Du Boys	Laursen	Toffaleti
Equizaroff	MacDougall	Waterways Experiment Station
Einstein	Meyer-Peter	Yang
Einstein-Brown		

Calculations illustrating the application of six of the above formulas:-

i. Einstein
ii. Einstein-Brown
iii. Engelund-Hansen
iv. Inglis-Lacey
v. Meyer-Peter
vi. Colby

are given as Examples A.4.1 to A.4.6, using the same data in each case. Results are shown on Figure 15.4.

In introducing the treatment of the Einstein method, the user is warned of the limitations of all formulas, especially for arid zone sedimentation problems.

The Einstein (bed-load function) method for prediction of total sediment transport is at present the only theoretical method which takes into account the full knowledge of flow hydraulics, including the theory of turbulence and laminar sublayer based on the work of Prandl, Von Karman, Nikuradse and Keulegan. It relates discharge conditions to the movement of sediment particles in which the motive forces are a combination of tractive force, due to shear stress of the flow on the particles, and the fluctuating turbulent component of the flow velocity. Einstein made a thorough analysis of all phenomena that occur under uniform flow conditions on an alluvial bed.

The probability theory is introduced in the method, as in the case of all theories relating to turbulence.

Einstein assembled knowledge of sediment transport processes and introduced three important original contributions, building a comprehensive and consistent theory describing all aspects of the phenomenon. He determined the numerical values of the constants by his experimental work in laboratory flumes and by analysis of water and sediment discharge data for natural channels.

The Einstein method is rather complicated and laborious to apply so that the hydrologist is tempted to disregard it. There are about 30 calculation steps that must be repeated for a combined series of water depths and sediment grain sizes. The Einstein theory is usually presented in a rather complicated way so that the inexperienced user does not know where to start and to finish the calculation, what basic data are necessary and the value of the result he can obtain. In addition it is often referred in the

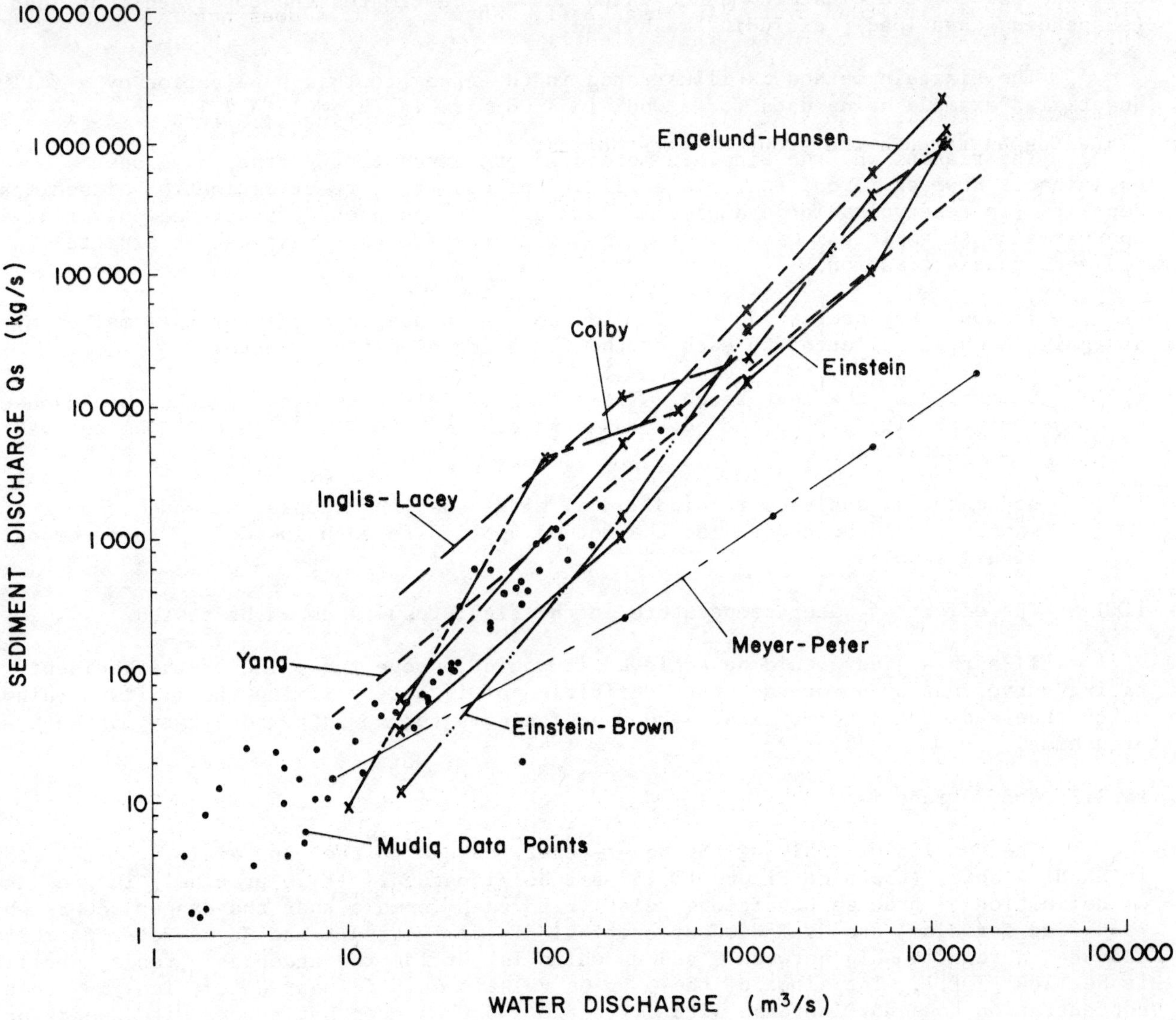

Fig. 15.4 Sediment transport formulas illustrated by data for Wadi Najran at Mudiq, Saudi Arabia

The figure illustrates the results of applying four sediment transport formulas to date for this location. The calculations for each method are shown by Examples A.4.1 to A.4.6. Sediment sample data, Q_s corresponding with discharge Q, which were used in the regression analysis shown by Figure 15.1, are also shown for comparison.

A wide range is revealed between Q_s predicted from Q by the various methods. This range is seen to be greatest (about an order of magnitude) at low Q values.

The figure is useful in demonstrating the uncertainty in estimating sediment transport and, thus, the useful life and economic viability of storage reservoirs in arid lands.

literature as "Einstein bed load formula" which creates confusion because bed load is only one part of the Einstein method which actually estimates the total sediment load (suspended + bed load), excluding wash load.

The Einstein method is illustrated in the Annex to this publication by a fully quantified example using data for a wadi in Saudi Arabia (Example A.4.1).

The reasons why the Einstein method is preferred to any other have been explained. However, proof that its application leads to correct estimation of sediment deposits (in reservoirs for example) is lacking. It has been demonstrated only that it approximates to empirical Q vs Qs or Q vs C with the accuracy that can be expected in sediment rate estimation.

It would be interesting to compute the confidence intervals of this method by assigning a margin of error to each of the following elements of data:

i) Selection of the test reach. The effect of selecting other reaches would check the stability of the method. The standardised cross-section could be derived differently.

ii) Bed material analyses for individual widely varying samples selected from all the samples could be chosen and the method applied to each instead of to a composite of all samples.

iii) The effect of water temperature in the final results could be tested.

If such a study could be achieved it would improve the value of the sediment rating curve, not by a more accurate definition of it but by fixing the limits within which true sediment rate may lie. The use of a computer would be necessary for such a programme.

15.2.2 Conclusions

The results of applying the Meyer-Peter formula to the case example of the wadi in Saudi Arabia, (shown on Figure 15.4) must be discarded. It is precisely in the lack of definition of precise conditions relative to each formula that the danger occurs when 'bed-load formulas' are used without critical discrimination. The Meyer-Peter formulas produces a relationship between Q and Qs which is outside the acceptable range specified in Section 15.1.2. The slope of the Q vs Qs relationship is near unity, implying that concentration does not increase with discharge and thus does not accord with theory or fact.

The Einstein-Brown and Engelund-Hansen formulas fit the measured data fairly well. The slopes of the theoretical Q vs Qs relationships (≈ 1.67) are somewhat greater than that of the straight-line regression derived in Section 15.1. Extrapolation from the actual set of data towards high values of sediment discharge using Einstein-Brown or Engelund-Hansen's formulas would be nearer to the expected range of exponents ($b = 2$ to 3).

It would be useful to test the sensitivity of these formulas to the variables involved. In particular, the bed material granulometry is represented only by the median diameter d_{50}. The effect of using d_s values at specified confidence limits attributed to this value could be assessed. The same procedure could be repeated for other variables such as velocity V, energy slope S, flow discharge Q. etc.

It would be desirable, as a conclusion, to give definitive guidelines for the use of certain formulas, given particular conditions. Unfortunately this cannot be done. Hydrologists must at present make their own decisions on the basis of available data and with a very critical state of mind.

The important points to emphasise are the wide disparity between predictions made using different formulas and the need to assess economic consequences of the risk taken in estimating the useful life of hydraulic structures.

The validity of two important basic assumptions should be considered when applying sediment load formulas to arid zone drainage channels:

- The assumption that the bed-material sediment supply is never limiting and that turbulence and thus transport capacity is fully developed,

- The assumption that stage-discharge relationships exist in conditions where the bed material sediment supply is unlimited for any discharge.

Thus, considerable error is implicit in mutually exclusive criteria necessary for use of these formulas.

In fact it is quite possible for channels incised in hard-rock upland areas to be scoured of all available bed material during rising flood stages, leaving unused transport capacity near the flood peak on the flood recession. Drainage channels in many such upland areas are often virtually devoid of transportable bed material. The area drained may produce large quantities of sediment which are transported by drainage channels and deposited downstream where energy is less but the assumption that the available energy in upland areas is fully utilised for sediment transport may be invalid.

For alluvial channels in which flow, velocity and sediment load cause changes of bed-form, cross-sectional area and shape during the passage of a flood, there is no such thing as a stage/discharge or depth/discharge relationship. In these conditions the notion of an equivalent bed-width to which a unit sediment discharge rate for a specified depth may be applied, is suspect. The fact that the mobile bed section is far from flat, with highly variable transverse velocity distribution suggests that formulas based on a theoretical mean vertical velocity distribution may produce misleading results.

For these reasons, empirical formulas based on measured reservoir sedimentation should not be discarded out of hand simply because all the variables remotely connected to sediment transport problems are not used. The fact that an existing reservoir has accumulated a measured quantity of sediment in specified climatic and geographic conditions is worth a great deal of speculation where water and sediment discharge data are lacking.

15.3 SEDIMENT YIELD AND DEPOSITION

15.3.1 Introduction

A common purpose of sedimentation studies for project design is the prediction of the temporal distribution of sediment deposits at the project site, e.g. in a reservoir, at an irrigation canal intake site, or any other water control structure and in estuaries where deposition may alter the direction and location of water courses serving agricultural needs.

Previous sections dealt with the establishment of one or several sediment rating curves for a given channel, providing an estimate of the instantaneous rate of sediment discharge in weight per unit of time as a function of instantaneous flow discharge.

In this section we consider the question of how to establish the amount, by weight and/or volume, of the sediments likely to be deposited at the project site. Integration of the "sedimentogram" (sediment discharge as a function of time) over a given time period will give the total weight of sediments.

15.3.2 Sources of Sediments

Unlike the hydrological cycle, the erosion-sediment-transportation-deposition process is one-way and non-stationary, since sediment material resources are not renewable.

The sedimentation process starts with the alteration and the degradation of the surface layers by heating and cooling, chemical and mechanical processes. This material is progressively eroded by the impact of rainfall and moved by runoff. Sediment is transported by wind and water to drainage channels and downstream by channel flow. This apparently simple procedure is in reality highly complex due to the great number of physical elements, functions of both time and space, that condition each step of the sedimentation process. In arid areas the complexity is greater than elsewhere due to the greater variability of these elements.

Extreme runoff events are of fundamental importance in the sedimentation process. For example a drainage channel may serve as a major sediment storage system during successive years with moderate runoff peaks causing a progressive rise in the longitudinal profile of the drainage channel system. The incoming sediments are delivered at a rate such that the ephemeral flows are incapable of removing them at a sufficient rate for a period of several years. When an exceptional flood occurs, a large proportion of the accumulated sediment may be removed and the bed may be scoured below its previous lowest level. The sedimentation process then starts again.

15.3.3 Erosion of the Watershed

Quantitative determination of the rate of erosion within a watershed is the subject of intensive research and data collection over recent years, particularly on small trial test plots. A number of equations and methods of computation have been developed which are based mostly on correlation procedures. Amongst them is the "Universal Soil Loss Equation" (Wischmeier and Smith, 1965) and its modified form (Williams and Berndt, 1972). The quantity of eroded material is usually expressed in weight per unit area per year and is a function of a series of factors including:

i) Climatic: rainfall characteristics, temperature variability.

ii) Geological: rock formations, degree of alteration, hydrodynamic properties;

iii) Topographical: slopes, relief, altitude etc.

iv) Vegetal cover: nature and density of species;

v) Human factors: agriculture, urbanization, engineering works.

all of which vary in time and space.

15.3.4 Sediment Delivery Ratio:

The sediment delivery ratio expressed over a long period of time is the ratio of the sediments transported by the watercourse network at a particular cross-section divided by the total quantity of erosion of the drainage area at this particular cross-section. Measurements undertaken in the United States in existing reservoirs located in non-arid areas show that the sediment delivery ratio decreases with increase of the watershed area. As an indication, the following ranges are found:

Catchment area km^2	Sediment delivery ratio %
0.1 to 2.0	20% to 90%
2.0 to 100	9% to 50%
100 to 1000	3% to 15%
Above 1000	No data

The remaining part of sediments not transported from the drainage area by the streams is deposited in channels, on flood plains, terraces and alluvial deposits of all kinds.

The present state of knowledge of the erosion process and delivery ratio cannot be quantified with sufficient precision for project design purposes. Sedimentation investigations in relation to water conservation and agricultural development projects must start with the study of sediment transportation in the particular watercourses themselves.

15.3.5 Trap Efficiency

The trap efficiency is the ratio of the volume of sediments trapped in a reservoir to the volume of sediment that enters the reservoir. This ratio is required to estimate the rate at which a proposed or existing reservoir will silt up given an estimate of the rate of water and sediment discharge (Q and Qs) of the drainage system at the site. Trap efficiency is a function of:

i) The sediment size distribution - the fall velocity w of suspended bed material increasing with particle size diameter - and

ii) The time during which each elemental volume of inflow water loaded with sediments will travel from the inlet to the outlet of the reservoir.

Three authors have proposed empirical relationships based on data from surveys of reservoirs at various times. None of these three relationships takes into account the particle size distribution quantitatively.

1. Brown, C.B. relationship. 1944 (Figure 15.5a). The trap efficiency is expressed as a function of the ratio of the reservoir capacity to the watershed area. The data used by Brown show a reasonable grouping of data points between the envelope curves.

The upper envelope curve should be used for arid zone conditions due to the combined effects of the following:

i) the temporal variability of inflow;

ii) the relatively long detention time of water;

iii) the coarseness of the sediment load;

iv) the position of the reservoir outlet at the dam does not allow deposited sediments to be easily removed.

2. Brune, G.M. relationship, 1953 (Figure 15.5b). Trap efficiency is expressed as a function of the dimensionless ratio of the capacity of the reservoir (m^3) to the mean annual water inflow (m^3). Two envelope curves delineate the empirical plots representing data from surveyed reservoirs. The upper curve is representative of coarse sediments and the lower one of fine sediments.

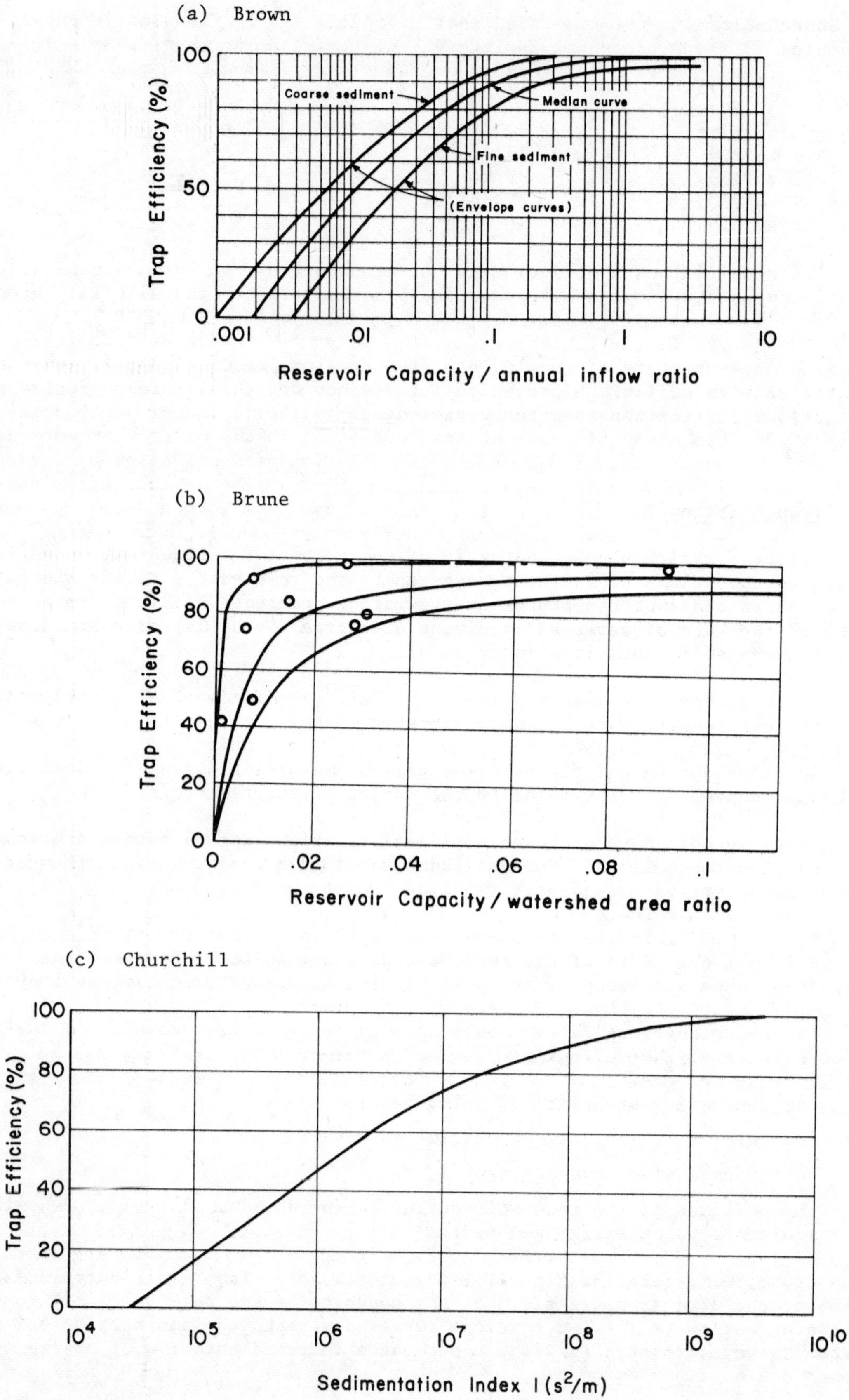

Fig. 15.5 Trap efficiency relationships

3. Churchill, M.A. relationship, 1947. (Figure 15.5c). Trap efficiency is expressed as a function of the "sedimentation index" I of the reservoir, defined as the period of retention T divided by the mean velocity through the reservoir. After simplification this index is

$$I = \frac{C^2}{Q^2 . L}$$

I : sedimentation index s^2/m
C : capacity of the reservoir (m^3)
Q : average daily inflow discharge (m^3/s)
L : length of the reservoir (m)

The best results are believed to be given by Churchill's relationship although no systematic verification has been made specifically for arid zones.

However, each of the three functional relationships are based on average conditions in time and space to derive a parameter of which the trap efficiency is a function. The various influences of the shape of the reservoir and inflow hydrograph and grain-size analysis, etc. are not taken directly into account.

In arid zones the sediments are often coarse with high settling velocities and it is known that the filling of existing reservoirs is extremely quick compared to temperate climate conditions. High trap efficiencies are therefore believed to be characteristic of arid zones.

The definition of a precise trap efficiency coefficient for a design purpose is not of major importance in relation to the wide uncertainty in the precision of estimation of sediment yield with available methods of prediction.

15.3.6 Fleming's Design Curves for Suspended Load Estimation

A rough preliminary method of estimating average annual sediment weight has been proposed by Fleming (1969) following studies of a collection of world-wide data. Such estimates should not be accepted as valid for design purposes, but only as giving a rough order-of-magnitude estimate.

Fleming assembled the data of annual sediment yield and flow volume for about 250 river catchments world-wide. He derived four relationships according to a qualitative classification of the catchment vegetation.

Regression analysis was used to derive the best statistical relationship between annual suspended sediment load and mean annual discharge. Two of the four curves refer to semi-arid and arid zones, with the definition:

- short grassland and scrub
- scrubland and desert

Figure 15.6 shows graphically these two relationships together with the points they are derived from. These curves should be used with caution bearing in mind that they would provide estimates of average annual sediment weight with a precision of, at best, about one order of magnitude.

Basic relationship: $Vs = AV^B$ or $\log Vs = B.\log V + \log A$

Vs : average annual sediment weight in 1000 tonnes/year

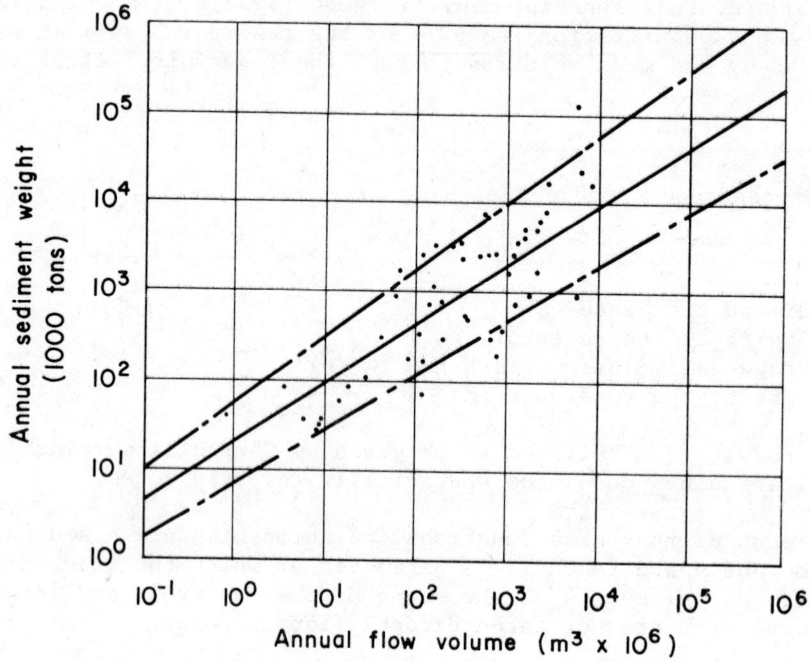

(a) Short grassland to scrub

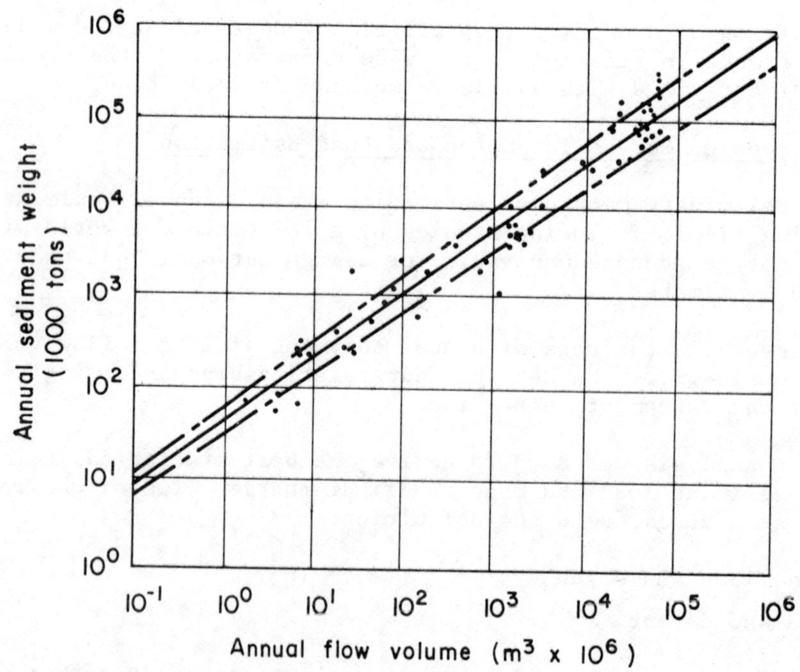

(b) Scrub land to desert

Fig. 15.6 Design curves for suspended load estimation (after Fleming, 1969)

V : average annual flow volume in Mm^3/year

Curve	Vegetational Cover	A	B
a.	Short grassland and scrubland	19	0.6523
b.	Scrubland and desert.	58	0.7158

In the case of the South Arabian wadi basin used for earlier example calculations, Fleming's curves would produce an estimated suspended sediment yield of about 0.6-1.7 M tonnes/year at 95% confidence level. The estimate may be compared with results obtained from the relationships established as described in Sections 15.1 and 15.2 of 1.75 M tonnes/year. This result suggests that Fleming's curves are suitable for first approximations in the absence of other data.

15.3.7 Specific Weight of Deposited Sediments

In order to transform a weight of sediments into the corresponding volume the specific weight of accumulated sediments must be estimated to predict the storage capacity that will be lost in a given period of time.

The specific weight of accumulated sediments is dependent on:

i) the specific weight of the individual particles, considered in practice as a constant:- = 2650 kg/m^3.

ii) the granulometry of the sediments: the finer the particle size the greater is the volume of voids available for water (including water adsorbed onto particle surfaces); with coarse but well graded sediments the voids in the coarser fraction are filled by finer sediments so that the effective voids ratio is smaller and the specific weight greater, than for uniform grain size.

iii) compaction of the sediment deposits, which increases with time and varies according to their granulometry.

Functional relationships of the specific weight of deposited sediments in reservoirs to the above parameters are given by Lane and Koelzer and developed by Miller to derive the following relationships (after conversion to the metric system):

$$\gamma_{10} = \gamma_1 + 0.67\ K$$
$$\gamma_{20} = \gamma_1 + 0.94\ K$$
$$\gamma_{30} = \gamma_1 + 1.09\ K$$
$$\gamma_{40} = \gamma_1 + 1.21\ K$$
$$\gamma_{50} = \gamma_1 + 1.30\ K$$
$$\gamma_{60} = \gamma_1 + 1.37\ K$$
$$\gamma_{70} = \gamma_1 + 1.44\ K$$
$$\gamma_{80} = \gamma_1 + 1.49\ K$$
$$\gamma_{90} = \gamma_1 + 1.54\ K$$
$$\gamma_{100} = \gamma_1 + 1.59\ K$$

γ = specific weight of deposited sediments in kg/m^3.

Subscripts refer to the number of years since sediment deposition. K is a function of the grain-size and of the degree of submersion of the deposited sediments.

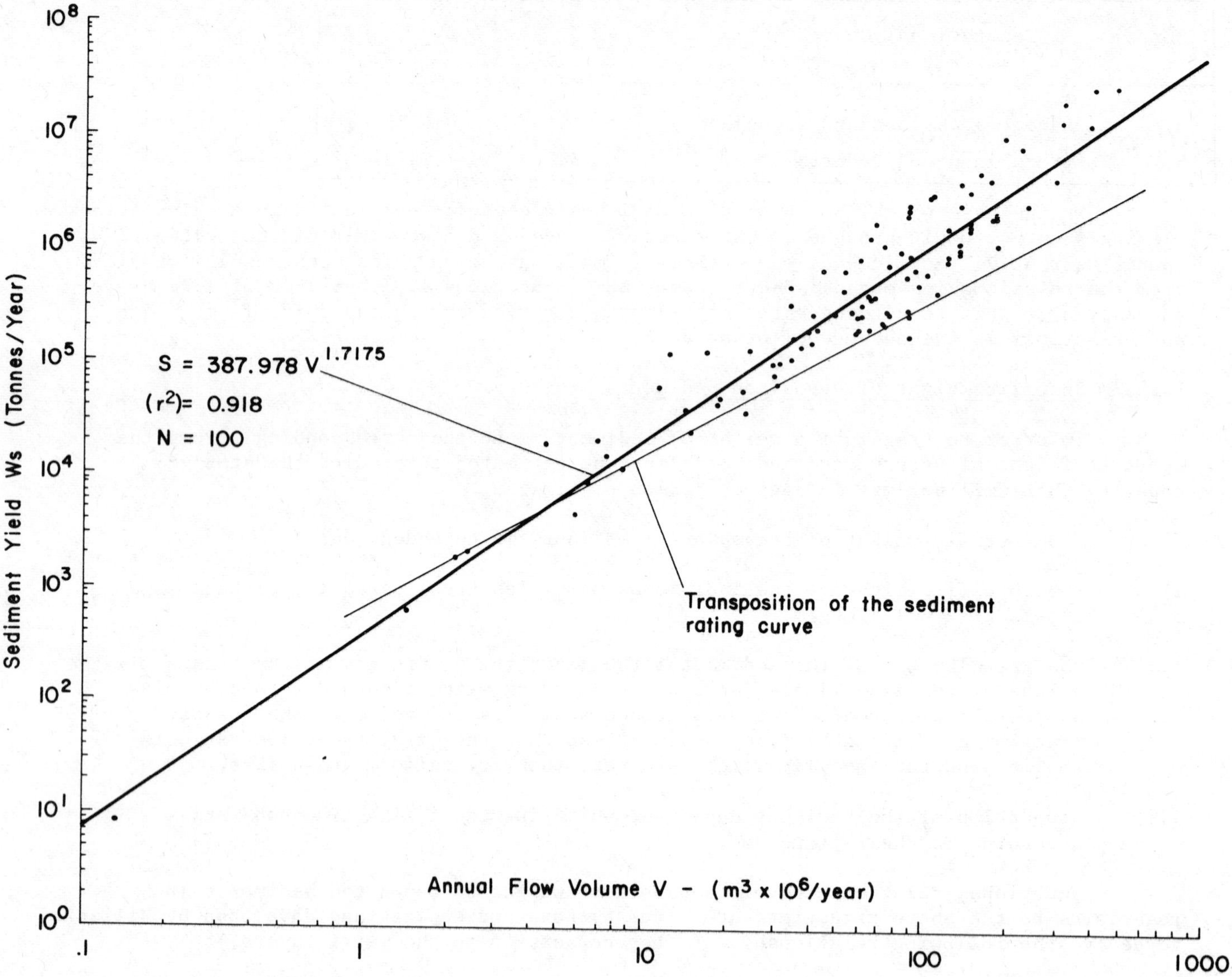

Fig. 15.7 Annual sediment yield vs. flow volume - computed from stochastic generation model

The figure shows the relationship between synthetic annual flow volumes and sediment yields generated using the stochastic model described in Section VII. It may be seen that prediction of annual sediment runoff weight from annual water runoff volume should theoretically be possible once a relationship of this type has been established, in this case 100 years (of synthetic data).

The adjusted regression equation $Q_s = 2.014Q^{1.3334}$ derived as shown in Figure 15.1 is also shown on the figure after converting Q-m^3/s to V $m^3 \times 10^6$/year and Q_s-kg/s to Ws-Tonnes/year. This demonstrates the impossibility of applying a regression equation based on instantaneous water and sediment discharge values to a mean discharge over a period of time. The functional relationships are completely separate.

Degree of submersion	Sand		Silt		Clay	
	γ_1 kg/m^3	K kg/m^3	γ_1 kg/m^3	K kg/m^3	γ_1 kg/m^3	K kg/m^3
(a) Sediment always or nearly submerged	1490	0	1041	91	481	256
(b) Normally a moderate reservoir drawdown	1490	0	1185	43	737	171
(c) Normally considerable reservoir drawdown	1490	0	1266	16	961	96
(d) Sediment in open air	1490	0	1314	0	1250	0

In practice, sediments are deposited continuously or intermittently over the life of the reservoir so that the specific weight should be integrated in time to include subsequent deposits and obtain an average specific weight of sediments at the end of the given time period.

Grain size analysis is required to define the percentage of each of the fractions, clay, silt and sand. The above equations are then applied separately to each of them, the average specific weight being their weighted mean.

15.3.8 Sediment Yield

In designing a water control structure the planning engineer needs to know, from the sedimentation study, the rate of storage loss by progressive sedimentation and what size of sand sluices or other means of reducing this rate should be provided in order to minimise the loss of capacity.

The best way of presenting the results of the sedimentation study is to prepare a curve, preferably with confidence limits indicated, of the rate of accumulation of sediments as a function of time.

Since Q_s is a non-linear function of Q the practice of applying the relation to monthly or yearly mean water discharge is as incorrect as would be the determination of stream flow from mean monthly or yearly water levels.

The correct method is to construct, point by point, the volume "sedimentograph" from the hydrograph, using the two relationships $Qs = f(Q)$ and $\gamma = f(t)$, and then integrating numerically or graphically.

Figure 15.7 illustrates the errors introduced by using monthly or annual discharge: the flow volume and sediment weight for each year are computed mathematically from a synthetically generated hydrograph covering 400 years and a relationship of the type $Qs = AQ^B$ (in weight/unit of time). It appears on the plot that for a time period of one year a wide range of sediment weights correspond with one value of flow volume. This is easily explained by the non-linearity of the Ws/V relationships when the algebraical concept of "function of function" is involved.

As an illustration, let us imagine two hypothetical annual hydrographs that produce the same flow volume in one year:

i) A constant base flow all through the year of discharge Q_1. The annual flow volume will be V_1 = 365 days x $Q_1.m^3/s$ days.

ii) A single flood in the year, of 1 hour (say) duration with a constant flow of discharge $Q_2 = 24 \times 365Q_1$. The annual flow volume being $V_2 = \frac{24}{24} \times 365Q_1 = V_1$ m^3/s.

When applying the relationship $Qs = aQ^b$ with b>1 the sediment discharge would also be constant as:

$$Qs_1 = aQ_1^b \text{ and } Ws_2 = \frac{1}{24}.a.24^b.365^b.Q_1^b$$

so that their ratio

$$\frac{Ws_2}{Ws_1} = (24 \times 365)^{b-1}$$

for b = 1.5 to 2.5, $\frac{Ws_2}{Ws_1} = 94$ to 820000

Taking a typical value of 2.0 for b: $\frac{Ws_2}{Ws_1} = 8760$

The annual sediment yield in the second hypothesis is therefore much greater than in the first one. In practice the sediment yield is greater for equal flow volumes when the annual hydrograph is characterised by a few high sharp peaks. Arid zone hydrographs have precisely these characteristics of high floods of short duration.

The simplest method of calculating sediment discharge weight or volume is as follows:

(a) calculate the instantaneous flow-duration characteristics from the instantaneous discharge hydrograph for the period considered.

(b) apply the derived equation $Qs = aQ^b$ to discharges corresponding with selected durations to produce an estimate of the instantaneous sediment discharge-duration characteristics.

(c) calcuate the sediment weight for the period by integration of the duration curve, either by planimetering the area under the curve or by multiplying the mean sediment discharge by the total duration.

(d) divide the total weight by the density of the sediments to estimate the dry volume.

Anything more sophisticated than this procedure is unwarranted.

The total sediment runoff weight or volume at a measurement site does not give any indication, other than qualitatively, of the useful life of a reservoir at the same point. For this we need either measured instantaneous hydrographs Q f(t) and convert these to sedimentographs Qs f(t) or we must generate realistic synthetic inflow sequences by computer modelling techniques, converting discharge rate mathematically to sediment inflow rate. The reservoir response must then be modelled to assess the sedimentation rate taking account of:

- reservoir capacity
- outlet capacity

- trap efficiency
- reservoir operation rules.

A reservoir designed only to detain water for as long as necessary to prevent flooding or enhance recharge could be expected to be empty for long periods and sedimentation would tend to occur in horizontal layers. On the other hand a reservoir kept substantially full for hydro power or agricultural purposes might create a sedimentation delta upstream with typical bedding dipping downstream towards the point of release.

15.4 CONCLUSIONS AND RECOMMENDATIONS

15.4.1 Conclusions

It should be clear from the foregoing chapters that sedimentation studies are a part of hydrology - and in particular of arid zone hydrology - where inevitably complicated and often tedious computations yield the least return in terms of precision and confidence in the results obtained.

There are very many experimental and theoretical formulas, most of which have some merit when applied to the normal run of conditions in the locality for which they were devised. They all tend to be of decreasing reliability in precisely those exceptional flow conditions which are generally assumed to be of most importance in transporting sediments: in fact even this assumption is based rather on subjective judgement than on rigorous proof. In humid conditions with perennial river flow and generally good vegetational cover the individual effects of exceptional flow conditions are almost always less marked while their overall effect on sediment yields is damped out. In arid conditions the mere existence of flow is exceptional and the only thing that is common to almost all flow events is their extreme flashiness and their extreme degree of variability.

The method of Einstein has been selected as being the one which most closely takes account of the physical processes in arid zone sedimentation. The large number of formulas which have been deliberately left out are found in the ASCE Sedimentation Manual which is thoroughly recommended for further reading.

15.4.2 Follow-up

It would advance the state of knowledge very considerably if the methods proposed in this section could be tested in arid zone ephemeral streams on which reservoirs already exist or are proposed. Simultaneous gauging of both water inflow and its sediment load (and possibly of the outflow and its sediment load) would be required and very careful topographic survey of the sediments deposited (by underwater sampling and sounding, or by normal methods when dry). If comparisons were made on an annual basis it might well be possible to detect the reasons for any serious discrepancies between calculated quantities, from inflow gauging, and quantities determined by survey.

It should be remembered that any refinement in calculation of total yields made possible by such a study would only be directly applicable to another stream having the same topography, geomorphology, vegetation, land use and climate. However these conditions of similarity would not be essential for verification of the relationships between water flow and sediment discharge in the stream itself.

Finally it must by now be clear that everything in this section from qualitative descriptions to attempts at quantification can, at best, only be guides to intelligent objective estimates backed by field experience.

GROUNDWATER

16. CHARACTERISTICS OF GROUNDWATER IN ARID ZONES

16.1 GENERAL

The contrast in the practice of groundwater hydrology between arid, dry, temperate, cold and hot zones is generally less spectacular than in surface water hydrology. Nevertheless, the hot arid regions as well as the cold arid regions have interesting characteristics. The rock types comprising the aquifers are as varied in one climatic region as another and therefore the geologic aspects of groundwater generally apply in most climates. The aquifer materials in many parts of the world have gone through similar geologic changes including changes in climate so the principles involved in determining how an aquifer reached its present state depend little on present climate at the surface.

Of course, present landforms do depend on Recent and Pleistocene climates but landforms are only one aspect of hydrogeology. The major difference in groundwater between arid and wet regions is in the surface water resources which become available to the aquifer.

Furthermore, in arid regions water itself is often more precious. Therefore small or poorly permeable aquifers, which would be considered useless in wet regions due to abundance of surface water, tend to have much higher value in arid regions.

Much can be said about extensive aquifers of the Arabian Peninsula and the North African arid zones, but similar aquifers exist in less arid and even in very wet regions of the world so the geologic principles are not essentially different.

The areas of Africa where the long, narrow, shallow alluvial aquifers are the major source of water are important, but very similar aquifers of equal local importance are found in many wetter regions of the world.

It will be evident that the major problem of groundwater hydrology in arid regions is simply that water is scarce, and is often used at rates exceeding recharge. Therefore more precise estimates of recharge are needed than in wet regions where the same type of aquifer may scarcely be used at all.

16.2 ROCK TYPES

16.2.1 Introduction

The emphasis on the discussion of rock types will be on typical problems in arid regions. Since most of these are similar in all climates only some aspects can be highlighted here. One of the few recent sources of general information is Unesco (1977).

16.2.2 Unconsolidated Sediments

Water-deposited sands and gravel aquifers are usually limited to relatively narrow river valleys, to deltas on the sea and on large lakes, to alluvial fans and to shoreline deposits related to beach-forming processes.

In many parts of the world old Quaternary river valleys with their gravel aquifers have been abandoned and sometimes can scarcely be recognized, yet may provide good water supplies. Alluvial aquifers in arid valleys surrounded by mountains are only recharged by river flow originating from the wetter high lands. Local rainstorms may be quickly concentrated in the alluvial channels.

While some of these aquifers are large (Nile Valley, Senegal River Delta, San Joaquin Valley, California) many hundreds are much smaller. Each, regardless of its size, can be of importance to those living on or near it in arid regions.

Modern wind-blown sands in most arid regions supply little water except where they are relatively thick and continuous and are recharged from underlying aquifers or from nearby streams. Where wind-blown sands fill old stream channels or other depressions in relatively impermeable rock they may provide useful storage for limited uses, since evaporation may be suppressed by the sand cover. Available research suggests that dune sands allow considerable infiltration of rainfall but that porosity is large and movement of water slow so that in high temperature regions much of the infiltrated water is lost by evaporation.

16.2.3 Consolidated Sediments

In general there is no fundamental difference between groundwater in consolidated sediments found in arid regions and those found in humid regions. Since water may be more valuable in arid regions the groundwater in consolidated sedimentary rocks of arid regions may be of more value than in humid regions. This can give the impression that these rocks are different but from a geologic point of view it is not true, particularly since many of these formations have been deposited when climate was completely different from the present.

The consolidated rocks generally have relatively low permeability and therefore groundwater movement will tend to be slow. This is a decided advantage where recharge is low, since the rocks will not drain rapidly. However, from a well production standpoint, the lower permeability tends to increase well and water costs and often limits the amount of groundwater that can be pumped to village and stock water use.

Of course, there are relatively permeable formations in this class due to limited consolidation and cementing. There are also some consolidated sediments which are essentially impervious except where fractured and faulted. If these fracture systems are extensive enough, major and important water supplies can be obtained if there is sufficient recharge or if the fractures drain a large region of low permeability but high storage capacity.

16.2.4 Carbonate Rocks

Carbonate rocks in hot arid regions generally are not as highly karstified as in colder and wetter regions. Past climates may have affected the carbonate rocks in a different way than present climates are affecting them, and therefore origin of carbonate rock permeability may have a very complex history. However, it seems to be relatively rare to have numerous and very extensive and large cave systems in most arid regions. Thus carbonate rock permeability in arid regions generally consists of relatively widespread small solution channels which form along joints and fractures related with tectonic activity (Jordan, Israel, Egypt).

Around the Mediterranean there are Quaternary coastal dunes which consist primarily of carbonate fossil shells. These also exhibit karstic features but are generally included as unconsolidated sediments and are relatively limited in extent.

Carbonate rocks in the Mediterranean and Arabian Region are often closely related with gypsiferous sediments as well as normal sandstones and shales. Some karstic features observed may be due to solution of gypsum as well as limestone (Qatar, Bahrain, Saudi Arabia, Iraq, Syria). Solution of gypsum may also result in fracturing of overlying limestones, thus allowing greater circulation and hence greater karstic development than where gypsum was not originally present.

Such relationships require extremely skilled and detailed geologic work to determine just what the permeability is, its distribution and how best to utilize the groundwater found there.

Where the carbonate rock is thin and inter-bedded with shales it will be very poorly permeable and not yield significant water to wells. Where the carbonate rock is massively bedded and even moderately affected by tectonics some degree of karstification, and therefore some permeability, may be found.

Carbonate rocks consisting of reef limestones may exhibit more karstification than nearby fine-grained limestones.

16.2.5 Volcanic Rocks

In most arid regions volcanic rocks are either associated with young volcanoes or are a part of an ancient volcanic system. In general, the ancient volcanic systems tend to be very poor aquifers except in fault zones. While some of the younger volcanic rocks are quite impervious, they can provide good water supply if rainfall is high enough. In very low rainfall areas, however, younger volcanoes generally are very permeable and therefore the groundwater drains away quickly to the toe of the volcano or to the sea, and sometimes into underlying aquifers.

Use of volcanic aquifers can be difficult due to costly and difficult drilling. As aridity increases the discovery of the spring outlets from the volcanic rocks is easier on land but if recharge is limited these springs may be dry much of the time due to the rapid discharge of the water of recharge.

16.2.6 Granitic and Metamorphic Rocks

Considerable regions of the land mass are underlain by these "hard" rocks which are practically impervious except where fractured and/or weathered. Fracturing of course depends on the tectonic conditions while depth of weathering depends very much on time and past climates.

The storage capacity of fractured rock is usually very limited so that in arid regions little of the rare recharge can be stored. Weathered rocks may have good storage capacity but low permeability. Therefore the ideal location for wells is in fractured rock overlain by saturated weathered rock.

Even in wet regions these rocks yield limited water and often at relatively high cost or with much labour (India, Africa, Australia). However, since such large areas are underlain by these rocks there may be little choice to inhabitants but to try to find water in them.

Fortunately, where weathering is extensive it seems that there usually is fractured rock below. However, in most hard rocks, fractures are rare below 50m and normally not found (except in highly faulted regions) below 100m.

In Saudi Arabia these rocks seldom yield water because of the limited rainfall, but they may yield some water when they are beneath wadi alluvium.

Where these rocks are very poorly permeable the small amount of water contained may be of very poor quality due to extremely slow circulation. Therefore fresher water usually indicates that there is circulation and therefore some replenishment and some permeability.

16.3 HYDROLOGIC BALANCE

16.3.1 Introduction

The principles of hydrologic balance (Unesco, 1977, Todd, 1959) are no different in arid regions than in other regions. The major difficulty is estimating recharge, while other items in the groundwater balance often are somewhat easier to estimate than in humid regions where interactions between surface and groundwater may be very much more complex in detail.

16.3.2 Recharge

As discussed elsewhere the reason for difficulty in estimating recharge in arid regions is the inadequacy of surface water measuring stations, high runoff in a short time, and irregular distribution of rain in time and space. Furthermore, the fact that all of the arid region may not be underlain by suitable aquifers complicates the spatial determination of recharge.

Where narrow alluvial aquifers are widely spaced they have relatively little surface area for direct recharge. Unless there is a large tributary area for runoff to reach the alluvium it may have very little recharge. If the alluvial valley has a mountainous tributary there may be more chance for seasonal and reliable recharge from the stream even though local rain on the alluvium may be very small.

In contrast to small river valleys is the problem of estimating recharge to very large and extensive aquifers. Direct recharge of rainfall can occur where there is an internal drainage system. While this situation is not the rule it is common in younger volcanic rocks and in some carbonate rocks (Qatar, Iraq, Saudi Arabia). Therefore it is important to note and determine the extent of such areas with some accuracy. A similar situation of course is sand dunes, where only major tributaries may be able to keep their channels clear. The work to date on recharge of sand dunes suggests that most of the water recharged is retained in the upper zones of large dunes where it eventually is evaporated due to high temperature. Furthermore, the sand dunes may be travelling over relatively impermeable deposits where recharge is not important.

In many arid regions however, there are extensive drainage systems which from time to time have runoff and therefore the possibility of recharge to alluvium in the stream bed and then into other underlying aquifers.

16.3.3 Groundwater Movement

Knowledge of groundwater movement is essential for determination of the water balances of aquifers. Problems of determining flow are little different in arid regions than elsewhere and depend, as usual, on aquifer size and on the precision desired for the water balance. Aquifer testing to determine parameters is not different in arid and wet zones (Kruseman and de Ridder, 1970).

Since recharge in arid regions is relatively small, the horizontal component of groundwater flow is usually important in the hydrologic balance and is usually the same order of magnitude as the natural recharge and discharge. In contrast, aquifers in extremely wet regions may have very large vertical components from recharge areas and to discharge areas, and vertical groundwater flow components may be many times greater than horizontal flow.

16.3.4 Groundwater Discharge

16.3.4.1 Natural Discharge

Natural groundwater discharge in arid regions often is found as natural springs, sometimes as swamps and more commonly as large salt flats called by various names in different countries. Natural discharge may also occur into streams, lakes and seas where it may not be so obvious and, when the outflow area is diffuse, may be impossible to detect (Unesco, 1977).

In much of the African and Asian arid regions, man has disturbed the natural springs for centuries if not for thousand of years, and evidence of such disturbance is not readily observable in most cases. This disturbance normally has been to dig into the spring in an effort to get more water. In some cases digging has been deep enough (Algeria) to disturb shallow impervious sediments, allowing the spring to flow strongly. As the spring flows vegetation grows, sand accumulates and raises the outlet, until a mound is formed large enough to slowly absorb the spring discharge. Work in Australia indicates that the size of the sand mound gives some indication of the spring discharge.

As aridity increases the number and discharge of natural springs normally decreases. In hot arid regions potential evapotranspiration is high; thus the amount of vegetation around small springs tends to be less than in cool arid regions. Estimation of total natural discharge is therefore difficult and imprecise. The estimation is further complicated where groundwater level is less than two metres from the surface. In theory, the amount of loss to the atmosphere depends on depth and type of sediment, the finer sediments having higher capillary rise. However, capillarity also depends on temperature which is another complex factor in these conditions. In shallow water table areas, phreatophytes further complicate estimation of natural discharge, since consumptive use data is available for only a very few phreatophytes (see Section 9.3).

16.3.4.2 Artificial Discharge

Artificial discharge from aquifer in arid regions presents more or less the same problems as in wetter regions. However, in addition to dug and bored wells, gently sloping tunnels are common in much of the Near East and North Africa. These tunnels, (called qanat, ghanat, falaj and other names) are usually constructed in the head of large alluvial fans to intersect the water table and then to carry it down by gravity to irrigate areas where the water table is deep. These tunnels when working properly discharge in a manner similar to springs with high discharge in the wet season and low discharge in the dry season. When there are sufficient numbers of such tunnels in a given area they act as drains, permanently keeping the high water table down. Measurement of the outflow of the tunnels is simple in principle but if there are large numbers of them it may be very difficult with limited resources.

Similarly, when there are many wells, estimation of discharge may be costly or impossible by direct measurement. Where pumps have electric motors sampling of power required applied to figures obtained from the power distribution agency is sometimes a practical solution.

Another way to estimate groundwater extraction, especially where there is no surface water available is to survey the land use and estimate the water requirements by land-use categories. In arid regions, use of air photography for this purpose can provide reasonably accurate results. However, in many warm arid regions, the cropping patterns may be quite complex so that land use at any one time may not be representative. Therefore sampling of land use over the growing season is generally attempted to increase the precision.

Even if land use could be precisely known, however, irrigation practices vary widely among farmers and it is often a difficult task to obtain water requirements which can be broadly applied to land use figures. Even if this were possible, the return of irrigation water to the aquifer must also be considered and this is subject to all the difficulties of estimating water requirements.

The problem of determining whether the return irrigation and rainfall will or will not return to underlying aquifers may be discussed briefly. In many arid regions the first wells drilled into an aquifer may flow as artesian wells, thus indicating that there is a confining layer above the aquifer. It is usually easiest to assume the confining layer is impermeable as long as wells flow. When the confined water table is lowered after heavy groundwater use, the assumption of zero vertical permeability often continues to be made. However, when the confined water table is pumped down until it is further below the shallow water table than it originally was above, a significant downward hydraulic gradient exists which can result in large vertical downward movement, even in relatively impermeable aquicludes. Such downward movement will tend to balance the hydrologic equation and can be very significant. Unfortunately, estimating the vertical permeability of an aquiclude is not easy. Leaky aquiclude permeability pump tests are the most commonly used method.

In spite of the difficulties involved in determination of natural and man-made discharge from groundwter, the techniques mentioned above have been applied with reasonable success to many basins.

16.3.5 Change in Groundwater Storage

In arid regions changes of groundwater storage are perhaps more variable and therefore of greater importance to aquifer use than in very wet regions. Of course, there is a wide range of groundwater level fluctuation and therefore change of storage.

On a regional basis those aquifers with little recharge and small permeability will exhibit small changes in water level. Similarly areas with high recharge and high permeability may also exhibit small water level changes. Thus, water level changes and change of groundwater storage depend on both the hydrology and the geology.

Determination of the storage coefficient in arid zones is just as difficult as elsewhere. Experience indicates that pump tests are seldom long enough to obtain reasonably true values of the storage coefficient. A good estimate of the storage coefficient of the materials penetrated by wells is ordinarily a better indication and requires a geologic sense of proportion. Some materials can of course be cored and the core tested in the laboratory. However, many cores may be required to reach conclusions on a large scale.

Groundwater storage is important in arid regions because it provides both seasonal carryover storage and long term carryover storage. In the event that extractions exceed recharge, the rate of water level drop will depend strongly upon the storage coefficient as well as other factors.

16.4 THE AQUIFER AS A TRANSPORT SYSTEM

The aquifer provides relatively free transport facilities. It transports water from recharge areas to areas of discharge, including man-made wells. In arid regions the transport system may be well below the surface and therefore not subject to evaporation losses as open canals would be. Thus, in analysing whether surface water should be recharged or whether it should be transported and applied directly, the transport character of the aquifer should be kept in mind.

16.5 THE AQUIFER AS A RESERVOIR

General aspects of the aquifer as a reservoir have been discussed above. It is essential to recognise that it is not feasible to prevent all losses from a groundwater basin just as it may not be feasible from a surface reservoir. However, in arid regions, reduction of evapotranspiration from groundwater compared to evaporation from surface water storage may be considerable.

It is generally desirable that groundwater extractions be controlled to equal long term recharge and to avoid serious damage such as saline water intrusion.

The level of operation of the groundwater basin should therefore be within certain limits - not so high as to cause excessive discharge or high water table, and not so low that poor quality waters intrude, that pumping levels become uneconomic, or worse, that the aquifer is dewatered and overlying owners deprived of a supply without recourse to a substitute supply.

In the western USA groundwater is usually legally overpumped only where a substitute supply is available, in order to avoid serious depletion or serious water quality deterioration.

16.6 GROUNDWATER MODELS

There is a wide variety of groundwater models which have been used for groundwater studies in all climates. From the point of view of the type of groundwater model, arid regions are no different than other climates, and the type of model used depends entirely on the preference and skills and resources available to the user. Kunkel (1973) presents a short discussion of data requirements for groundwater models in arid regions.

While there is a very large amount of literature on groundwater models there are no single source which gives the uninitiated guidelines on which model is best for a particular application. Instead, users tend to remain faithful to the particular kind of model they first learn about, amending it as they use it for different conditions. It is clear that, regardless of the mathematical solution used, models can be more precise in calculation than field data applied to them. Therefore the calculation processes are no longer an issue. Pickett (1975) summarises the types of models available.

Accuracy of data used and detail which is desired in the model will always be important issues in deciding validity of model results. One of the most obvious examples is modelling groundwater quality. It is well known that individual chemical analyses of water from nearby wells, properly collected and analysed, may fluctuate widely both in time and space. A quality model therefore may be mathematically precise but will not be able to duplicate complex details because geologic, hydrologic and water quality parameters may not be available with sufficient precision.

It cannot be over-emphasized that the obvious use of a model for prediction is often overshadowed in importance by use of the model simply to understand what data need to be obtained in order to better understand the basin being studied.

Groundwater models may be rather simple or they can be very complex. The more complex a model is the more data are needed. For a first attempt on a particular basin or aquifer, it may be better to resolve overall problems of water balance and boundaries with a relatively simple model. The results can then be used to speed up verification of a more complex second-stage model.

In very large aquifers, some of which are in arid regions, the time for small changes to propagate may be very long and this is perhaps one of the few differences in use of groundwater models in arid regions. Determination of the time for a major hydrologic change to affect a large aquifer is important in order to put man's own activities and planning scale into proper perspective. In very large, thin, unconfined aquifers, hundreds of kilometres in extent, one portion of the aquifer might be essentially dewatered, yet it may take hundreds of years for the effect to propagate across the basin. When saline water bodies are also involved, the time for significant particulate movement may also be very slow relative to man's planning horizon. There is some evidence that the saline water wedge in the Nile alluvial deposits is not yet in equilibrium with the present fresh water balance even after several thousand years. Similarly, movement of saline water in the Indus Plain alluvium appears not yet to be in equilibrium after 70 years or so of extensive recharge by surface water applied for irrigation. Groundwater models can provide some insight into the time scale of such problems.

Groundwater models can be of particular value for studying the various alternatives for planned extractions which exceed recharge, a common situation in arid regions.

16.7 GROUNDWATER DEVELOPMENT

16.7.1 Introduction

Differences between arid and humid zones in the development of groundwater for irrigation and other uses are few. Again the major difference is the frequent availability of surface water for at least part of the year in wetter areas where groundwater may only be a supplemental supply.

In arid regions, groundwater may be the only supply and therefore its reliability must be determined. When water extraction is only a small proportion of recharge there is of course little problem. As extraction approaches and exceeds recharge, the precision of the water balance becomes more important for future planning.

Thus, when groundwater extraction is less than mean recharge it is said that the "safe yield" is not exceeded. However, in many arid zones only limited extraction will result in "overdraft" (will exceed safe yield). Then it is a matter of decision of how rapidly to use the groundwater in storage.

16.7.2 Extraction Methods

In general, extraction methods do not depend as much on aridity as on hydrogeologic conditions.

16.7.2.1 Well design in arid areas must take into account the possibility of lowering of the static water level. This may be done by hand calculations of the water balance or in more detail with simulation models. Other problems affecting well design, such as corrosion, occur in all climatic zones (Campbell and Lehr, 1973).

Upper limits of well discharge are set generally by the physical nature of an aquifer. The actual development limits may be set by economic and other factors. Well design for irrigation must take into account a host of factors such as farm sizes, cropping patterns, type of on-farm irrigation and distribution, reliability of supply needed, etc. even before well discharge is chosen. On the other hand, depth and permeability of the aquifer, depth of water table and drilling and material costs and operation costs will also determine the most economical well discharge.

Reliability in arid regions is of course a major factor in the development of groundwater.

16.7.2.2 Other extraction methods (Huysmans, 1972) include buried galleries, excavated tunnels or qanats and others requiring different engineering approaches. In sand rivers simply burying a pipe at the base of the sand may suffice, but construction of a partial subsurface dam or barrier may also be needed. In less permeable rocks where well discharge is constrained, small storage may be needed to provide adequate water during the working day.

16.7.3 Sub-surface Dams

In semi-arid regions an important source of water is the intermittent stream which may have several floods per year of relatively short duration and high peaks, followed by a certain amount of low flow in the hydrograph recession. Often this recession flow may continue for some time in the streambed below the surface. In areas where the volume of gravel or sand fill under the stream channel is significant, and this alluvial fill is underlain by some impervious strata both at the bottom and sides of the valley, it is possible to create a storage reservoir below the surface. The method consists of crossing the valley at a carefully selected site by an impervious wall or barrier which extends to the tight basement. The volume of storage created above this underground dam depends upon the cross section of the valley, the depth to the impervious layer, the longitudinal gradient of the channel and the porosity of the alluvial fill. Where these site conditions are favourable, and the predicted intermittent flow is adequate, it is possible to store water in a manner which is often much more economical than a surface reservoir of equivalent volume. Other advantages are much less evaporation loss, no loss of volume due to trapping of silt, little loss of land, no relocation of works or property in the valley area, no breeding ground for mosquitoes, and no costly spillway requirements.

Underground dams may be constructed in several ways depending upon the depth required, nature of the overburden, availability of local materials and equipment. Small depths can be achieved by trench excavation and plastic membranes. Greater depths require more sophisticated engineering methods such as piling or slurry walls. The stored water can be extracted by shallow pumping from wells or collection galleries above the membrane, or by gravity if sufficient slope is available below the dam.

16.7.4 Recharge

Recharge works in many arid areas may be constructed to reduce evaporation losses somewhat and to augment groundwater storage. In richer countries large complex recharge works have been installed. However, smaller scale but effective recharge is possible in many stream channels and on alluvial fans near their apex (Agricultural Research Service, 1970; Bize et al, 1972).

16.7.5 Groundwater Quality

One of the major aspects of use of groundwater in arid zones is its quality. Prior to large, long-term investments the quality aspect must also be studied. With increased extraction the natural discharge of water from the basin, which usually also carries out the dissolved salts, may be diminished or stopped, so that salts may accumate in the soil and in the groundwater.

16.7.6 Drainage

Development of groundwater for irrigation ordinarily will not require extensive drainage. However, import of surface water in large amunts to an arid area where groundwater is not used may result in sometimes spectacular rise of the groundwater table.

In some parts of the world, very serious soil salination has occurred requiring extensive drainage and soil reclamation. While similar phenomena occur in wet regions, the rainfall often is high enough to flush the soil salts downward and eventually into natural and artificial drains. (U.S. Bureau of Reclamation, 1972).

16.8 GROUNDWATER DATA NETWORKS AND EXPLORATION IN ARID ZONES

There is little technical difference between establishing groundwater data networks in arid and wet zones. The obvious differences are generally due to local features. If anything, data collection in arid zones tends to be easier, particularly in areas where groundwater is already being utilized. Here there will normally be all-season roads, wells useful for measuring, drillers' well logs, etc.

In less developed groundwater areas, there may be good access to surface features but few wells and well data. In these regions, electrical restivity surveys used properly with good judgement often provides a reasonable base for location of more costly boreholes (Zohdy, et al, 1974; Astier, 1971).

17. SPECIAL TECHNIQUES USEFUL IN ARID ZONE HYDROLOGY

17.1 REMOTE SENSING

17.1.1 Introduction

Aerial photography has been an invaluable tool for both surface and groundwater hydrology for many years. With the advent of satellites remote sensing has expanded for two major reasons: first, imagery is available for much larger areas than can be obtained from most aircraft; second, satellite imagery can be obtained repetitively for the same place at relatively low cost.

At the same time, there has been a proliferation of ways to handle the large amounts of imagery now being obtained, including new electronic-optical means to interpret the imagery. Many of these new techniques can be applied with great advantage to aerial photographs or even ordinary photos taken on the ground with a hand camera.

Simultaneously there has been a considerable advance in cartograhic and related techniques for improved utilisation of the imagery.

Of first and most obvious value: some of the satellite imagery can provide a directly usable base map up to a scale of 1/250,000.

Landsat imagery is taken in four spectral bands so that some features show better in one than in another. Water, for example, shows up best in band 7, while clear water a few cm deep will not be evident on band 4,5 or 6. Green vegetation shows up best on band 4. Careful superpositioning of the four images using colour filters allows the interpreter to bring out a variety of features not easily seen on one image alone.

Large flooded areas can be determined by using an image before (or after) and during the flood stage, superimposing the two with appropriate colour filters so that the flooded area becomes easily visible. The scale of the Landsat imagery is such that features smaller than perhaps 100m to 140m in size can seldom be distinguished. However, vegetation along stream channels, irrigated land etc. are usually visible in arid regions.

Large-scale geological features are sometimes seen best on Landsat. However, conventional air photography is still necessary for detailed work.

None of the remote sensing techniques can detect water more than a millimetre or so below the ground. Therefore groundwater studies must use the same interpretative analyses as with air photos.

Since Landsat imagery is repeated each 18 days, it is sometimes possible to detect vegetation growth resulting from rainstorms in otherwise arid regions.

There remain a wide variety of applications not discussed and an infinite variety of possible future applications. The most comprehensive manual on Remote Sensing is by Reeves, et al (1975).

17.1.2 Rainfall Estimation

Another way to refine water resources estimates in arid zones is to improve the estimate of rainfall. In many arid regions rainfall occurs mostly from convective storms triggered by orographic or air mass instability effects.

In the past the only means of improving areal rainfall estimates has been to increase the number of raingauges. In most arid regions this is impractical since some storms cover only a few square kilometres and raingauge spacing of the required density on a national or basin-wide scale would be uneconomic.

Very detailed work on individual thunderstorms indicates that one gauge per three or less km^2 is needed for a 10% precision estimate of areal rainfall. It is unlikely therefore that simply increasing raingauge density will ever be a practical solution for estimating storm rainfall over large arid areas.

In general, the more arid the area, the fewer storms are recorded by a particular gauge. However, storms normally occur nearly each year somewhere in a region. Furthermore, the intensity of rainfall is usually high enough to cause local runoff and perhaps recharge to aquifers.

There remain two alternatives to improving rainfall estimate in arid regions. One is by using weather radar, the other is by using meteorological satellites.

1. Weather Radar

Weather radar requires costly and sophisticated equipment and, unless the equipment is needed for other purposes, generally is beyond the resources available for water resource studies in most arid regions. Weather radar operates only within line of vision and is therefore limited to a radius of about 150 to 200 km.

To detect and calibrate 99% of the 24 hour rain greater than 2mm rainfall, a gauge density of about one per 50 km^2 is required. Even this gauge density is difficult to maintain in many arid regions (Herndon et al 1973).

It is likely that, where weather radar is available, some combination of radar and gauges with meteorological satellite imagery interpretation may become a practical and economic method of estimating areal rainfall on large areas.

2. Meteorological Satellite Interpretation

A brief description of satellites which are intended for meteorological purposes is given below:

1. The US polar orbiting NOAA series of weather satellites have a recurrent coverage every 12 hours. The best ground resolution is about 1km in both the visible band (0.4 to 0.7m) and in the thermal infra-red. Many APT stations receive only at very low resolution (about 8-10 km). The maximum range of receiving stations is about 6,000 km and the Near East for example is covered by the receiving station at Lannion in France. FAO studies related to desert locust survey and control (Barrett, 1977) indicate that under arid/semi-arid conditions the relatively high resolution imagery (VHRR) is useful for locating areas of rainfall and estimating the severity of the rainfall using grid cells of 25 x 25 km and 50 x 50 km.

2. The US polar orbiting DMSP weather satellite has ground resolution of about 0.6 km, which gives recurrent imagery every 12 hours in a visible and thermal band. Data is usually transmitted directly to stations in the USA, but ground facilities also exist elsewhere in the world.

Table 17.1 summarises available information on the rapidly expanding US satellite programmes being developed for launching in the next few years. In addition there are the West European and USSR geostationary weather satellites to be stationed at 35,400 km altitude (see table), which can provide valuable information on rainfall over catchments. The USSR Soyuz and Salyut programmes with relatively high resolution images may also be expected to continue. The USSR is prepared to provide imagery of a country on request.

TABLE 17.1

PRESENT AND NEAR-FUTURE UNITED STATES SATELLITE REMOTE SENSING PROGRAMMES

Year	Satellite	Sensors	Remarks
1975-78	Landsat-2	RBV(0.505-0.750 μ *) MSS(0.5/06,0.6/07, 0.7/0.8, 0.8/1.1 μ	Polar orbiting - 18 days frequency. Direct transmission and via tape storage. Spatial resolution about 0.45ha. Effective transmitting range about 2,500 km.
1977-?	NOAA (-5)	Visible, thermal IR (2 channels)	Weather data, polar orbiting, 12 hour frequency. High Resolution 1km and 2.5 km; but most stations can receive only standard imagery (e.g. 100 km).
1977-?	DMSP (followed by Block 5D	Visible, thermal IR (2 channels)	Weather data, polar orbiting, 18 days frequency. Few receiving facilities. Resolution 0.6 km.
1977/78	Heat capacity mapping mission	0.5/11 μ, 10.5/17.5 μ	To study thermal inertia. Resolution 0.5 km.
1978	Seasat	Includes SLR radar (20/25m?). Precision radar altimeter. Microwave radiometer	Polar orbiting for oceanographic and coastal studies - reception US, possibly Europe. A European station would probably cover part of the Near East.
1978 (early)	Landsat-C	RBV(0.505-0.750 μ)	Polar orbiting, 18 days frequency; similar to Landsat-2; but spatial resolution 40m in RBV. Also thermal scanner.
		MSS(0.5/06, 0.6/07 0.7/0.8, 0.8/1.1, 10.4/12.6 μ*)	Transmission range about 2,500 km
1978	Tiros-N	Visible, thermal, micro-wave (4-channels)	Weather data. Polar orbiting; frequency 12 hours
1978?	Nimbus-G	6 channel MSS with narrow band passes	To test advanced NASA sensors
1978	GOE System	Visible and thermal IR	Weather data, geostationary. Near East covered by Meteosat (resolution 2.5 and 4.0 km) (European Space Agency) and satellite (GOMS) over Indian Ocean. Direct data transmission exceeds 10,000 km. Frequency every 30 minutes to stations in France, USSR and Germany (and elsewhere if installed).
1980 ?	Space shuttle		To provide a manned link between satellite and earth. Funding partly approved.
1981 ?	Landsat-D (Landsat-4)	MSS similar to "C" but a thermatic mapper (resolution 30m? 6 bands)	Polar orbiting - 9 days frequency, but at lower altitude than Landsats 2 and 3. Change to X-band data transmission.

* μ = Microns

Follansbee (1973) describes a method to estimate rainfall and Barrett (1973, 1977) presents a more useful method using the measured rainfall at gauges to calibrate the satellite data (Annex A.5). The meteorological satellites receive the thermal infra-red band from which surface or cloud top temperature can be estimated, which in turn can be used for rainfall estimation. Barrett (1977) simply by-passes the temperature stage by using measured rainfall. The study area is divided into a grid of 16, 25, 49 or 64 km squares. Grids squares having rain gauges are called gauge cells ("gacells") and those without gauges are called satellite cells ("satcells").

For extensive desert areas it is obvious that large cells are easier to use but larger cells result in less precision since the rain gauge data becomes less reliable. A major problem with polar orbiting satellites is the infrequency of daily sampling. Stationary satellites will resolve this problem but increase the labour involved in analysing data. It seems clear that automatic data processing of meteorological satellite data will eventually allow it to become a standard rainfall estimating device. A detailed discussion of a technique of estimating rainfall from raingauge and satellite imagery will be found in Annex A5.

17.2 USE OF ISOTOPES IN ARID ZONE HYDROLOGY

17.2.1 Introduction

The use of isotopes is a comparatively recent addition to the range of tools available to the hydrologist. Among the United Nations organizations the International Atomic Energy Agency is the one responsible for this field of work. This Agency has a special unit, the section of Isotope Hydrology, devoted to research into and guidance on the use of isotopes in hydrology. Their first major publication on this subject was in 1962 and their "Guidebook on Nuclear Techniques in Hydrology" (1968) is still the most complete work on the subject, though there have been several publications on advances in specific fields since that time and reference should be made direct to the Agency for the most recent knowledge. In this Paper the intention is only to give a general background to the subject so that the reader can see where it fits into the overall field of study of arid zone hydrology.

The IAEA gives a description of isotopes in the following words:

> "Isotopes of an element are atoms or nuclides having the same number of protons in the nucleus, and thus the same atomic number, but differing in the number of neutrons, and hence in their atomic mass, from other atoms of the same element. Whenever a nucleus of an isotopic species is unstable, it undergoes a single nuclear transformation, or a series of them, leading to a stable state in the form of a different element or isotope. The changes are accompanied by emission of radiation known as radioactivity. These are radioisotopes, in contrast to stable nuclides or stable isotopes".

Thus there are two classes of isotopes: the stable, which do not change with time although their concentration is affected by other processes (e.g. the evaporation of water) and the unstable, which decay with time. There is also another major grouping of isotopes into environmental, i.e. those which occur naturally, and artificial, which are created by man. These two groups will be discussed in the following paragraphs.

17.2.2 Environmental Isotopes

For the purposes of this Paper, the definition given by the IAEA is useful: "... environmental isotopes are defined as those isotopes whose natural abundance variations may be used for hydrological studies, for example heavy stable isotopes of hydrogen and oxygen in water molecules, cosmic-ray produced tritium ^{14}C (carbon-14) and ^{32}Si (Silica-32), and members of the uranium and thorium families.

"Included in this group are also man-made radioisotopes, such as tritium and ^{14}C, released to the environment through the testing of nuclear devices, reprocessing of nuclear fuel and waste disposal. Although these are artificially produced radioisotopes, the researcher has no control of their production and mode of release to the environment".

17.2.2.1 Stable Isotopes

The IAEA description says "The principle heavy stable isotopic components of water are $HD^{16}O$ (deuterium) and $H_2{}^{18}O$ (oxygen-18). They occur in natural waters in concentrations of about 320 ppm and 2000 ppm respectively. The varying proportions of these components in terrestrial waters can be measured by a mass spectrometer".

The value of deuterium and oxygen-18 to the hydrologist lies in the fact that the processes of evaporation and condensation alter not only the absolute concentration of these isotopes in the water molecule, but also their proportion to each other. Thus the concentration and proportion of these isotopes vary from one water mass to another and can be taken as a kind of isotopic "signature" of that water mass, which will be constant until changed by mixing with other waters with different signatures or by evaporation.

17.2.2.2. Unstable or Radioactive Isotopes

By definition, unstable isotopes decay with time and hence the "half life" is a useful descriptive parameter. The most common unstable environmental isotopes used by hydrologists are tritium (^{3}H) and carbon-14 (^{14}C). The half life of the former is 12.26 years and of the latter about 5600 years. The IAEA gives the following summary of the characteristics of tritium and carbon-14: "Tritium is the radioactive isotope of hydrogen ... Its occurrence in precipitation originates from two causes. The first, a natural one, is production by the interaction of cosmic high-energy radiation with atmospheric components. Estimates of the concentration of tritium in precipitation resulting from this source vary, but seem to be of the order of 10 TU. Tritium Unit (TU) is defined as one atom of ^{3}H per 10^{18} atoms of ^{1}H, or 1TU is equivalent to 7.2 dpm (disintegrationsper minute) per litre of water, or 3.2 pCi/l (pico-curies/litre). The second source is man-made and is derived, since 1952, principally from the detonation of thermonuclear devices. This production has swamped the former by interjecting periodic pulses into the atmosphere, with the result that precipitation has been labelled with an amount of tritium which can be relatively easily measured" ... "It follows that the occurrence of tritium in precipitation results in an overall labelling of water in the hydrologic cycle. Tritium is therefore found to varying degrees not only in precipitation but also in surface waters, groundwaters and the oceans. However, before one can interpret tritium values obtained from samples of surface or groundwater, it is necessary to have information on the concentration of tritium in precipitation in the area under study."

"Carbon-14 is a radioactive isotope of carbon ... (and) is produced in the upper atmosphere by cosmic-ray produced neutrons reacting with nitrogen. The carbon is oxidized to carbon dioxide and mixes with the atmospheric carbon dioxide reservoir. Testing of nuclear devices added a considerable amount of ^{14}C to the atmospheric reservoir.

"The ratio of ^{14}C to ^{12}C is constant in any living material which uses carbon dioxide in its life cycle and in substances, such as water, which react with carbon dioxide, and is referred to as 'modern ^{14}C content'. The year of reference is 1950. The modern ^{14}C content remains constant as long as the material is in equilibrium with the atmospheric reservoir, but when it is removed from the reservoir, as when water leaves the aerated zone, the amount of ^{14}C will decrease according to the law of radioactive decay. In principle, therefore, the ^{14}C content of the water is a measure of the time elapsed since it left the aerated zone.

"Since 1952, due to thermonuclear detonations, a considerable amount of ^{14}C has been added to the atmospheric reservoir. The present levels of ^{14}C in the atmosphere are thus larger than 'modern'."

Details of other unstable (radioactive) isotopes which may be useful under certain conditions will be found in the IAEA literature.

Although in principle the use of radioactive isotopes to date the origin of water masses is simple, in practice there are several complicating factors (see the IAEA publications) on which research is continuing. Moreover the sampling and analysis of the radioactive isotopes pose problems not found with the stable isotopes due to the danger of contamination from other sources and to the low levels of energy emitted and the low concentrations.

17.2.3 Artificial Radioisotopes

Artificial radioisotopes are a subclass of "unstable" isotopes. They are prepared by a bombardment of single elements with nuclear particles. Due to their radioactivity they constitute a health hazard, however small, and so their use must take into account considerations of safety. Their most common use in hydrology is as tracers of the movement of water, sediment, etc.

"The application of radioactive tracers to problems involving water movement is, in effect, a logical extension of the well-known tracer techniques in which dyes and salts are used as tracers for water. Inherent in the use of radiotracers are the advantages of very high detection sensitivity (hence, very low required concentrations and elimination of density effects), a choice of a variety of nuclides alien to the geohydrological system (hence, unique identification and low background) and the technical possibilities of tagging large volumes of water. The drawback of radiotracers lies in the relatively expensive and specialized field equipment, necessity for specialized training and compliance with health and safety regulations."

The choice of tracer is influenced by many factors - for example, the duration of the experiment, for ideally the tracer should have a half-life of about this length. The type and energy level of the emitted radiation will determine to a large extent the required concentrations, instrumentation for detection, and decision on laboratory or field analyses. This is a specialized field for which specialized advice should be sought. And a comparison with other types of tracer - such as dyes and salts - which may be simpler to use should be made before a decision is taken.

17.2.4 The Use of Isotopes in Practical Hydrology

Basically the characteristics of isotopes useful to the hydrologist are "signature" and "decay". These characteristics can be used, singly or combined, to assist in answering such practical questions as the source of a given water, its relationship (e.g. mixing) with other sources, and its age. The signature characteristic, either using naturally occurring isotopes or artificially applied isotopic "labels" can also be used to study the rate of movement of groundwaters, water in the unsaturated zone, and sediments.

In terms of arid zone hydrology, one of the most useful applications of isotopes has been to assist in the assessment of groundwater recharge. This aspect is illustrated by a Case Study in Section 18. Separation of income from capital, i.e. water which is being recharged into the aquifer at the present time compared with "palaeo water" which had its origin in a different hydrological environment thousands of years ago, is an important distinction for the planning of water resource exploitation. Moreover, where modern day recharge is thought to occur, the mechanism and frequency by which this happens is of great interest. Environmental isotopes can be particularly useful here. As an example, in one limestone desert area there are many shallow internally draining depressions with catchment areas from a $0.25 km^2$ to $45 km^2$ in size. In the lowest point of the depression, mudflats or fine sand are often found, which are flooded by rain from the surrounding catchment after storms. These ponds may exist from a few days to a week or more. The question was raised as to whether the rise in groundwater levels after storms represented recharge through these ponds or whether the water was lost by evaporation only. Isotopic studies of the groundwater showed no enrichment of oxygen-18 in comparison with deuterium. This meant that the recharge water had not been exposed to any significant evaporation loss and hence that it was storm water that had infiltrated quickly after the storm in fissures or solution channels and was not water that had been ponded on the mudflats.

In many arid regions rainfall occurs in mountain areas and may reach lower levels either as surface or subsurface flow. Deuterium and oxygen-18 are especially useful to differentiate water sources by altitude of the source area.

Other examples of the use of isotopes are given in relation to sediment load measurements, recharge and resource assessment the IAEA Guidebook and in the Papers of international conferences on the subject. However, to conclude this section two points should be made. The first is that the interpretation of isotopic signatures is often very imprecise. The use of isotopes may raise as many new problems as it solves, due to apparent inconsistencies resulting from the interplay of many factors which affect the isotopic content of a particular water at any particular time. The use of isotopes in hydrology is a fascinating subject but the further that research proceeds, the more complexities are revealed. It may often be useful but is seldom conclusive on its own.

This leads into the second point, which is that isotopes are only one additional tool available to the hydrologist, to complement the well established uses of hydrochemistry and hydraulics. It is very seldom if ever that the use of isotopes alone can solve a specific practical hydrological problem (an exception may be the use of isotopes as tracers; but the use of tracers is in fact no new technique; only the use of an isotope as the tracer is new). It is to be expected that a new tool such as isotopes may initially have a kind of mystique, particularly as their use requires different specializations (measurement, handling, safety, etc). Some isotope specialists may indeed have encouraged such as air of mystery in the early days. Now, however, it is generally accepted that the isotope specialist should be one of a team of experts working on any hydrological problem, and, as is a normal requirement for good results from any team, each contributor's role cannot be successful unless integrated into the efforts of the team as a whole.

17.3 USE OF CHANNEL GEOMETRY FOR ESTIMATING STREAM FLOW

The estimation of streamflow in ungauged catchments in arid areas is a most difficult problem. The use of runoff coefficients is likely to lead to very big errors in areas where the coefficient may vary from 1 or 2 percent in flat, sandy or limestone areas, to 50 or even 80 percent in bare rock mountains. In such cases, the study of the shape of stream beds can be helpful. It is clear that stream channels are formed by the flow which they carry and that therefore the two are related. Many studies of this relationship have been made over the years, e.g. Moore and Donald O., (1968), I.A.S.H., (1968). Dincer (1979), working in Saudi Arabia, has found quite a good relationship of the form $Q = a \times W^n$, where Q = mean annual discharge in Mm^3

a is an empirical coefficient
W is the width in metres of the channel between sand bars and
n is an empirical exponent.

This worker studied about 20 sites where mean annual discharge had been estimated for about 10 years. He found a = approx. 1/40 and n = approx. 2.

Clearly these relationships are of a regional nature and cannot be extrapolated within non-homogeneous areas. The error of estimate may be large and may include systematic errors of unknown magnitude if the runoff data from which constants are derived are subject to the uncertainties discussed in Section 12.2. The estimate is likely to be highly sensitive to W, which is seldom well-defined. However the method may be better than immediately available alternatives.

It is likely that a relatonship of the same form can be found between flood peaks of a given frequency (e.g. once in five years) and channel width. It should be noted that the width in question is not the full width of the flood channel, but the width between bars which carry medium and low flow and that this method applies only to alluvial channels. A comparison of mean annual discharge at different points along the length of a river channel calculated in this way has been used to obtain first estimates of recharge to channel bed aquifers.

Other workers have established relations between channel geometry and bankfull discharge. Difficulty is experienced however when it is proposed to transpose regime equations of this type to channels with bed materials other than unconsolidated sediments.

18. RESOURCE ASSESSMENT METHODS

18.1 INTRODUCTION

Occurrences of surface water flow in the water-courses of arid lands are unpredictable, of short duration and highly variable. The instantaneous discharge-duration and discharge-volume percent curves shown by Figure 18.4(a) and (b) demonstrate the problem of controlling runoff resources having these characteristics. Regulation of the natural sporadic discharge distribution by means of surface reservoirs presents many problems among which the most important are:

- The high ratio of storage to mean annual runoff volume required to produce the degree of control necessary for economic agricultural development. Table 18.4 illustrates the problem in relation to a study carried out in Saudi Arabia.

- The high evaporation which tends to make over-year storage uneconomic and even seasonal storage wasteful of precious resources. Table 18.4 shows the effect of evaporation on a hypothetical scheme designed to control discharge to a constant maximum value compared with a scheme to regulate by temporary detention storage using an uncontrolled outlet. This comparison shows clearly that serious losses may result from surface storage schemes designed to control flow for the convenient operation of irrigation schemes based on canal distribution systems. Regulation to make it possible to divert a high proportion of total runoff by traditional spate irrigation techniques is likely to be more economically productive.

- Sedimentation of surface storage reservoirs may make their useful life uneconomic.

Groundwater storage evens out the effect of runoff variability. Hence, groundwater schemes are generally to be preferred, since they are normally less susceptible to evaporation and sedimentation. The available storage is often many times greater than mean annual flow and recharge may be increased by a degree of surface water regulation required only to improve the amount and distribution of recharge.

Surface and groundwater resources, particularly in arid zones, are inseparable. Change in groundwater storage may be the largest element in the water balance even when a period of several years is considered.

The case studies described here produced resource assessments for purposes ranging from simple reconnaissance level estimates to design of major engineering works.

18.2 SURFACE WATER ESTIMATION AND USE

18.2.1 General

The collection of reliable records of runoff in arid zones is expensive and is only justified on ephemeral water courses where some indication exists of abundant resources.

Project runoff gauging sites should be limited to points of particular significance for development. It is not feasible, during project hydrological investigations, to determine rainfall with the precision required for reliable studies of rainfall-runoff processes and project investigations should not normally include the establishment of rain-gauges for this purpose.

Runoff records in arid zones are typically short and unreliable. Little confidence can be placed in the direct use of simple monthly or annual runoff statistics for project planning. Even daily runoff data can produce misleading impressions of water availability. Instantaneous hydrographs are necessary for design purposes.

18.2.2 Surface Water Estimation Methods

Where rainfall is very variable in space the use of deterministic rainfall runoff models is generally impracticable. Since it is surface water which is the fundamental resource, either for direct use or for groundwater recharge, it is usually preferable to concentrate financial resources on runoff measurement. The selection of station location may be on the basis of immediate need: the water resources nearest the water demand for agricultural development are studied first. The apparent size of the resource and the representativeness of the site for transposition of data may also be important factors.

There is usually the need to extrapolate to unmeasured locations. However, on either a storm or an annual basis such simple formulas as $R = a(P - b)$ where R = runoff, P = rainfall and a and b are constants, can be highly misleading unless based on long (and accurate) rainfall and runoff data.

The main factors affecting the values of these constants are the slope, the surface retention and infiltration capacity of the soil or rock surface and the rainfall intensity. Mean annual rainfall is unrelated to the size of individual storm runoff events. However, annual rainfall amount and distribution clearly affect land use and, where it is sufficient to support dry farming, the construction of terraces and other works to delay runoff may have a major influence in increasing the runoff threshold value b (or in preventing nearly all runoff to downstream channels) and also in increasing the possibilities for groundwater recharge.

The relation $R = aP$, which omits the loss constant b, is seldom of any value.

Runoff coefficients on a storm basis vary widely. A study in the highlands of East Jordan gave values of 0.1% to 68.5%, with an average of about 5%. In Qatar, with very small catchments and no channel losses, a mean figure of 35% was obtained. A similar figure of 25 to 35% has been suggested for hard rock mountain catchments in Oman.

Another simple relation is of the form $R = aP^b$. On two small hardrock catchments in Oman, very limited data gave values of a = 0.033 and b = 1.5 for in millimetre units. Elsewhere in the same region values were vastly different for the same study period. This form of non-linear relationship is more likely than the assumption of linearity.

Simple rainfall-runoff relations of the type illustrated here can be quite misleading, particularly on a short-term basis. However, if used carefully to extrapolate under conditions that are known to be similar they may be of some use to estimate the order of magnitude of runoff.

It is important here to stress the effect of basin and sample size on constants a and b in the empirical relationships discussed above. On Figure 18.1 are shown the P v R relationships for a few drainage basins. The constants a and b are derived from least squares regressions assuming a linear relationship. The outer limit curves indicate the range within which sample data points lie at 95% confidence. The inner limit curves indicate the range of uncertainty about runoff predicted from P at 95% confidence. The effect of an increase in sample size on the outer curves would be random because we could expect the same degree of scatter in the population of annual rainfall and runoff values. The effect of an increase in sample size on the inner curves is to reduce the apparent tolerance on runoff predictions made from rainfall. This is true only insofar as the mean runoff predicted from an annual basin rainfall series is concerned. The tolerance on predictions for individual years would remain approximately as indicated by the outer limit curves.

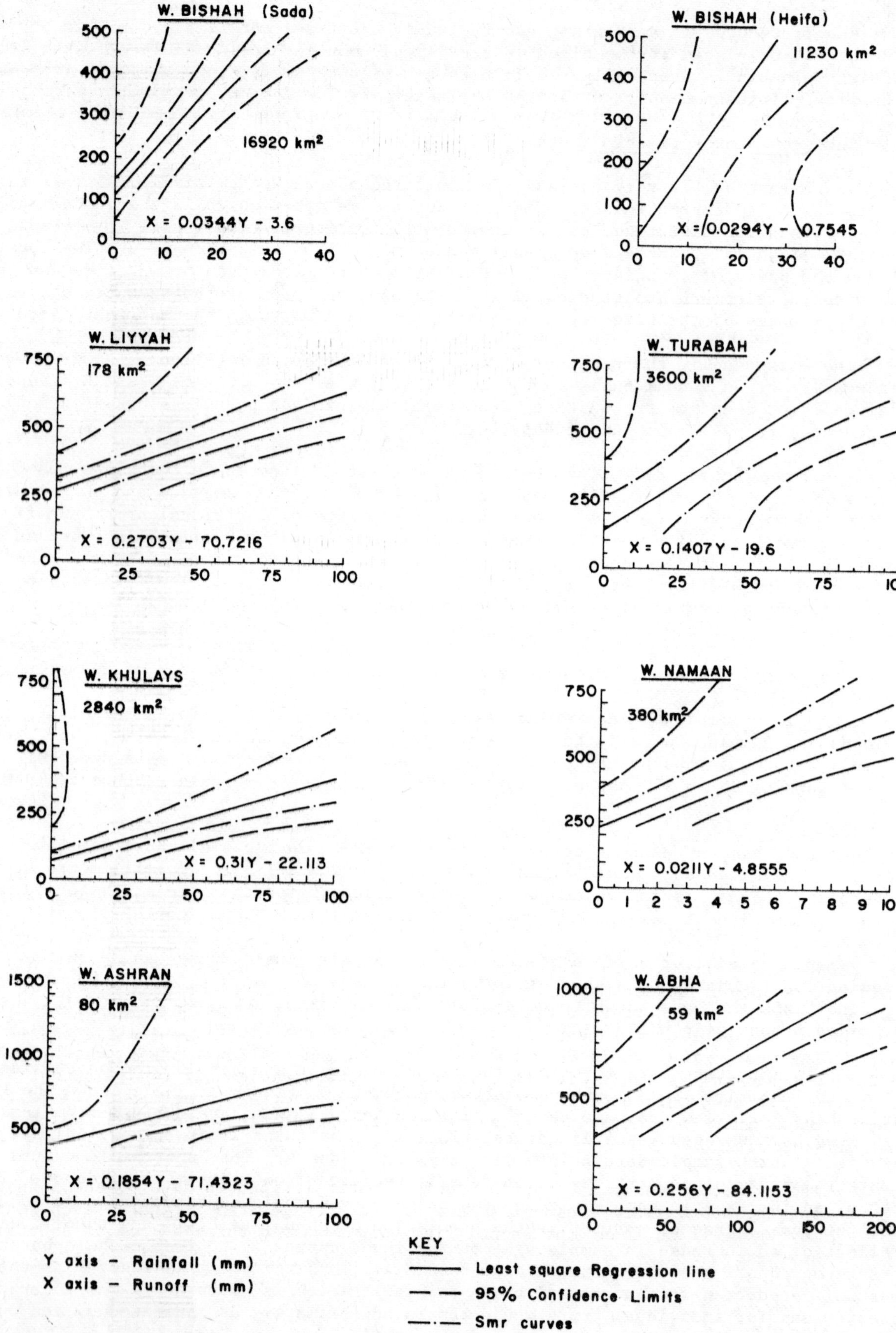

Fig. 18.1 Annual rainfall - runoff distributions, Saudi Arabia

Medium-term rainfall and runoff figures reported by Davy et al (op.cit) for drier tributaries of major rivers in the Sudano-Sahelian zone of West Africa are interesting. They show the wide range between minimum and maximum runoff values in this part of Africa and the low percentage runoff typical of medium to large basins in semi-arid climates (Table 18.1).

TABLE 18.1

Rainfall and Runoff for the Sudano-Sahelian Zone of West Africa

Basin	Tributary	Drainage Area km^2	Record Period	Long-Term Basin Rainfall $\bar{P}$ mm	Annual Runoff Mean mm	%$\bar{P}$	Max mm	Min mm
Niger	Goroual	7500	1961-72	470	34	7.2	59	19
	Maggia	2525	1954-72	500	16	3.2	41	2.4
Niger	Goulbi de Maradi	5400	1956-72	700	36	5.1	79	14
Chari	Bahr Azoum	80000	1953-74	800	12	1.5	22	3.1
	Bahr Keita	14000	1954-67	1000	90	9.0	220	5.1

18.2.3 Estimation of Mean and Variability of Annual Runoff

Where the annual runoff data for arid zone basins are available for a number of years they show the following properties:

- the coefficient of variation is of the order of 1.
- the distribution is positively skewed with mode < median < mean,
- increasing frequency of zero annual runoff with distance downstream of the main runoff-generating zones, especially below alluvial aquifers with large recharge and infiltration and storage capacities.

Normally the value of the mean annual runoff $\bar{R}$ at a gauging station is the statistic which may be estimated with the highest degree of accuracy from a given sample. However, a number of short-term records for stations whose annual runoff volumes are independent may provide a more reliable indication of the coefficient of variation CV than of the mean. The runoff data for selected stations in Saudi Arabia shown in Table 18.2 illustrate this point. With this amount of data, mean annual runoff cannot be estimated with the confidence resources for major engineering investment. If we take other factors into consideration we may be able to narrow the range within which $\bar{R}$ lies. Such factors are:

a. The existence, size and type of surface water irrigation development downstream and the approximate consumptive use of the average area irrigated each year. We may know little about the annual flow volume wasted through lack of control however.

TABLE 18.2

ANNUAL RUNOFF DATA - (SAUDI ARABIA) Mm^3

STN. / YEAR	J-401	J-402	Ta-401	Ta-402	Ta-406	B-402	B-406	A-401	A-402	B-404	N-401
1966	30	x	(0.6)	(1.2)	x	0.4	-	-	x	(-)	(15)
1967	18	TR	(1.0)	0.4	x	41	27	1.4	x	(4.3)	(80)
1968	26	1.5	(4.4)	9.8	176	157	114	7.6	3.2	(2.9)	56
1969	138	1.6	(2.0)	8.2	160	69	13	1.2	0.2	(2.7)	155
1970	18	TR	(0.3)	0.9	90	121	85	0.2	0.2	2.3	31
1971	17	0.2	2.4	TR	11	33	23	1.1	0.3	1.2	32
1972	27	-	1.8	1.1	71	21	81	4.7	2.2	12.1	42
$\bar{R}$	39	0.5	(1.8)	3.1	102	63	49	2.3	1.2	3.6	59
$\hat{\sigma}_R Mm^3$ (1)	43.9 113	0.8 160	1.4 78	4.1 132	68 67	57 90	44 90	2.8 122	1.4 117	4.0 111	47 80
$SE_R Mm^3$ (2)	16.6 43	0.3 60	0.5 28	1.5 48	30 29	21 33	16 33	1.1 48	0.6 50	1.5 42	18 31
$SE_R 0.95$ (3)	41 105	0.8 160	1.2 67	3.7 119	83 81	51 81	39 80	2.7 117	1.7 142	3.7 103	44 75

Notes:

1. σ_R is the standard deviation of the sample values expressed in absolute and percentage values.
2. $SE_{RO.95}$ is the standard error of the sample mean at the 95% confidence level expressed in absolute and percentage values.
3. The average value of the coefficient of variation CV = $(\hat{\sigma} / \bar{R})$ = 1.01. The approximate average value of $SE_{RO.95}$, expressed as a percentage is ± 100% for the average number of record years.

b. The existence of groundwater irrigation upstream. The calculation would be very roughly the same as for (a) since the difference in overall efficiency between exploitation of the water resources by spate irrigation before infiltration and recharge and by pumping from shallow aquifers afterwards, is small by comparison with the uncertainty in the estimate of surface runoff escaping either system.

c. The likelihood of the runoff record coinciding with a period of above- average or below-average rainfall may give an idea in which part of the range of possible $\bar{R}$ the true value may lie. This does not imply a rainfall runoff relationship - but is a qualitative indication only.

Both (a) and (b) require estimates of ETa losses from irrigated crops and natural vegetation, often including extensive infestations of phreatophytes drawing water from alluvial aquifers. These losses may represent the largest component of the water balance in some areas.

For some feasibility studies, often leading to large-scale expenditure on dams and irrigation works the assessment of annual runoff and its variability have been based wholly on short-term relationships between annual rainfall and runoff. However well rainfall and runoff are measured, regressions based on short-term data can be highly misleading and should never be used as the only basis for conclusions as to potential resources.

Table 18.3 shows mean and median annual runoff values for a number of gauged basins in Arizona, in which diversions for irrigation and water supply are minimal. The generally large proportional differences between mean and median are an indication of the positive skewness of the annual runoff distribution. Not all the basins listed can be described as arid, but the skewness characteristic of annual runoff distributions in arid and semi-arid climates is present.

18.2.4 Stochastic Runoff Generation

18.2.4.1 General Concepts

Stochastic methods of extending hydrological records, seek to overcome the problem of short historical data series by generating synthetic sequences. Although generated series are neither actual historical records nor forecasts of future events, they may represent equally likely sequences of the hydrological variable of interest. They may in some circumstances be used to estimate the frequency distributions of extreme events.

Stochastic generation cannot in any sense manufacture new information, but is essentially a technique for extracting the maximum amount of information from the original data. As such, it provides a powerful tool for the analysis of arid zone runoff data; it allows the limited data available to be analysed on an event-by-event basis, thus avoiding the inevitable loss of information associated with the use of aggregate statistics. A monograph by Fiering and Jackson (1971) would serve to introduce the interested reader to the subject of synthetic streamflow generation. A paper by Weiss (1977) gives a theoretical description of the so-called "shot noise" class of models found useful for representing the arid zone runoff process.

Shot-noise models are used to generate runoff data, by the superposition of individual synthetic flood events to form compound hydrographs similar to the historic series.

These may then be integrated over any desired period to estimate the statistics of discharge volumes. An important feature of this approach to model formulation is that it requires the instantaneous discharge hydrograph to be studied. As the mechanics of flood wave propagation, recharge by bed infiltration and sediment transport in arid zones

TABLE 18.3

SUMMARY OF RUNOFF DATA FOR SELECTED STATIONS IN ARIZONA

River Basin and Station	Drainage Area km^2	Annual Runoff Mean m^3/s	Mean mm	Median m^3/s	Median mm	Extremes Qp m^3/s	Years of Record
Gila							
Rillito Creek	2280	0.46	6.3	0.23	3.1	680	64
Santa Rosa Wash	4560	0.44	3.0	0.28	1.9	1500	18
Sycamore Creek	11.5	0.014	4.4	0.011	3.1	55	11
Dry Beaver Creek	364	0.91	78			750	19
Red Tank Draw	126	0.161	40	0.137	33	300	15
Rincon Creek	115	0.137	38	0.105	29	270	20
Sabino Creek	91	0.34	118	0.232	80	220	47
Tanque Verde Creek	110	0.23	67	0.16	46	87	13

can only be understood on the basis of instantaneous hydrographs, development of shot-noise models helps in engineering studies of these processes.

The synthetic sequences generated by stochastic models may be used in conjunction with deterministic simulation models to examine alternative project designs. Engineering studies for agricultural development in arid zones may require estimates of the impact of the proposed works on flood peaks and on the distribution of infiltration and flow volumes at various locations down the course of a main channel. An approach to these studies is to establish a digital simulation model of the flood wave propagation and bed infiltration processes to be modified by the proposed works and then to "feed" this deterministic model with series drawn from stochastically generated data sequences. The series should be long enough to produce assessments of the risks associated with alternative proposals.

Because of the rapid fluctuations in discharge in arid zones, simulation models of flood wave propagation must be operated with short incremental time steps. In practice, limited data on channel hydraulic charactistics and the available time and money, may dictate that a hydrological (or storage) method of flood routing be adopted for such studies. Modified pool-routing methods have been found to be reasonably reliable for studying flood wave decay and bed infiltration as flows traverse channels typical of arid zone alluvial plains. Such models must be calibrated and tested by comparison with observed hydrographs covering a range of discharges in the study reach. Detailed observation of a series of flood waves should, therefore, have high priority in planning field investigations if this approach is to be applied.

18.2.4.2 A Conceptual Basis for a Stochastic Model of Ephemeral Runoff

Examination of arid zone hydrographs suggests that flow periods may be decomposed into a series of quasi-instantaneous rises in discharge followed by negative-exponential recessions. Figure 18.2 shows the decomposition of a typical hydrograph. The durations of no-flow periods are denoted by TAU and the number of discrete rises in each flow period by N. The individual hydrograph rises are termed "flood events" and their magnitude is denoted by QP. The time lag between the rises within each flow period is denoted by DT.

Diskin and Lane (1972), concluded that the four elements defined above may be considered to form series of random variables. Studies of decomposed hydrographs indicate that recession segments associated with the rise of individual flood events may be represented by relatively simple deterministic recession functions. These may be established by deriving two recession parameters from a plot of QP and the associated runoff volume V taken from the record.

The conceptual decomposition of the runoff data described above may be viewed as a stochastic model of the runoff process. If the statistical distributions of the four random variables, TAU, N, QP, and DT, are estimated as well as the value of the two recession parameters, Monte Carlo methods (Fiering and Jackson, op.cit) may be employed for the production of synthetic data series (See schematic flow chart - Figure 18.3).

This model of the runoff process implicitly assumes that no correlation exists between the various values of the four stochastic variables and that individual values of each variable are independent of all antecedent values of the same variable. Historic series are usually too small for rigorous statistical tests to detect such correlations. However, consideration of the physical processes involved suggests that no strong cross - or auto-correlations dominate the statistical behaviour of the runoff process, except possibly the duration of a no-flow period. This means that we do not expect short-term persistence to be present in the runoff pattern either between consecutive QP, DT, N and TAU values or between Qp and DT etc. We assume the probability of a small DT value to be the same regardless of the magnitude of the preceding QP or DT values. Where marked seasonality exists, the procedure used to model the duration of no-flow periods should obviously take this into account.

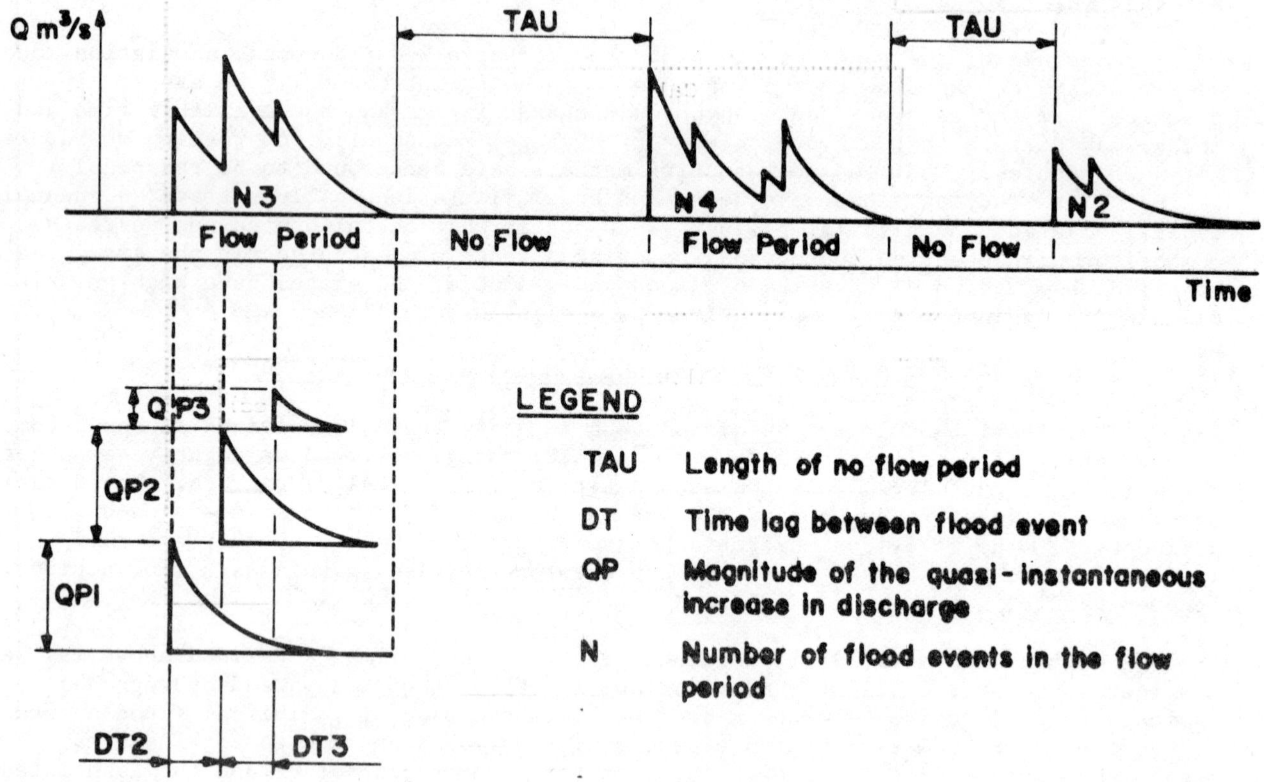

Fig. 18.2 Conceptual basis of the stochastic runoff model

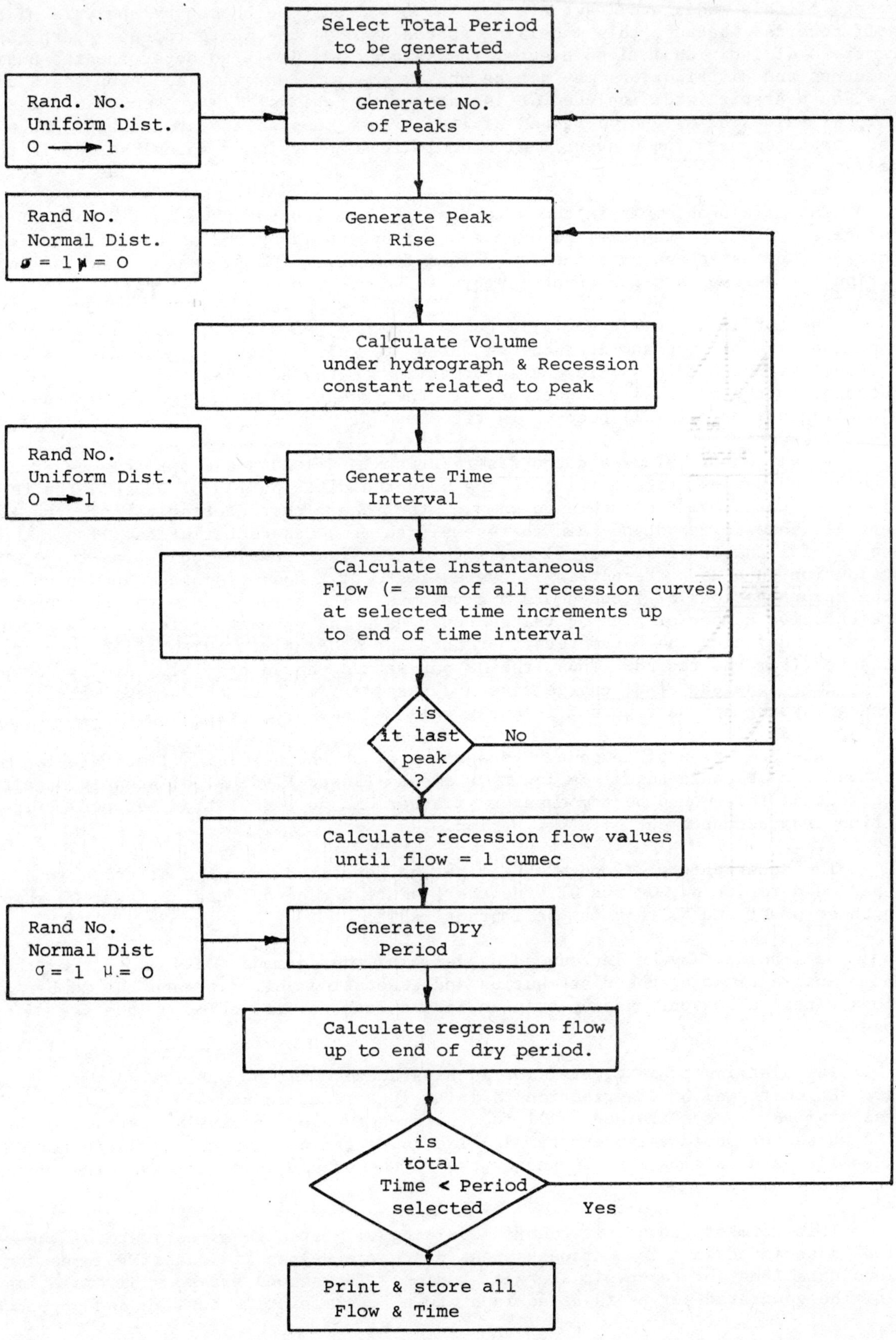

Fig. 18.3 Stochastic generation model - schematic flow chart

In practice, the probability distributions from which the four stochastic elements of this model are drawn are not known and must be chosen by analysis of the runoff records. Because this modelling approach uses the instantaneous hydrograph parameters of individual flood events, the samples available to determine the basic model parameters and distributions may not be unduly small. For example, seven years of data for a South Arabian wadi included 61 isolated flow periods, 239 values of QP, 178 values of DT, 61 values of N, and 60 values of TAU. These samples are at least as large as those sometimes used for conventional hydrological analysis of annual series in temperate zones.

The main problem is in the selection of distribution type to fit each of the four parameters. Clearly, the limited experience represented by the sample size and the errors in the instantaneous discharge values from which the recession function and QP distribution derive, are important factors to be considered.

The total TAU sample normally comes from a mixed distribution since it includes dry periods in both wet and dry seasons which are seldom clearly definable. A very large TAU sample might contain sufficient information to provide a basis for a seasonal adjustment to the probability function. In the absence of long periods of record, conjunctive use of rainfall records is recommended.

Total runoff volume and the distributions of flood peaks and volumes produced by the generated series are sensitive to the selected distribution of QP. If the record period does not provide an unbiased sample, the runoff characteristics inferred from the generated sequence cannot be said to resemble the true characteristics. Thus, the presence of a number of 50-year floods within a 10-year sample could bias the QP distribution such as to result in an overestimate of runoff resources and an underestimate of the return periods of flood peaks and volumes. The absence from the record of a representative number of high QP values could have the opposite effect. The problem is also one of defining the parameters to reduce the subjectivity with which they are abstracted from the record. Thus, the QP parameters should possibly have two components, that of the magnitude of discharge rise and that of time between successive rises. The discharge hydrograph must be sufficiently well defined for this purpose.

Interpretation of compound hydrographs of which the rising limbs are composed of more than one QP could result in fewer QP events (lower N value) of greater magnitude. The effect of this would be compensatory if there is no correlation between QP frequency and time from commencement of a wet period.

The occurrence of flood events might be generated from the distributions of TAU, wet period duration DW, QP and DT. However, since a record contains fewer DW samples than those of DT and N, this may be impracticable.

Seasonality may be introduced by defining the seasonal date as that of the first flood event, estimating the distribution and generating this date and the number of wet periods within a season. Again, this implies a long record period or the use of rainfall records.

The distributions selected for each parameter must be realistic over the entire range. Log-extremal QP distributions are likely to lead to unrealistic generated values, unless arbitrarily constrained. All four parameters are, basically, partial-duration series whose threshold value and sample size is determined by the discharge threshold between flow and no-flow periods and by the degree of smoothing involved in definition of hydrograph shape.

This form of stochastic runoff modelling will produce an estimate of the return periods associated with instantaneous peak discharge values and runoff volumes for various durations for reservoir storage studies. The second and possibly more important use of the generated series is as an input to a deterministic recharge and groundwater model of the type discussed above, for which short-duration mean discharge values are required. A further purpose of stochastic generation is to provide an estimate of the sediment load, for which instantaneous discharge hydrographs are required.

18.2.4.3 Case 1. Application of a Stochastic Runoff Model to a Design Problem

The generation functions for the parameters TAU (length of dry period in hours), N (number of hydrograph rises in a wet period), DT (time interval in hours, between successive hydrograph rises within a wet period) and QP (hydrograph rise m^3/s) were calculated from the sample, as follows:

1. $Tau = 10^{(2.319 + 0.78605R)}$ hours.
corresponding with a Log-Normal distribution.

2. $N = (Ln\ (1 - P)/ -0.5)^{1/0.67}$
corresponding with a Weibull distribution.

3. $DT = -24Ln\ (1 - P)$ hours
corresponding with an exponential distribution.

4. $QP = 10^{(1.3086 + 0.7805R)} m^3/s$
corresponding with a Log-Normal distribution.

Where: R is a Uniformly distributed random number between 0 and 1.

P is a Normally distributed random number with mean 0 and variance 1.

The recession function was evaluated from the data sample as

$Q_t = QPe^{\alpha . t}$

Where t is the time in hours from the hydrograph rise QP and α is a negative exponential function of QP, e.g.:

for QP = 100 m^3/s $\alpha = -0.1076$

for QP = 500 m^3/s $\alpha = -0.126$

for QP = 1000 m^3/s $\alpha = -0.132$

For example: if QP = 1000 m^3/s; $Q_{24} = 100\ e^{(-0.132 \times 24)} = 42\ m^3/s$.

If QP = 100 m^3/s; $Q_{24} = 100e^{(-0.1076 \times 24)} = 7.6\ m^3/s$

Figure 18.4(a) shows a typical generated wet period runoff sequence. Figure 18.4(b) shows the instantaneous discharge-duration relationship for a 10-year simulated flow sequence and Figure 18.4(c) shows the relationship between instantaneous discharge and percentage of flow volume.

Water losses by evaporation from a hypothetical reservoir were calculated by applying the Penman Eo estimate, expressed as an evaporation flux from unit area, to the model-calculated water surface area. These losses are shown in Table 18.4.

The study requirements were to determine the quantity of water which could be diverted for irrigation both with and without reservoir storage regulation. In the absence of regulation the volume of water capable of being diverted for irrigation is represented by the area of the instantaneous flow-duration curve for the natural flow hydrograph lying below the diversion rate as shown by Figure 18.5. With storage regulation this volume is represented by the area lying below the instantaneous flow-duration curve of the regulated out-flow.

Table 18.4 shows the results of routing a 10-year synthetic hydrograph sequence through storage with different uncontrolled outlets or at controlled release rates.

Figure 18.5 illustrates the exercise described above and is an example of the use of synthetic hydrograph sequences to assess the value of storage and the degree of control provided by different project configurations. On the basis of the 10-year sequence it would seem that, for the case illustrated by the Figure, a storage volume of up to about $40Mm^3$ with an uncontrolled outlet of one square metre would permit diversion via a 10 m^3/s canal of 84 Mm^3 per annum out of the total runoff of 101 Mm^3/s.

The effect of reservoir regulation on recharge processes in downstream aquifers may be studied by models of this type. It may be necessary to assess the additional recharge induced by detention storage regulation. In the example given, the effect of storage in one case is to reduce the peak discharge rate from 3350 m^3/s to 290 m^3/s, thereby enhancing the recharge opportunity downstream.

Studies of the benefits of reservoir regulation in flood alleviation may similarly be studied with synthetic flow sequences.

18.2.5. Deterministic Rainfall-Runoff Models

By definition, a rainfall-runoff model requires a reasonably precise measure of basin rainfall input in order to predict the output at the point of interest with acceptable precision. Since runoff response is highly dependent on rainfall intensity and surface interception storage characteristics and since both are highly variable in space, the possibility of predicting runoff response at the basin outlet would be remote, even if the areal and temporal rainfall input pattern could be expressed with exact numerical precision. The input to, storage change within and output from a small representative basin in an arid zone is capable of being modelled. However, the results obtained may be transposed to larger basins only on a probabilistic basis; that is, the long-term mean and variability of runoff from any part of the catchment may be estimated but the long-term spatial distribution of the runoff responses and storage levels are such that estimation of their effect at a major basin outlet presents severe problems.

To illustrate the problem, attempts to relate runoff response to rainfall input in an arid environment are cited below.

18.2.5.1 Case 2. Wadi Liyyah - Saudi Arabia

Estimated monthly runoff data for the seven-year record period are given in Table 18.5. Figures 18.6 and 18.7 show daily and monthly rainfall and runoff regressions to illustrate the initial degree of scatter before attempting more sophisticated techniques. Intercepts on the X (rainfall) axes are about 7mm and 16mm respectively. The relatively high correlation coefficients, 0.87 for daily and 0.81 for monthly regressions, are very highly significant in the statistical sense. However the degree of scatter, as indicated by the outer (95% confidence) curves shows that little reliance can be placed in individual daily or monthly runoff estimated from the regression equations.

If the mean annual basin rainfall and the distribution of daily and monthly basin rainfall amounts can be estimated, mean annual runoff may also be estimated, as follows:

- calculate the mean annual number of rainfall occurrences above specified depth thresholds (exceedances),

- calculate the runoff depth corresponding with each rainfall depth threshold from the regression equations,

TABLE 18.4

ANNUAL YIELD AND LOSSES - COMPARISON OF STORAGE CAPACITIES AND TYPE OF RELEASE (mean of 10 years routing)

Storage (Mm^3)	Release Type	Orifice Area (m^2)	Release Rate (m^3/s)	Total = Release Rate (Mm^3)	Total Spill (Mm^3)	Total Evap. (Mm^3)	Annual Duration = Release Rate (Hours)
0	Controlled	-	2	27.8	71.2	0	737
0	Controlled	-	10	39.7	60.3	0	367
43	Controlled	-	2	52.3	41.1	5.7 (1)	5812
43	Controlled	-	10	92.8	5.4	1.8	2409
43	Uncontrolled	1	< 17	96.0	3.1	0.9	-
43	Uncontrolled	5	< 85	98.9	1.0	0.1	-
86	Controlled	-	2	63.4	24.3	10.7 (1)	8766 (2)
86	Controlled	-	10	97.7	0	2.3	2517
86	Uncontrolled	1	< 18	99.0	0	1.0	-
86	Uncontrolled	5	< 87	99.9	0	0.1	-

(1) There is residual storage at the end of routing.
(2) Assumes mean year = 365.25 days.
(3) Total inflow 1008 Mm^3 (101 Mm^3/a)

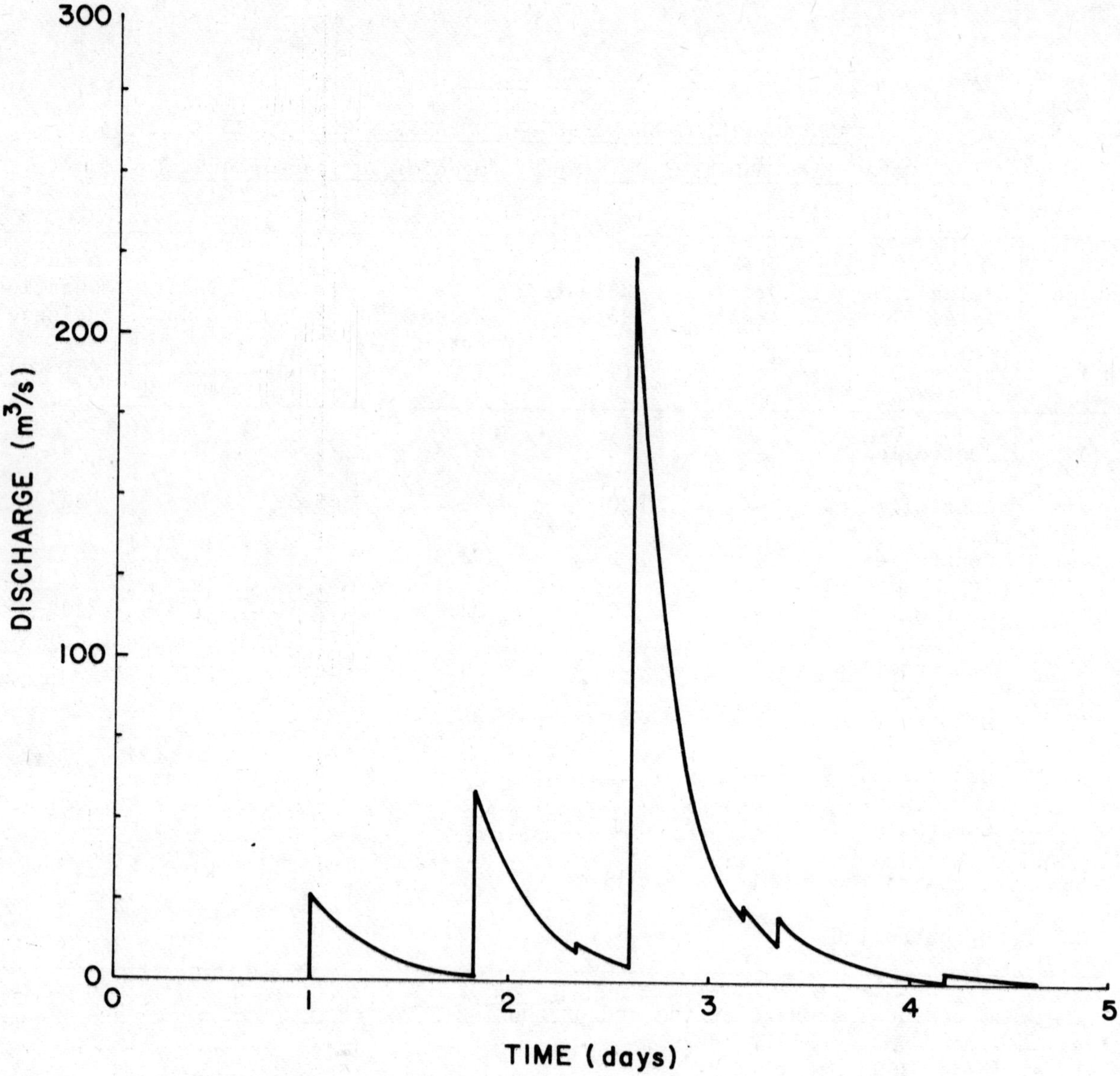

Fig. 18.4(a) Typical generated instantaneous discharge hydrograph

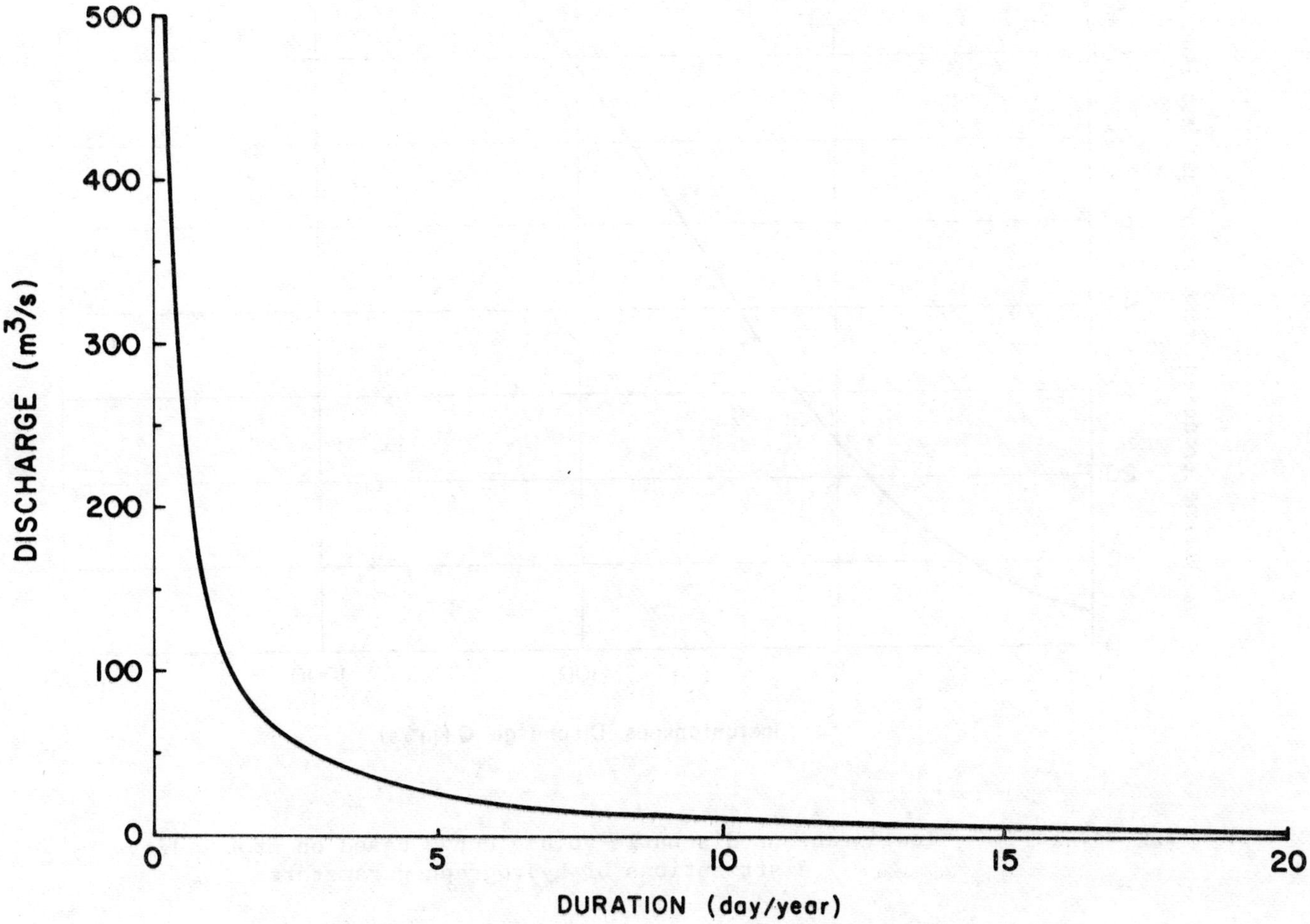

Fig. 18.4(b) Instantaneous discharge - duration curve, 10-year generated sequence

The figure shows the duration of discharge at instantaneous rates greater than indicated. Duration is expressed as mean days per year in which discharge is equalled or exceeded. This duration may be converted to hours or second by applying the appropriate factor. The area under the curve is the total runoff volume

For an irrigation diversion structure without storage at this location, the volume of water capable of being diverted may be visualized as the area below a horizontal line corresponding with the capacity of the diversion canal. This type of curve is also useful in illustrating the problem of gauging in arid zones. The shape of the curve shows the need for gauging stations capable of accurate measurement up to high discharges to reduce errors in estimating total resources.

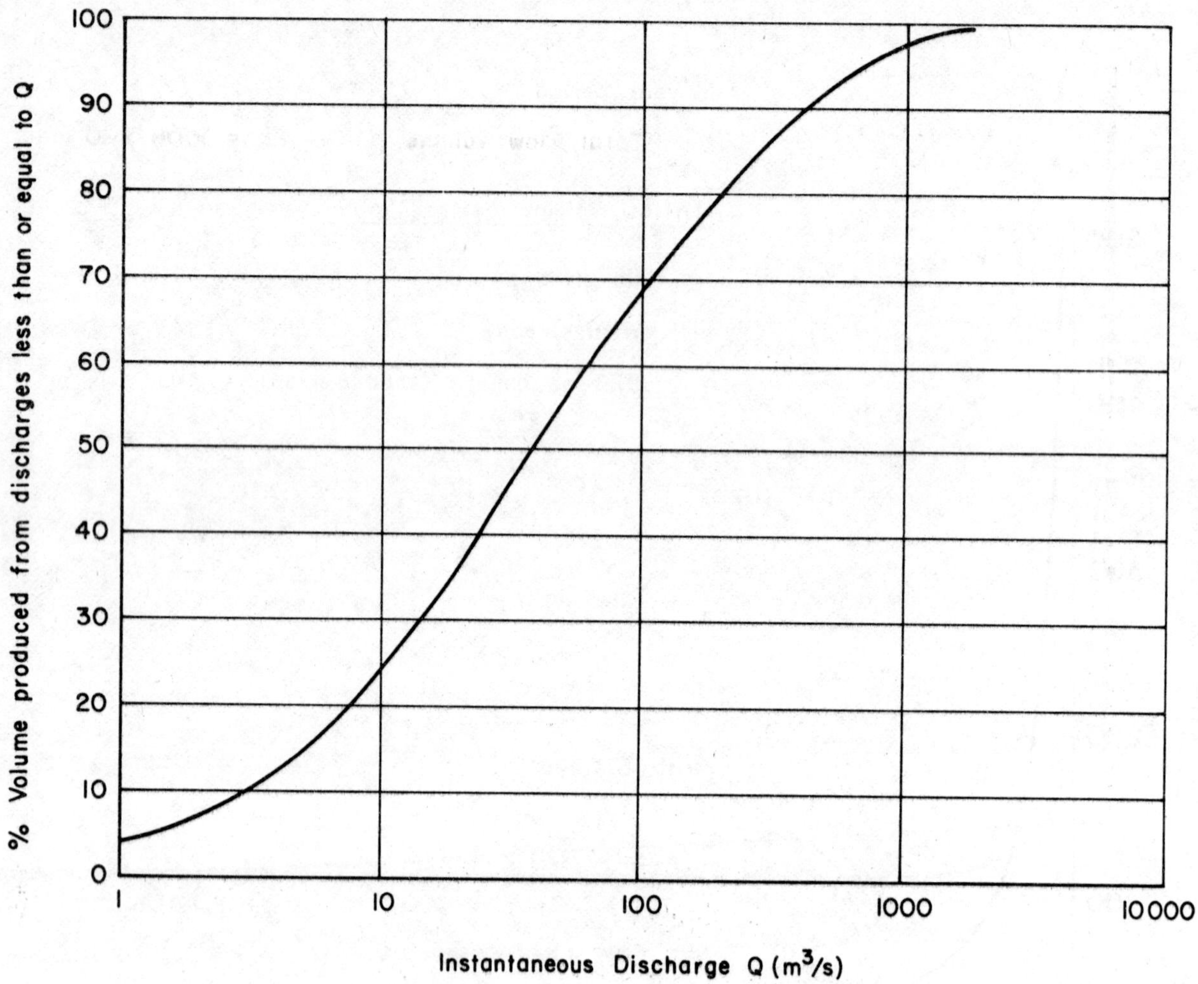

Fig. 18.4(c) Instantaneous discharge-volume curve based on probability distributions of hydrograph parameters

This figure effectively results from plotting the area under the instantaneous discharge-duration curve (Figure 18.4(b) against discharge rate.

The volume of runoff, as a percentage of mean annual volume capable of being diverted for a given canal capacity is shown directly. Volumes corresponding with this capacity will vary highly between years.

The temporal distribution of diverted irrigation water can only be estimated from detailed statistical analysis of hydrograph shape and of the random and systematic distribution of flow and no-flow periods.

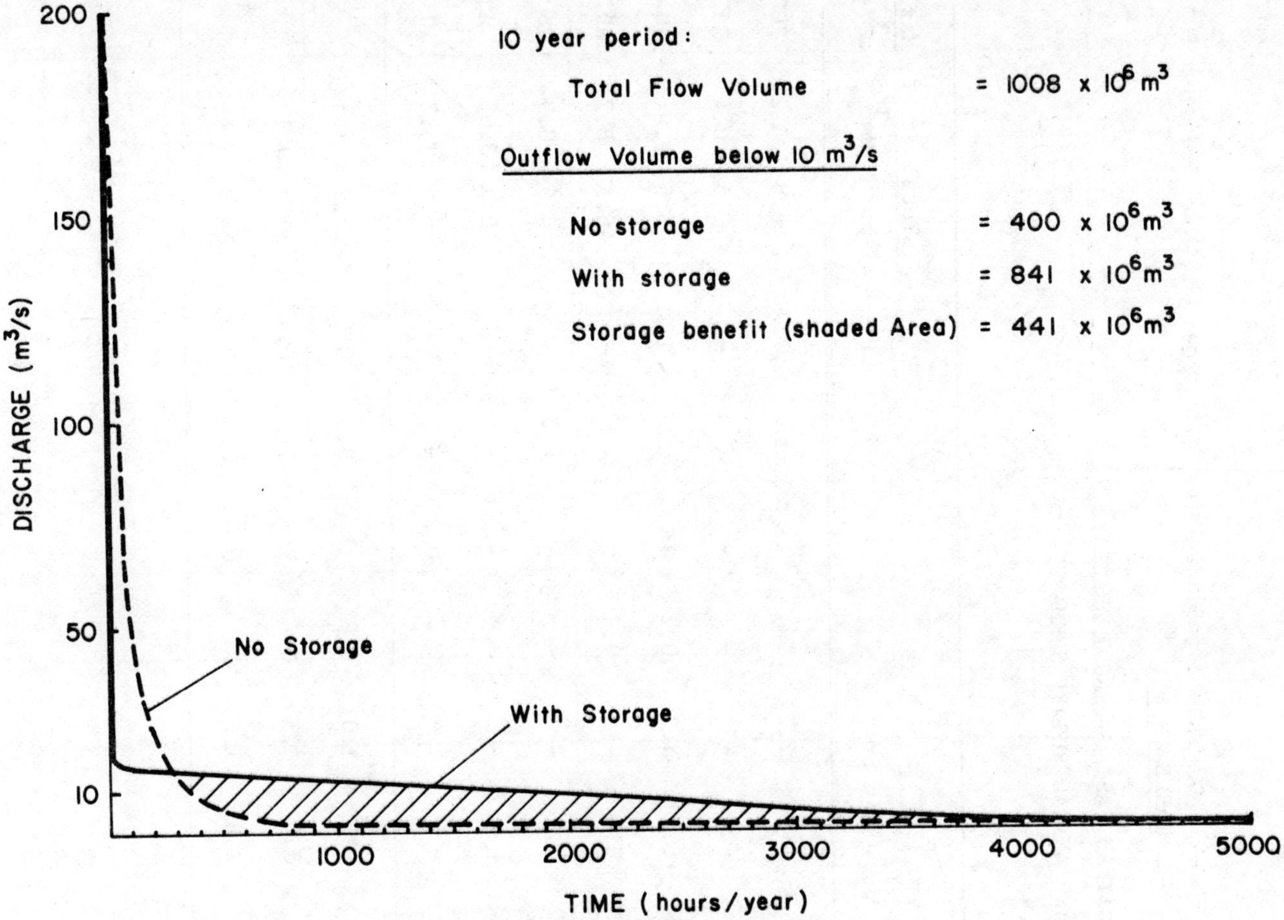

Fig. 18.5 Instantaneous flow-duration curves for generated wadi inflow to detention storage and routed outflow

The figure shows the benefit of storage where the required diversion rate is > 10 m^3/s.

TABLE 18.5

WADI LIYYAH BASIN - SAUDI ARABIA

MONTHLY AND ANNUAL RUNOFF SUMMARY - mm

YEAR	P/R	J	F	M	A	M	J	J	A	S	O	N	D	Total mm	Total Mm^3	Max Disch. Qp m^3/s	Max Disch. Qp DATE
1966	R	-	-	-	3.4	-	-	-	2.2	TR	TR	1.0	0.2	6.8	1.2	20	14.4
1967	R	0.1	-	0.6	-	-	1.7	-	-	-	-	0.1	-	2.5	0.4	21	7.6
1968	R	-	-	-	26.7	9.3	6.1	8.0	3.2	0.2	TR	1.0	1.0	55.5	9.8	240	17.4
1969	R	22.8	9.7	9.0	2.4	1.8	0.1	0.3	-	0.1	0.1	-	-	46.3	8.2	100	5.1
1970	R	1.6	-	0.1	2.6	-	-	0.3	-	-	-	0.3	-	4.9	0.9	18	25.4
1971	R	-	-	-	-	-	-	-	-	0.1	-	-	-	0.1	0.01	1	8.9
1972	R	-	-	-	0.1	1.6	3.1	-	1.4	0.2	0.1	-	-	6.5	1.12	40	9.6
MEAN	R	3.5	1.4	1.4	5.0	1.8	1.6	1.2	1.0	0.2	TR	0.3	0.2	17.5	3.1		
$\hat{\sigma}$	R	8.5	3.7	3.4	9.7	3.4	2.3	3.0	1.3	0.4	X	0.5	0.4	23.1	4.1		
SE95	R	7.9	3.4	3.1	8.9	3.1	2.1	2.8	1.2	X	0.4	0.3	0.3	21.3	3.7		

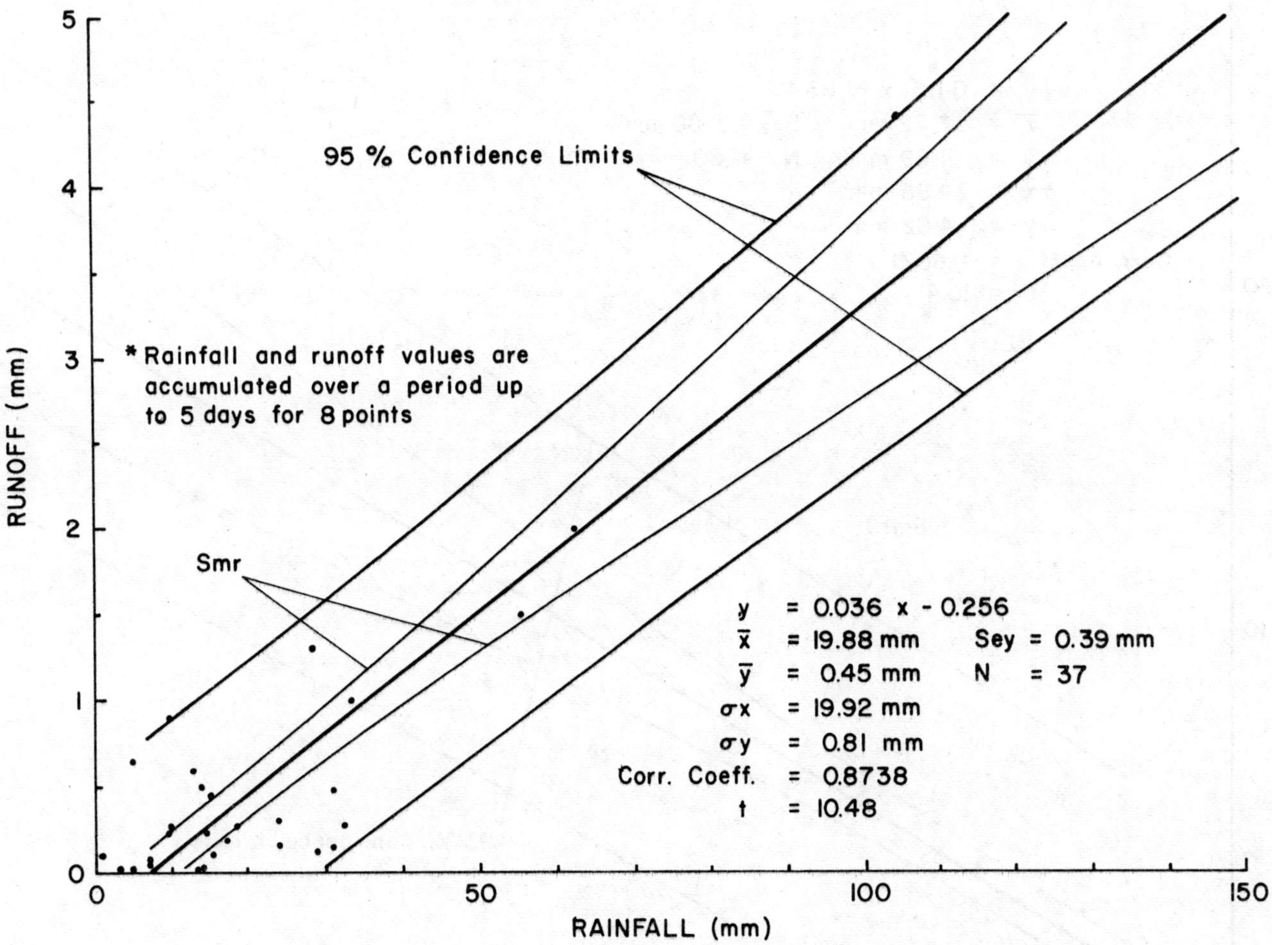

Fig. 18.6 Liyyah basin runoff vs. rainfall (daily values)

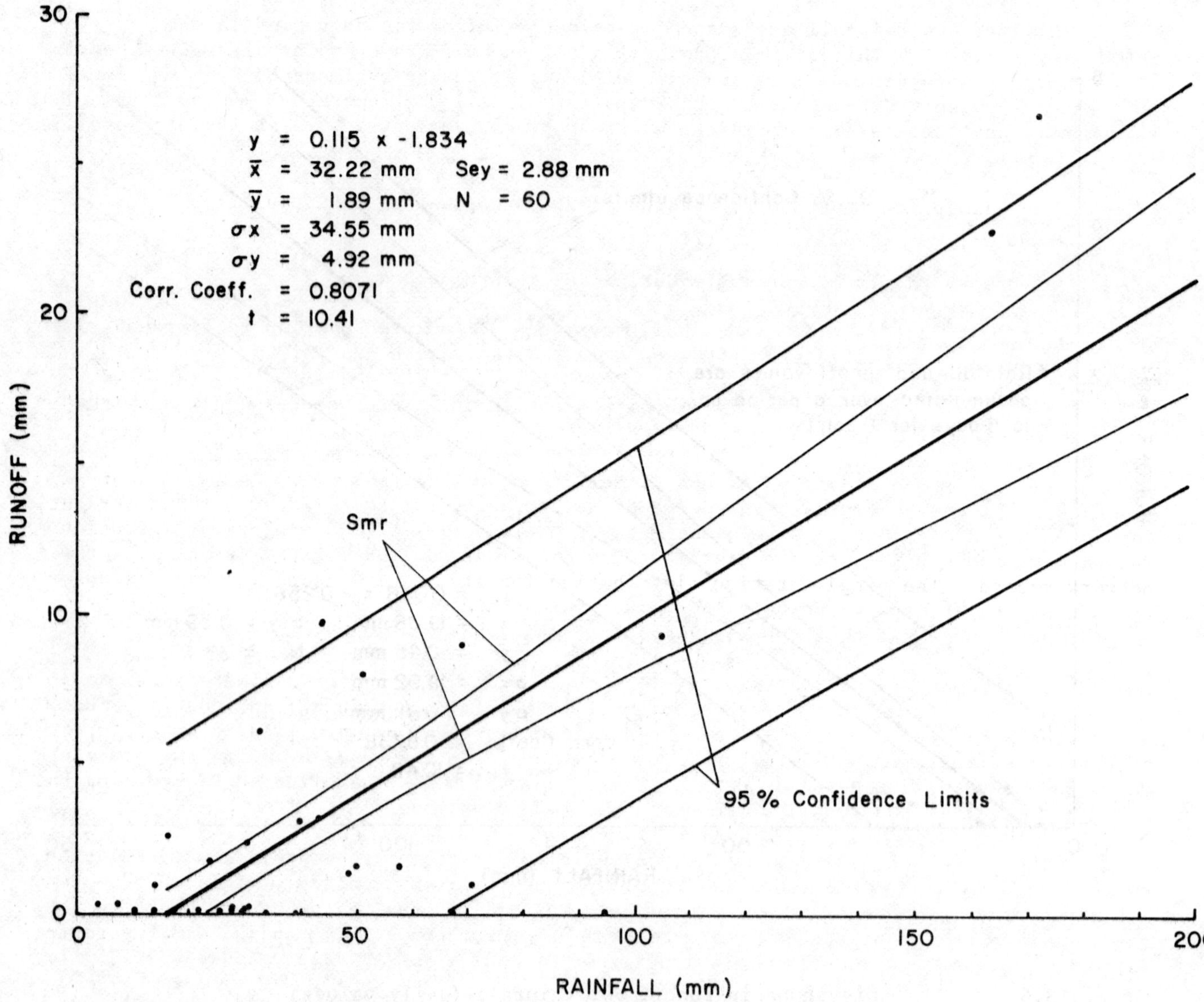

Fig. 18.7 Liyyah basin runoff vs. rainfall (monthly values)

- plot number of occurrences against runoff depth to produce, effectively, a runoff-duration curve (Figure 18.8).

- calculate the area beneath the curve.

This produces an estimate of mean annual runoff. In the case of Wadi Liyyah the calculation is as shown by Example 18.1.

Daily basin rainfall exceedances were derived from the data used in the regression analysis. This assumes basin rainfall to be the average of six raingauges in and adjacent to the basin. The resulting mean annual runoff estimate of 2.55 Mm^3 is sufficiently close to the measured mean runoff (3.1 Mm^3) for the record period (Table 18.5) to suggest that there are no large calculation errors. Uncertainties in the estimate lie in:

- the assumption that the regression equation applies to the populations of rainfall and runoff,

- errors in runoff measurements during the record period,

- the fact that rainfall and runoff have been accumulated over up to five days for eight of the thirty six values used in the regression analysis.

- the assumption that the distribution of basin input exceedances applies to the long-term.

Of these, the first is the most likely source of large error. An adjustment may be made to the basin rainfall exceedance distribution by comparing it with the distribution for any long-term raingauge within the network or in a similar meteorological situation. The network and single- station distributions are first derived for the duration of the network record. The single-station distribution for its longer record is then derived and factors calculated to correct the network exceedance values to their long-term equivalent. These are then the basis for the runoff calculations.

If the mean rainfall:runoff regression constants are a good estimate of those which would be derived from the population of long-term mean rainfall and runoff values, and if the estimated distribution of basin rainfall exceedance is similarly accurate, the resultant estimate of long-term runoff must be good. In any case, by using the procedure described above, a degree of smoothing takes place which has the effect of decreasing the weight of rainfall and runoff extremes in the record period.

The Laurenson (1964) runoff model, used in an attempt to improve the correlation between measured and predicted daily runoff volumes, was unsuccessful. This was due very largely to the low precision of the estimated rainfall inputs to the basin sub-zones for the maximum duration of one day required to model the catchment response. The theoretical number of gauges required for basin input precision of $\pm$ 5mm at 95% confidence is 250!!

Further studies were undertaken to discover if the precision of the rainfall input estimate improves with increased rainfall amount, as might be expected. The area within which the study basin is situated (containing 12 recording raingauges) was stratified according to mean annual isohyetal zones and the standard error of daily rainfall input to each stratum was calculated and used to estimate the overall input error to the area covered by the network ($\simeq$ 700 km^2). Daily rainfall input $\bar{P}_w$ was calculated as the sum of the stratum mean gauge catches weighted by the stratum area. $\bar{P}_w$ was then plotted against the associated areal variability expressed as the percentage standard error. The resulting scatter diagram was smoothed by calculating the mean of the percentage standard errors for $\bar{P}_w$ intervals to produce the curves shown on Figure 18.9. The graph, which also shows the weighted mean standard deviation of point rainfall for each storm, illustrates the relative increase in the standard deviation of the sample point values and relative decrease of the corresponding standard error of the mean, with increasing rainfall amount.

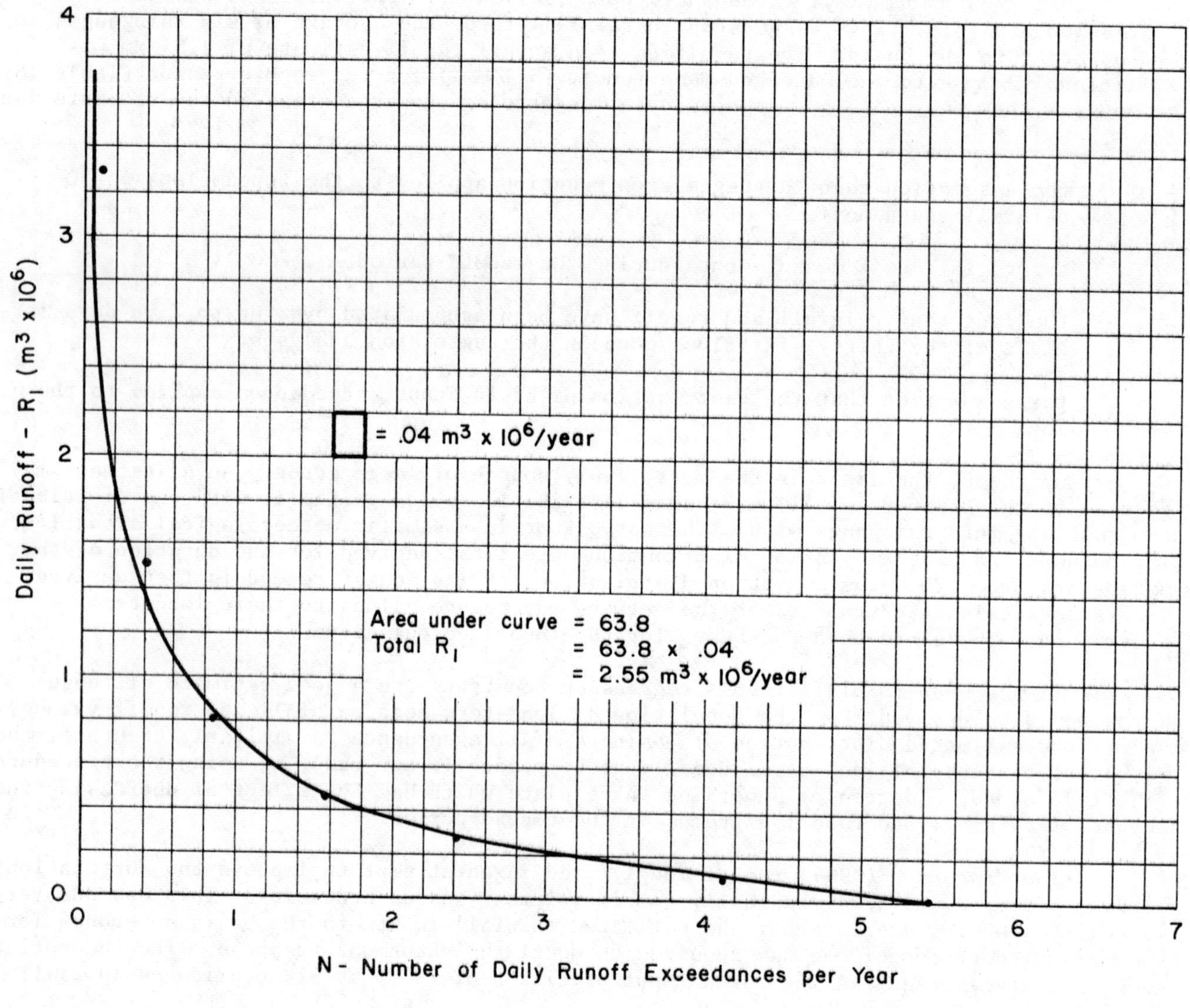

Fig. 18.8 Estimated distribution of daily runoff occurrences above specified thresholds - Wadi Liyyah, Saudi Arabia

Example 18.1

Estimation of Mean Annual Runoff - Wadi Liyyah, Saudi Arabia

1. Distribution of mean basin daily rainfall; assumed to be the average of the available records for six stations.

P - Class From mm	To mm	N - Number of Occurrences in in 7 years	Rainfall Threshold Exceedances	R-Runoff(1) Mm^3	Runoff Exceedances per year
0	7	22	60	0	8.57
7	10	9	38	0	5.43
10	15	12	29	0.10	4.14
15	20	6	17	0.28	2.43
20	30	5	11	0.46	1.57
30	50	3	6	0.81	0.86
50	70	2	3	1.52	0.43
70	100	-	1	2.23	0.14
100	150	1	1	3.30	0.14
	Total	60	60		

(1) Calculated from $R = -0.256 + .0356P$

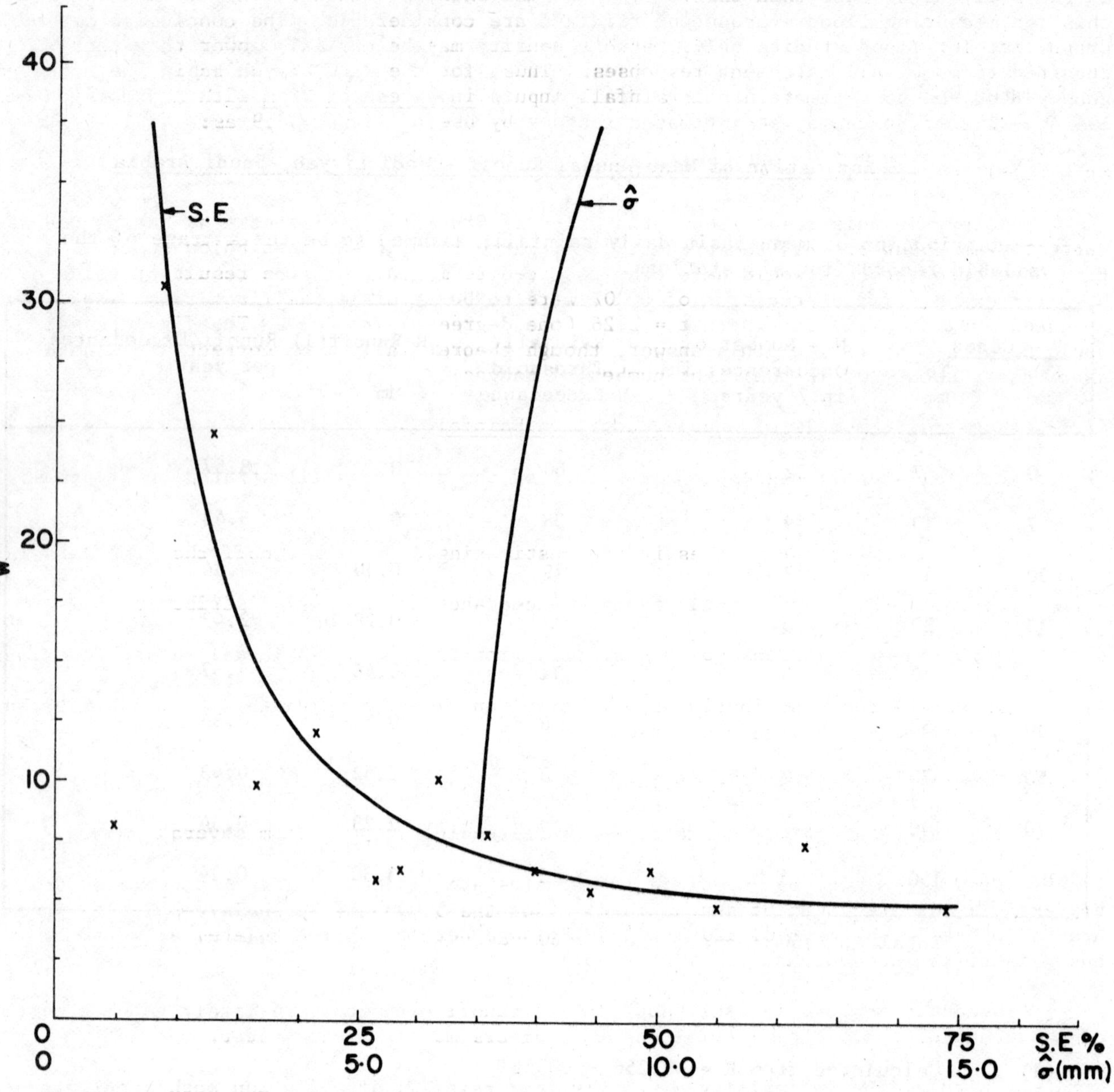

Fig. 18.9 Taif area - areal variability of daily rainfall

The curves show the relationship between weighted mean daily rainfall input ($\bar{P}_W$) and the associated percentage standard error (SE%) and the standard deviation ($\hat{\sigma}$).

This finding is important in showing that the areal variability of major storms is generally much lower than that of more normal rainfall events. The implications of this for measuring flood - producing rainfall are considerable. The conclusion may be drawn that for flood studies only, network density may be markedly lower than that required to model all catchment responses. Thus, for the Wadi Liyyah basin the number of gauges required to estimate areal rainfall inputs in excess of 25mm with a precision of ± 10% at 90% confidence may be estimated roughly by use of Figure 18.9 as:

$$Nr = \left(\frac{to}{SE_r}\right)^2 = \left(\frac{2 \times 8}{2.5}\right)^2 \simeq 40 \text{ at } 95\% \text{ C.L.}$$

Note that this result shows the value of Student's t statistic as having been set correctly at about 2. Had the predicted raingauge requirement been below about 20 a trial and error process would have been required to match t and the resultant value of Nr. For example, if a precision of ± 20% were to be required the first trial would have produced a value of 10 for which t = 2.26 (one degree of freedom). The final result would have been Nr. = 12. This answer, though theoretically more correct would have shown that "about 10" is the right number of gauges.

18.2.6 Regression Models of Annual Runoff on Rainfall

Rainfall-runoff regressions for four of the data sets illustrated on Figure 18.1 are shown in Table 18.6.

To use the selected regression for estimating long-term runoff the procedure is:

- estimate the basin annual rainfall exceedance probability distribution.
- apply the regressions to this distribution to produce an annual runoff distribution.
- integrate the distribution by plotting runoff exceedance against probability and calculating the area under the curve.

Example 18.2 illustrates the procedure.

The inaccuracies of rainfall:runoff regressions result from several causes.

Firstly, annual basin rainfall estimates are subject to large standard errors . For example the precision of annual rainfall on the 5,000 km^2 W. Khulays basin in Saudi Arabia with a 7-year mean of 120mm and a 15-gauge network was calculated as ± 75% by analysis of variance of annual point values.

Secondly, errors in individual annual runoff estimates are likely to be generally of the order of ± 30%, in alluvial channels errors may be much greater.

Thirdly, the variability of within-year rainfall distribution both areally and temporally would result in poor relationships even between precisely-measured inputs and outputs.

As an example, in 1969 very severe flooding resulted from heavy rainfall in the north of the Khulays basin which was inadequately gauged by the basin network. The runoff evidence in this part of the basin and in the adjacent basin to the north showed that much higher rainfall had occurred than was recorded. 1969 annual runoff was about 140 Mm^3 for a basin input estimated from the available records as 460 Mm^3 representing 30% runoff. Monthly figures for the flood month (January 1969) were; rainfall 200 Mm^3, runoff 120 Mm^3 - representing 60% runoff, much too high a figure.

It might be argued that despite the inadequacy of rainfall data, this form of estimate is the best available. This is not necessarily so. With seven years of runoff data, analysis of the individual hydrographs (or even daily runoff data) would be likely to reveal more stable statistical relationships (as described in Section 18.2.4) from

TABLE 18.6

Regression Analysis - Runoff on Rainfall (Annual Totals) Saudi Arabia

Station No.	Station Name	Drainage Area Km^2	No. of Years	Mean Input Mm^3	Mean Runoff Mm^3	Regression Constants a	Regression Constants b	r^2	SE Mm^3	t'(3)	Significant at 95% (t.05,5=2.27)
A-401	Abha										
	(1)	59	7	30	2.3	- 4.963	0.256	0.613	7.9	6.29	Yes
	(2)					8.03×10^{-9}	5.571	0.920	0.89	7.58	Yes
Ta-402	Liyyah										
	(1)	178	7	60	3.1	-12.600	0.270	0.625	2.50	6.45	Yes
	(2)					5.38×10^{-17}	9.234	0.706	2.59	3.47	Yes
J-401	Khulays										
	(1)	2840	7	340	39.3	-62.801	0.310	0.437	32.9	4.41	Yes
	(2)					.004	1.540	0.430	38.8	1.94	No
B-402	Bishah										
	(1)	11230	7	2500	90	- 8.472	0.029	0.148	52.4	2.08	No
	(2)					1.463×10^{-13}	4.241	0.438	72.5	1.98	No

NOTES:

(1) Linear Fit
(2) Power Fit
(3) $t' = \frac{r \sqrt{(n-2)}}{\sqrt{(1-r^2)}}$
(4) Note that derived b values may be unrealistically high, producing runoff estimates greater than rainfall when extrapolated beyond the range of the data sample.

EXAMPLE 18.2

Estimation of Mean Annual Runoff - Wadi Liyyah, Saudi Arabia. Catchment Area 178km^2

1. Distribution of basin annual rainfall. Only 7 values are available so assume a Normal distribution and calculate using the method of Beard (1962).

Year	Basin Rainfall Pa mm	Mm^3
1966	330	58
1967	290	51
1968	380	68
1969	420	75
1970	260	46
1971	240	43
1972	360	65
$\bar{P}a$	330	58
$\hat{\sigma}$	65	12

p% =	0.25	1	5	10	15	50	90	95	99	99.75
K =	-4.62	-3.36	-2.08	-1.54	-1.21	0	+1.54	+2.08	+3.36	+4.62
$K\hat{\sigma}$ mm	-300	-218	-135	-100	-79	0	+100	+135	+218	+300
$\overline{Pa} + K\hat{\sigma} = Pa_p$	30	112	195	230	250	330	430	465	548	630
Pa_p Mm^3 =	5	20	35	41	45	59	77	83	98	112
R_1 =	0	0	0	0	-0.4	3.3	8.2	9.8	13.9	17.6

Notes:

(1) K is a frequency factor for the Normal distribution.
(2) $\hat{\sigma}$ is the best estimate of the population standard deviation.
(3) $\bar{P}a$ is the sample annual rainfall mean.
(4) Pap is the annual rainfall estimate of probability P.
(5) R_1 is the estimated runoff corresponding with Pap: = 12.6 + 0.27 Pap (Table 18.6).

2. The non-exceedance probabilities P% are plotted against R_1. The area under the curve represents mean annual runoff and is calculated as 3.8 Mm^3, as shown on Figure 18.10.

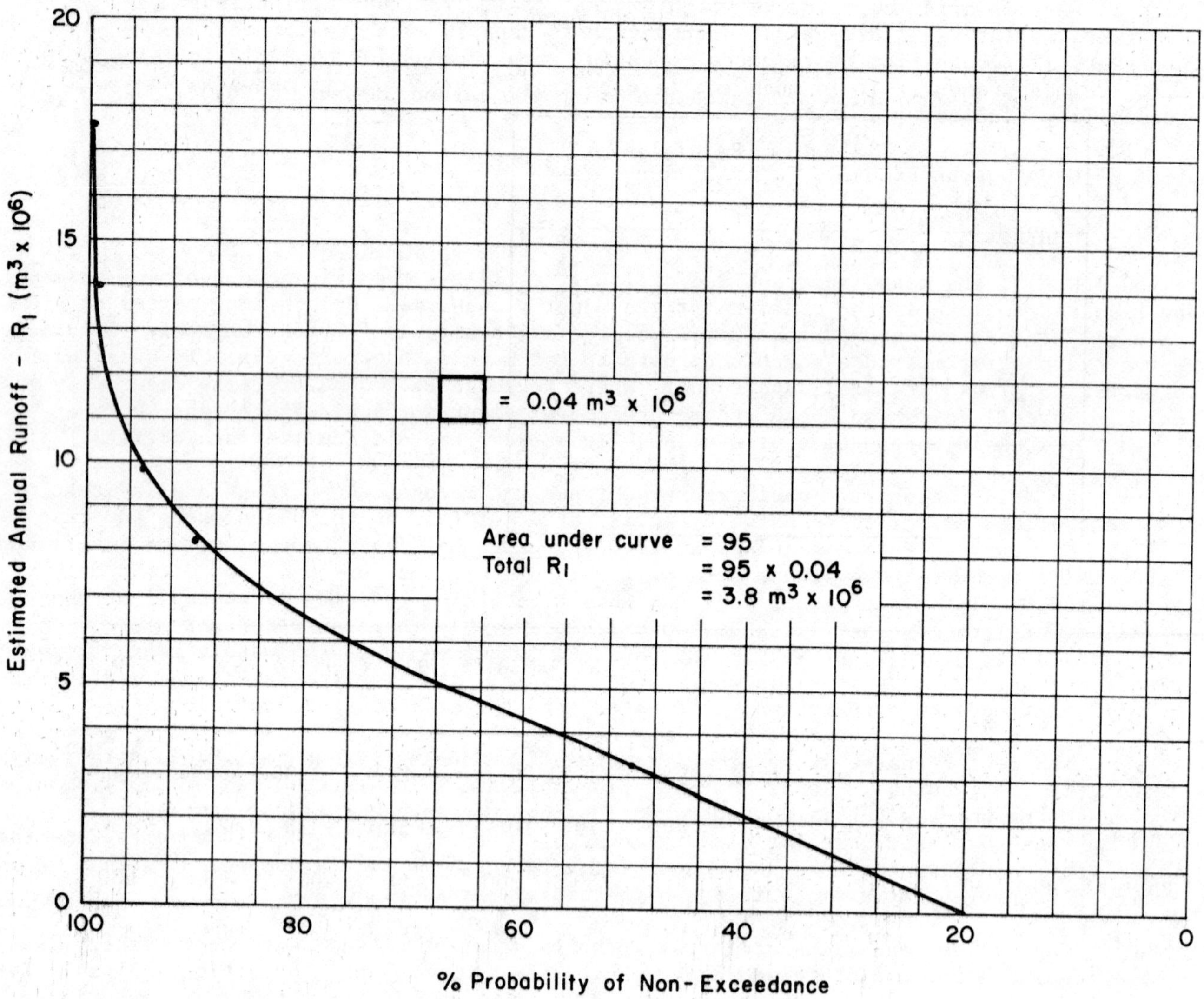

Fig. 18.10 Estimated probability distribution of annual runoff Wadi Liyyah, Saydi Arabia

which better estimates of the mean and variability of annual runoff would emerge. Such detailed studies would also lead to more realistic design of hydraulic structures. It is worth restating here that runoff data collection is not only the most useful hydrological activity in arid zones, it is also the most difficult. A long record of accurate runoff data is rare in arid zones and is worth a great deal. The cost of intricate (and sometimes worthless) hydrological speculation is usually much greater than would be necessary if recognition of resource potential were accompanied by early investment in resource measurement. It is much too late to install hydrological networks during feasibility studies.

18.3 GROUNDWATER RECHARGE ESTIMATION

18.3.1 General

One of the most important and difficult estimates the arid zone hydrologist may be required to make is the estimate of groundwater recharge. Unless long series of rainfall, runoff and groundwater level records are available - a most unlikely situation - the hydrologist must use short-term data to estimate a long-term mean. In areas with high variability of rainfall and runoff, where recharge to groundwater may occur only once in several years and where groundwater development for agriculture represents a rapidly increasing proportion of available resources, the difficulties are obvious.

In arid regions two basic recharge types are recognized - direct and indirect. The term Direct refers to recharge by infiltration of rain at the place where the rain falls. By contrast Indirect refers to recharge which has first become surface runoff and later infiltrates through stream beds or the floors of local areas of concentration, depression or flats. In either case the moisture-holding capacity of the soil at the point of infiltration must first be replenished and only then can water percolate downward to the groundwater body. The physical mechanism is approximately the same, but direct recharge is only possible where storm rainfall less surface retention, surface runoff and immediate evaporation, is greater than the soil moisture deficit.

Indirect recharge is possible from runoff-producing rainstorms when a relatively large area contributes to local concentration of water, in a stream bed or depression. The area over which water is concentrated for indirect recharge is, of course, much smaller than that available for direct recharge. Thus it may be that in some regions the higher the rainfall, the more important direct recharge may be in comparison with indirect recharge, for example in the high-rainfall areas of Jordan.

Depressions and flats are areas in which water-borne sediments accumulate. In these areas the soil moisture storage capacity is high, producing conditions suitable for vegetal growth and higher evapotranspiration losses. In addition, the finer materials characteristic of drainage depressions may produce greater potential for evaporation from groundwater and soil moisture storage.

Examples of direct and indirect groundwater recharge estimation in arid zones are given as Case Studies in this Section. Recharge equations may be expressed simply as follows:

a. $$DREC_{\Delta t} = P_{\Delta t} - R_{\Delta t} - ETa_{\Delta t} \pm S_{\Delta t}$$

Direct groundwater recharge during time interval Δt = rainfall - runoff - evapotranspiration from soil moisture ± change in soil moisture storage. ETa from groundwater is omitted from the equation since, by definition, it has already been recharged.

b. $IREC_S = QA_t - QB_t - E_t$

Indirect recharge rate at time t from a steadily flowing stream = flow at A - flow at B - evaporation from stream and bed.

c. $IREC_{NS} = VA - VB - E$

Indirect recharge between two sections A and B produced by the passage of a flood wave = volume flowing past A - volume flowing past B - evaporation from stream and bed during the time interval over which the volumes are measured.

d. $IREC_D = H_{\Delta t} - E_{\Delta t}$

Indirect recharge from an internal drainage depression during time interval Δt = change in water level - evaporation from water and wetted surfaces during Δt.

Recharge by underflow is not taken into account in the above.

For estimation of recharge by small differences between measured flows the flows should preferably be constant in time and the differences must be appreciably greater than measurement errors. Indirect recharge estimates may be made under unsteady flow conditions if the total flood hydrograph at A and B are measured and reliable estimates made of evaporation from the wetted surface of the channel, from the stream itself and from soil moisture storage.

In addition to these methods which are based on the calculation of water balances, groundwater recharge may also be estimated by groundwater models, by measuring groundwater discharge through springs and by estimating the baseflow component of the total runoff hydrograph.

The estimation of recharge using groundwater models is based on the principles of continuity and of flow in aquifers. To use this method, measurements of groundwater levels and estimates of the aquifer hydraulic properties of transmissivity and storage are necessary. The time interval for the model may be shorter than a day if instantaneous water level data are available. In such cases the arrival of recharge events - particularly those due to floods in ephemeral stream beds - and their gradual dispersion through the aquifer can be modelled. In the case of direct recharge, it is more common for a monthly, seasonal or annual time interval to be used.

Estimation of recharge to a groundwater reservoir by measuring the discharge from the reservoir, if possible, avoids the complexities of simulating the recharge process but is only applicable where aquifer leakage is small or can be estimated by other means. Where aquifers discharge through springs and as stream baseflow these should be measured to provide data on the storage and depletion characteristics of the groundwater resource in addition to producing data on the amount and variability of the surface water resources more immediately available for use downstream.

18.3.2 Direct Recharge

Direct recharge from rainfall and subsequent drainage beyond the effects of evaporation is more difficult to estimate than indirect recharge. The main reason for this is the fact that in arid zones the amount of recharge is often only a minute proportion of rainfall, although recharge volume over an extensive surface may be enormous.

Direct recharge may be estimated by:

- Calculation of the water budget within the zone of evaporation from the surface layers of the soil. This involves a modelling approach in which the rainfall input is routed through one, or possibly several, storages. Evaporation from storage is assumed to follow a predetermined, empirically-based rule. Recharge is the residual in the balance (rainfall = runoff - evaporation + change in storage + recharge).

- Establishing networks of soil moisture measurement stations in typical recharge zones and determining the moisture capacity and drainage characteristics of the recharge medium. Soil moisture measurement methods may include electrical resistance sensors set at intervals, over the depth range from which evapotranspiration may take place and neutron scattering probes with access tubes installed to similar depths. Water in excess of that corresponding with the field capacity of the recharge medium beyond the root zone is assumed to be recharge. The recharge rate is estimated from moisture gradients in the soil profile.

- Radioisotope method in which the isotopic (usually tritium) activity is measured over the soil depth profile and related to the temporal pattern and isotopic history of rainfall in the region. If the isotopic profile, in the depth zone below the reach of evapotranspiration, can be correlated with rainfall incidence, mean annual recharge is the total soil moisture content divided by the number of years represented by the identifiable isotopic pattern. The method relies for its validity on the fact that water molecules draining through the zone of aeration displace those absorbed onto the soil particles, with a "piston-type" effect. The usefulness of tritium as a tracer is due to the very large and sporadic artificial introduction of tritium into the atmosphere since testing of thermo-nuclear devices began in 1953. The effect of these tests was to "label" the rainfall for those years in which they took place and for those in which they did not.

The water budget method is a typical example of widely used "book-keeping" procedure applied to direct recharge problems. Uncertainties in the results produced lie in evaporation and soil moisture storage and capacity estimates, and in spatial rainfall variability. Measurement and estimation errors in these items may be much larger than the difference between them.

In arid zones all such methods are suspect unless supported, at least qualitatively, by groundwater data. 10-day rainfall on recharge zones can only be estimated within wide tolerances. Unless the recharge medium is coarse sand, infiltration amount cannot possibly be spatially uniform. Errors due to the assumption of spatial uniformity of rooting depth must be large where vegetation is predominantly located where runoff concentrates and infiltration is correspondingly greatest.

Evaporation from surface and soil moisture storages may be calculated from lysimeter experiments but results are virtually impossible to extrapolate to large areas.

Attempts are sometimes made to estimate direct recharge on an areal basis by interpolating rainfall depths at the node points of a grid system from rainfall at gauging stations. This approach is useful where it is required to study the piezometric head response and aquifer flow patterns on a regional basis. However, it is worth considering the precision with which interpolation between gauged falls is possible in arid zones, before developing an interpolation model. Section 5.1 discuss methods for calculating errors in areal rainfall estimation which could as usefully be applied to deterministic groundwater recharge models as to those which attempt to predict runoff from basin rainfall. It may be found in some areas and for some durations that an arithmetic average of gauged rainfall is as precise an estimate of areal input as would result from complex mathematical "surface" fitting techniques.

To estimate long-term direct recharge, which by definition refers to a point rather than an area, it is necessary only to use point rainfall as input to the water balance model. In this way the recharge to a particular surface type may be expressed in the probabilistic sense rather than deterministically. Thus, an actual or synthetic rainfall record may be used to determine the amount and temporal distribution of point recharge though it would first be necessary to confirm the recharge model empirically. Point recharge processes may be simulated in this way but the aquifer response cannot.

18.3.2.1 Case 3 - The Southern Desert of Jordan

A UNDP investigation of the sandstone aquifers of East Jordan included a report by Lloyd (1969) on the hydrogeology of the Southern Desert. The report described the method used to estimate direct recharge by calculation of the water budget in the top soil layers of selected recharge surface types.

The method used was:

- Classification of recharge surface types. Nineteen distinctive range types were recognised, of which six are extensively distributed. These range types were grouped according to their apparent recharge characteristics. Bare sandstone outcrops were considered to be too indurated to be capable of acting as a recharge medium.

- Two recharge factors were determined for each recharge surface type. C, a drainage factor defined as the available water (depth in mm) within the influence of vegetation rooting systems, calculated as 50% of the field capacity of the soil (water weight as a percentage of the weight of soil) and applying this value to the bulk density and depth of soil in the root zone. D, the maximum soil moisture deficit determined by field measurements at the end of the dry season.

- Water was assumed to be evaporated at the potential ETp (Penman) rate when the soil mositure deficit SMD was less than C. When SMD was between C and D evaporation was assumed to occur at 10% of the ETp rate.

- Evaporation was calculated using Penman's method with an albedo $r = 0.15$ and using McCulloch's (1965) altitude correction.

- A running balance of accumulated SMD was calculated for 10-day time intervals. Negative SMD was assigned to recharge.

18.3.2.2 Case 4. Recharge Through Sand Dunes in Saudi Arabia

In an investigation carried out in the Dahna sand dune region of Saudi Arabia, Dincer et al (1974) studied the infiltration and recharge properties of desert sands. Observations of temperature gradients, moisture content, particle size distribution of the sand and the tritium, deuterium and oxygen 18 content of soil moisture showed that a significant proportion of 80mm mean annual rainfall percolates through the sand dunes and may be an important source of recharge to underlying limestone aquifers.

The method was as follows:

- Sand samples were collected from different depth intervals by augering.

- The tritium content of water extracted from the samples was measured.

- The tritium content of rainfall measured at Bahrain was tabulated together with the rainfall for Bahrain, Riyadh and Khurais, the latter two stations being those nearest the study area (Table 18.7).

- The tritium depth profile was compared with the tritium content of rainfall measured at Bahrain, taking account of the radioactive decay time of tritium (12.26 years).

TABLE 18.7

Precipitation (Bahrain, Khurais, Riyadh) and Environmental Isotope Data (Bahrain)(1)

Year	Month	Monthly precipitation (mm)(2) Bahrain	Khurais	Riyadh	Tritium (T.U.)	Error ±	Deuterium D^{o}/oo	Oxygen-18 o O^{o}/oo
61	11	41					16	0.6
61	12	4			250	25	11	-1.7
62	02	2			810	40	12	2.4
62	03	7			-	-	29	-
62	11	1			300	25	24	3.1
63	03	1		no data	-	-	34	6.3
63	04	4			1420	30	14	1.4
63	05	10			-	-	10	0.0
63	10	2			470	45	16	0.4
63	11	16			-	-	2	-3.5
63	12	2	no data		575	25	25	2.3
64	01	32		69(4)**	480	25	-28	-5.3
64	02	27		20(13)	720	25	17	1.5
64	04	1		0.7(1)	-	-	19	1.6
64	12	53		72(7)	126	13	-22	-4.9
65	01	17		6(4)	105	20	4	-2.2
65	04	8		38(10)	315	20	20	2.6
66	02	21		12(4)	-	-	9	-0.6
66	04	14		10(5)	-	-	19	1.4
66	12	1		0.1(1)	-	-	27	-
67	01	2		0.4(2)	-	-	38	5.4
67	02	8	1(1)	4(3)	-	-	35	4.2
67	03	5	7(5)	18(6)	-	-	31	4.6
67	04	6	26(5)	36(9)	-	-	29	3.9
67	05	3	0 -	0 -	-	-	19	-
67	11	4	20(4)	17(10)	-	-	- 3	-1.9
68	02	53	37(8)	31(5)	52	5	-	-
68	04	12	19(8)	44(7)	111	5	- 3	-2.3
68	11	3	4(1)	2(2)	-	-	9	-0.2
68	12	5	7(2)	5(3)	42	6	10	0.8
69	01	88	100(20)	85(17)	29	1.4	-0.1	-
69	02	10	34(5)	3(2)	79	8	-	-1.2
69	03	1	4(5)	8(5)	47	6	31	-
69	04	49	11(7)	20(6)	58	3	-10	-
69	10	9	5(1)	0 -	-	-	13	0.7
69	11	9	3(4)	2(4)	-	-	19	0.6
70	01	7	5(7)	7(4)	-	-	-	1.2
70	03	1	5(3)	5(4)	73	3	-	3.9
70	12	1	1(1)	0.3(1)	122	8	-	1.1
71	01	3	1(2)	0 -	71	3	-	-
71	02	8	1(2)	0 -	53	3	-	-
71	03	2	6(5)	16(8)	113	4	-	-
72	01	20	20(10)	27(6)	28	1.4	-	-
72	02	10	0.4(2)	0.5(1)	37	2	-	-
72	03	42	37(9)	49(8)	34	4	-	-

(1) Bahrain data from the International Atomic Energy Agency Publications and files; Khurais and Riyadh data from the Ministry of Agriculture and Water files.

(2) Values in the parenthesis show the number of rainy days.

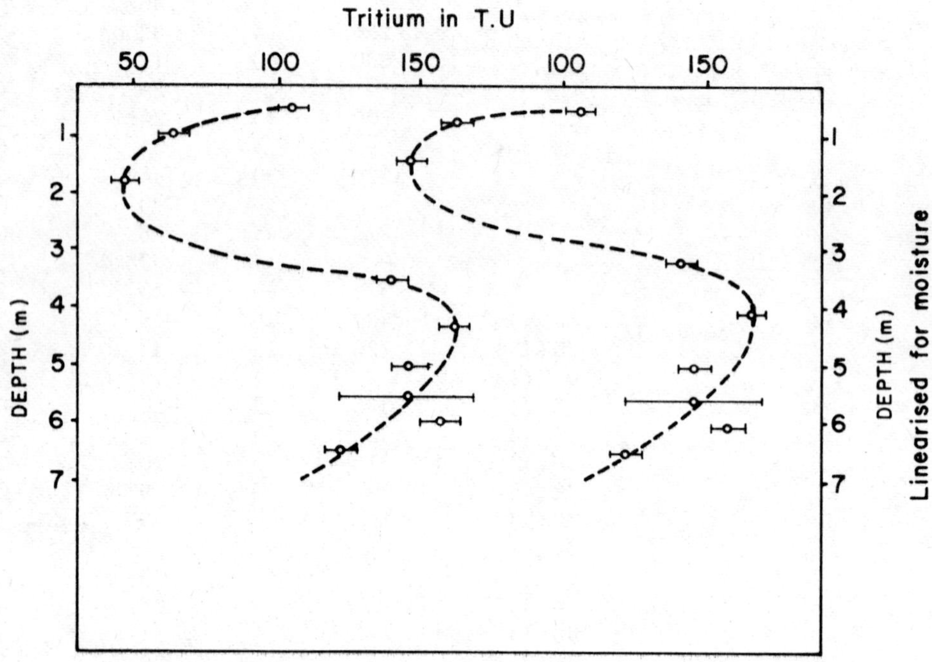

Fig. 18.11 Tritium content of the sand moisture at 114 km on the Riyadh-Dhahran highway

As it is known that the tritium content of rainfall is uniform over large areas, it was possible to extrapolate from Bahrain to the study area without introducing large errors. However, the possibility of identifying the major tritium concentration peaks in the soil profile is conditioned by the amount of rain falling on the study area. The correlation coefficient r between Bahrain and Riyadh monthly rainfall is 0.754 and between Bahrain and Khurais 0.799; both values being highly significant. However, daily rainfall would be expected to be much more variable spatially and the depth of percolation is much more dependent on daily than on monthly rainfall depth. So dissimilarities could be expected between the tritium profiles of sand dunes in Bahrain and the study area.

Comparing the tritium profile shown on Figure 18.11 with the tritium content of Bahrein precipitation shows that the tritium content at a depth of 2m is contributed by 1969 rainfall and the peak tritium content at 4m to have been that contributed by 1964 rainfall. Multiplying the depth (minus surface dry layer) with the mean moisture content of the sand in the depth zone above 4m, the annual mean recharge was estimated as 23mm.

The relative lack of recharge since 1969 was explained by the temporal distribution of rainfall in subsequent years when falls were generally light, although annual amounts were near normal.

18.3.3 Indirect Recharge

The physical processes involved in indirect recharge may be highly complex, for example where it takes place through the bed of an ephemeral alluvial channel as the result of rare and sporadic, highly complex flood waves. In these cases the depth, width and duration of the recharge stream varies with time and infiltration rate may also be a function of sedimentation processes. Indirect recharge may occur from concentrations of runoff in local depressions too small to quantify.

The case studies described below each illustrate typical situations in which it is required to estimate recharge as a result of these complex processes. The one factor which makes the task easier than that of estimating direct recharge, is the size of the recharge water source and recharge amount in relation to the area in which recharge takes place.

The case studies described below range from preliminary resource estimates to feasibility studies leading to water resource system design.

18.3.3.1 Case 5. Groundwater Recharge in Qatar (Pike et al, 1975)

1. Introduction

Direct and indirect recharge from rainfall to groundwater is the sole natural water resource of Qatar. Quantitative estimates of its magnitude and frequency are of fundamental importance to any assessment of available groundwater resources. Almost the entire land surface of Qatar is made up of some 850 contiguous depressions of interior drainage with catchments varying from 0.25 km^2 to 45 km^2. A large percentage of these depressions have a circular pattern and are a few metres to 20m below the surrounding erosion surface with colluvial soils of up to 1m in depth covering the floor of the depression. While direct recharge from rainfall undoubtedly takes place after significant storms, the major recharge mechanism is indirect with runoff from the surrounding catchment ponding in the depression floor. Part of the ponded water is evaporated and the remainder recharged to groundwater.

2. Recharge Area

The catchment areas of individual depressions were delineated on accurate 1:50,000 maps, their areas computed and indexed. Large flat areas with no drainage were excluded as being inactive. In this way 850 individual active internal drainage systems were identified and found to have a total area of 6,942 km^2 or 61% of the total land area of Qatar. An external drainage area of 2,676 km^2 forms a narrow belt around the periphery of the peninsula where runoff usually terminates in a series of playas.

The peninsula was divided into ten major basins delineated on the basis of low topographic divides across whose boundaries groundwater movement may or may not take place.

From rainfall data and runoff estimates based on post-flood surveys of depressions, it was deduced that runoff from individual storms of over 10mm per day amount to between 15% and 25% of the rainfall, although two surveyed occurrences in 1972 and 1975 illustrate the importance of storm duration and intensity characteristics on runoff variability and the errors that are likely to occur in making broad assumptions.

3. Recharge Amount

Recharge resulting from individual storms was estimated from observations of water level decline in depressions less Penman Eo estimates.

From the results of these studies recharge was estimated over the whole country as follows:

- Isohyetal maps of total storm rainfall in excess of 10mm ("effective rainfall") were prepared and the effective rainfall calculated for each basin.
- 2% of effective rainfall was assumed to be recharged directly to groundwater.
- 35% of effective rainfall was assumed to be infiltrated to the depression soils, the areas of which had been determined.
- Soil moisture storage capacity was calculated and deducted from infiltration with an allowance made for interception by vegetation.

The mean recharge for 1971-76 was estimated in this way as 32 Mm^3 representing 15% of effective rainfall (as defined) and 9% of total rainfall.

18.3.3.2 Case 6. Kairouan, Tunisia

Besbes et al (1978) considered four methods for estimating groundwater recharge to an alluvial aquifer system in Tunisia.

a. Flow in the unsaturated zone.

$$Rec. = L \times \bar{W} \times I \times D$$

in which L is the length of streamflow over which recharge takes place, $\bar{W}$ is the mean width of the stream, I is the infiltration rate and D is the duration.

Difficulties in applying this method arise from:

- Estimation of infiltration rate, its variation with time, with depth to the phreatic surface, and in space.

- Removal of surface fines from the alluvial channel during the passage of a flood exposes the coarser materials with higher infiltration rates. Experimental determinations of I in a dry alluvial channel are thus underestimates.

- Evaporation from channel bed and water surfaces is difficult to estimate.

b. Water Table Recovery

The recharge volume may be estimated as:

$$Rec = A \bar{H} S$$

where A is the area of saturated alluvium, $\bar{H}$ is the mean groundwater level rise over the area and S is the specific yield of the aquifer material in which H occurs. This method depends for its accuracy on the configuration of the phreatic surface, on the density of observation wells and on the precision of the S estimate.

c. Model Calibration

Recharge is estimated by optimisation of the transmissivity T and storage S parameters in a digital nodal network model. The T and S calibration is achieved by a trial and error process in which both historic steady state piezometric surface and the groundwater level hydrographs at individual observation wells by time controlled simulation are reproduced.

d. Deconvolution

If transmissivity is time invariant, the equations of flow in a saturated aquifer are linear. The relationship between recharge and water table recovery can be written as a convolution integral

$$s_i(t) = \int_0^t r(t) \, \phi_i (t - \tau) \, d\tau$$

where $s_i(t)$ is net recovery of the water table in observation well i as a function of time t, r(t) is recharge through the stream bed at time t and $\phi_i(t)$ is the linear impulse response of the system (saturated aquifer plus unsaturated zone between water table and stream bed) relating recharge to recovery in observation well i . τ is a time shift which is constant for a particular case. $\phi(t)$ is the observed or computed response for unit recharge. Here, $s_i(t) = h_i(t) - d_i(t)$ is the average over the time-step of the difference between measured high $h_i(t)$ and assumed level $d_i(t)$ in the absence of recharge, i.e. the level corresponding with natural groundwater decline. d(t) may be estimated from a digital model or by fitting a simple exponental recession coefficient to measured well hydrograph recessions during recharge-free periods. Recharge through the stream bed at time t, r(t), is taken at the time it leaves the stream. It is averaged over a time step generally longer than the duration of one flood when discretizing the convolution integral. Thus the recharge is the volume infiltrated during a time step.

Two simplifying assumptions are required:

- transmission through the unsaturated zone is a linear process.

- the spatial distribution of recharge retains the same pattern; i.e. for any given flux r(t) at time t a constant percentage of r infiltrates through a given section of the bed. If this assumption is invalid, e.g. small floods disappearing in the upper reach and larger ones being more evenly distributed over the recharge zone, another approach should be used.

The recovery S_i is obtained from the hydrograph of observation wells i. If the impulse response ϕ_i for each well is known, recharge may be computed by deconvolution (matrix inversion techniques). The method is comparable with the unit hydrograph method in surface water flood estimates. The unitgraph is equivalent to ϕ_i, the hydrograph to r_i and the net rainfall profile to r(t).

Besbes et al found that the results obtained by the deconvolution method were similar to those obtained by the water table recovery method using an automatic piezometric surface mapping technique. The validity of recharge estimates may be checked by testing the digital model, using recharge estimates derived by deconvolution to simulate groundwater level response. Results of applying automatic piezometric surface mapping techniques allied to specific yield estimates and the deconvolution technique combined with aquifer modelling to estimate recharge to two alluvial aquifers during two separate periods, were as follows:

			Recharge Mm^3	
			Fall 1969	April 1970 to April 1974
1.	Piezometric Surface Mapping			
	Basin			
	a.	Zerouad	116	
			74	
	b.	Merguellil	120	
			100	
2.	Deconvolution and Modelling			
	basin			
	a.	Zerouad	82	
			54	
	b.	Merguellil	60	
			50	

The Zerouad basin is 9,000 km^2 and the Merguellil basin 1,100 km^2. The surface area of Quaternary alluvial deposits forming the aquifer is about 3,000 km^2.

18.3.3.3 <u>Case 7. Estimation of Recharge to Desert Basins of Nevada</u> (Watson et al, 1976)

The Maxey-Eakin (1949) method of recharge estimation was investigated to determine whether or not a statistically rational basis exists for its use. This method calculates recharge from specific percentages of precipitation estimated to fall in 6 different elevation zones. These zones are delineated on maps, their areas determined and mean zonal rainfall input calculated. Multiple regression equations are developed to obtain the best fit of zonal rainfall with recharge. It was assumed that hydrologic equilibrium exists, i.e. that recharge equals discharge in the closed basins. In fact, several valleys were not in equilibrium due to agricultural development.

Multiple and simple linear regressions were developed, the former using five independent variables (the zonal rainfall inputs) and the latter, the separate zonal inputs. The dependent variable discharge (= recharge) was that reported from 63 groundwater studies in the Basin and Range province of the Western United States.

The results obtained are shown in Tables 18.8(a) and (b) which show the regression coefficients as percentages and the 95% confidence interval about these for the 63 data sets. Watson et al concluded that although the predictive value of both multiple and simple linear regression of discharge on rainfall is slight, the simple regression coefficients were more stable. They felt that the basic approach is the only practical

TABLE 18.8

(a) Results of multiple linear regressions showing variability of the coefficients; 63 Observations

Precipitation Zones (in)*	Coefficients (%)	95% Confidence Intervals (%)
20	20	± 15
15-20	19	± 16
12-15	- 1	± 6
8-12	4	± 2
8	0	± 1

(b) Results of simple-linear regressions showing variability of the coefficients

Precipitation Zones (in.)*	Number of Observations	Coefficients (%)	95% Confidence intervals (%)
20	23	63	± 14
15-20	44	36	± 5
12-15	57	15	± 2
8-12	63	9	± 2
8	51	3.4	± 2.2
Total precipitation	63	3.4	± 0.6

* 1 in = 25.4 mm.

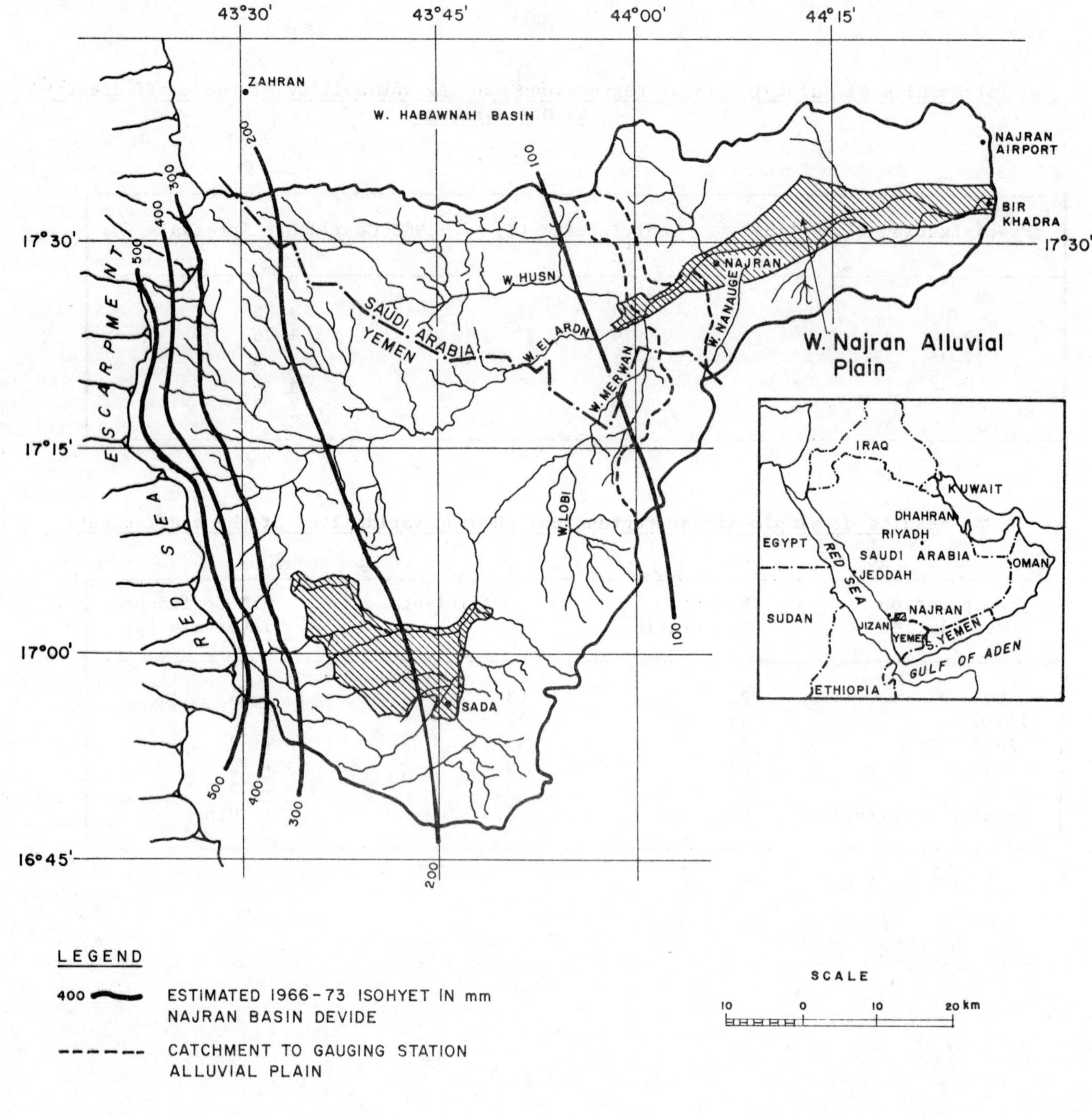

Fig. 18.12 Drainage network and rainfall - Wadi Najran

method of assessing recharge, despite the fact that it does not take account of the many variables, other than rainfall, affecting recharge.

It should be noted that few arid zones have sufficient data on which to base regression models. Normally a more complex model, based on short-term data, is necessary.

18.3.3.4 Case 8. Wadi Najran, Saudi Arabia (Figure 18.12)

A feasibility study was conducted to determine the water resource and flood protection benefits to irrigation agriculture in the Najran alluvial plain, provided by the construction of a reservoir. The hydrological and hydrogeological studies were required to produce quantitative assessments of the existing water resource system represented by the alluvial basin and of the effect of imposing a new system of storage and regulation of natural inflow sequences on the natural recharge mechanism downstream.

The aquifer system in the study area consists of 370 km^2 of alluvial sediments which have accumulated on the crystalline basement complex at the outlet of a 4800 km^2 basin draining eastward from the Red Sea escarpment. Estimated long-term mean basin rainfall is 170mm ranging from 500mm at the escarpment to 100mm at Najran. Seven years of runoff records of variable quality were available for a station located within the alluvial plain. Two new stations were installed for the purposes of the study, one in a stable channel reach at the head of the alluvial plain and the other at a constriction in the plain some 45 km^2 downstream. The existing station had no control and discharge hydrographs were based on current meter measurements and, to a small extent, slope-area surveys. Cross section surveys and ad hoc current meter and slope-area measurements were made at intervals between the upper and lower stations and the existing station was upgraded.

Previous work had shown that the stochastic flow generation technique would be applicable despite uncertainty as to the quality of the runoff data. There was no effective alternative to this approach since it was necessary to model the infiltration and recharge characteristics under existing conditions and following possible reservoir regulation. These characteristics, and sediment transport rate, are sensitive to instantaneous flow rates.

A flood propagation and wadi bed infiltration model was developed and integrated with a groundwater model. The flood wave propagation model used a modified pool-routing technique based on surveyed cross sections. Infiltration rate was initially estimated from measured discharges and planimetric survey of flow areas during steady flow conditions and was considered to be time-invariant. The routing interval was two hours. Infiltration was calculated for each recharge node of the groundwater model. Initial model calibration was against a well-monitored single-peak flood wave (Figure 10.1) and was achieved by varying infiltration rate F and Manning's n. The best fit was achieved initially with F = 10 cm/hour and n = 0.02 (Figure 18.13). The model was later re-calibrated against more representative multipeaked hydrograph data for the upstream and downstream stations. Calibration criteria were to preserve the recorded flow volumes and volume ratios at the downstream stations. This was achieved with F = 4.2 cm/hour and n = 0.038. Different F and n values would have been required to meet different calibration criteria (e.g. flood peak simulation).

The historic 7-year hydrograph sequence was then routed through the recalibrated flood wave propagation and infiltration model. Table 18.9 summarises the results which indicate a mean annual outflow of 7 Mm^3 representing 11% of inflow at the upstream station. Two 30-year synthetic sequences were then routed through the model with the results shown in Table 18.10. These indicate a mean surface outflow of 13 Mm^3 representing 14% of the estimated long-term mean inflow to the alluvial basin.

TABLE 18.9

SIMULATED RESPONSE TO HISTORIC FLOW SEQUENCE

(Summary of Results)

YEAR	INFLOW AT EL SHEFA	WADI BED RECHARGE	BIR KHADRA OUTFLOW	
			VOLUME	% OF EL SHEFA INFLOW
	Mm^3	Mm^3	Mm^3	%
1967 (from July	75.6	68.6	7.0	9.3
1968	55.6	53.6	2.0	3.6
1969	143.3	112.9	30.4	21.2
1970	26.8	25.5	1.2	4.6
1971	40.2	38.5	1.8	4.4
1972	42.1	40.5	1.6	3.8
1973	39.6	36.8	2.8	7.1
1974 (until April)	50.8	46.4	4.4	8.6
TOTAL	474.0	422.7	51.3	-
MEAN (based on 7 years)	67.6	60.3	7.3	10.8

TABLE 18.10

Simulated Response of Najran Alluvial Basin
To Synthetic Flow Sequence

Year	Inflow At El Shefa Mm^3	Bar Khadra Outflow Volume Mm^3	Bar Khadra Outflow % of El Shefa in-Flow %	Year	Inflow At El Shefa Mm^3	Bar Khadra Outflow Volume Mm^3	Bar Khadra Outflow % of El Shefa in-Flow %
1	51.6	3.45	6.7	31	66.7	4.76	7.1
2	375.4	88.42	23.6	32	0.1	0.00	0.0
3	141.2	9.23	6.5	33	64.1	6.26	9.8
4	86.3	3.33	3.9	34	109.8	4.77	4.3
5	4.5	0.26	5.8	35	474.0	194.16	41.0
6	85.9	4.60	5.4	36	66.8	7.60	11.4
7	71.4	2.36	3.3	37	62.3	3.97	6.4
8	85.4	8.42	9.9	38	23.5	1.46	6.2
9	30.4	1.13	3.7	39	72.0	3.24	4.5
10	55.0	3.62	6.6	40	119.9	6.99	5.8
11	86.7	9.31	10.7	41	88.1	6.81	7.7
12	14.5	0.09	0.1	42	119.3	7.29	6.1
13	76.4	4.83	6.3	43	106.5	18.18	17.1
14	33.4	1.20	3.6	44	134.4	25.44	18.9
15	225.0	16.42	7.3	45	92.0	7.10	7.7
16	72.3	3.15	4.9	46	33.5	2.43	7.3
17	23.0	0.51	2.2	47	118.8	8.94	7.5
18	59.4	2.36	4.0	48	303.4	24.55	8.1
19	69.9	2.60	3.7	49	44.9	2.27	5.1
20	14.0	0.47	3.4	50	48.1	3.04	6.3
21	64.1	2.68	4.2	51	141.7	17.91	12.6
22	103.5	10.14	9.8	52	36.0	1.67	4.6
23	41.4	2.06	5.0	53	38.3	2.53	6.6
24	140.2	7.93	5.7	54	187.6	85.05	45.3
25	289.8	86.28	29.8	55	132.5	9.17	6.9
26	32.3	1.55	4.8	56	17.1	0.92	5.4
27	161.5	12.14	7.5	57	60.9	4.30	7.1
28	153.9	28.30	18.4	58	5.7	0.00	0.0
29	185.4	13.47	7.3	59	100.2	5.97	6.0
30	70.0	3.03	4.3	60	18.2	0.51	2.8
30 yrs to - Total	2903.8	333.70	11.5%	30 yrs Total	2886.4	467.29	16.2%
Mean	96.8	11.12		Mean	96.2	15.58	
				60 yrs Total	5790.2	800.99	13.8%
				Mean	96.5	13.35	

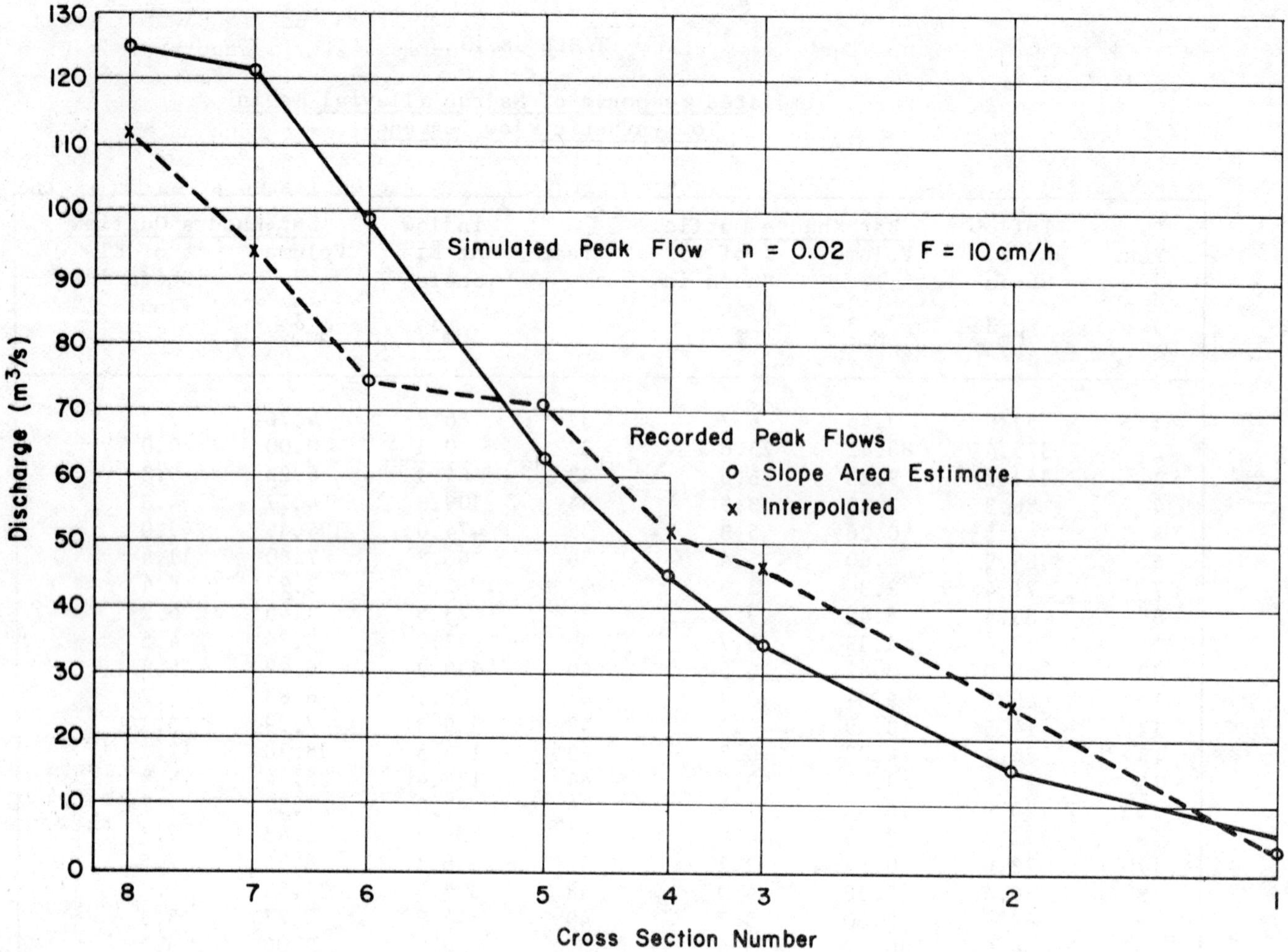

Comparison of simulated and observed peak discharge along the course of the wadi

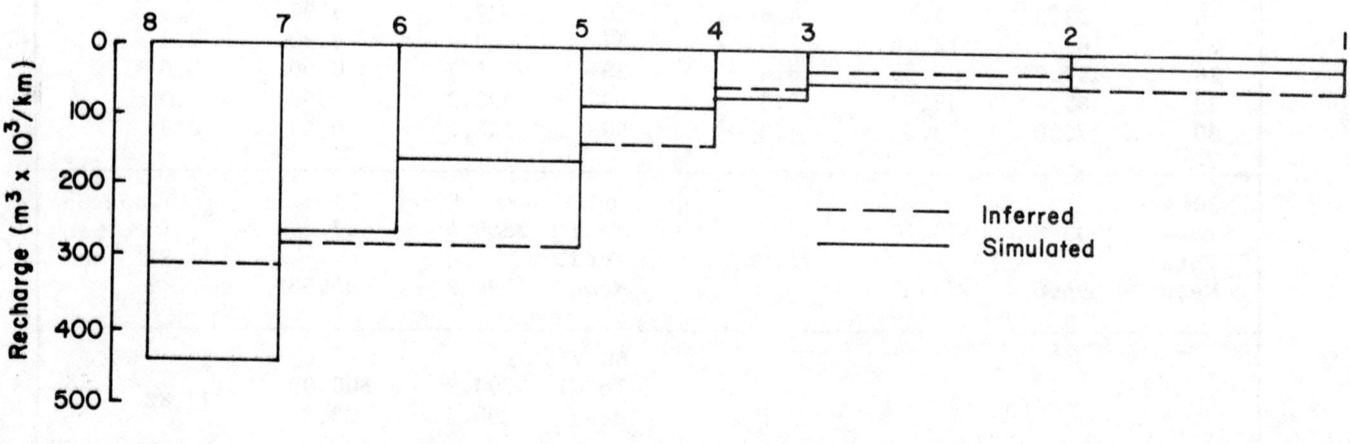

Comparison of inferred and simulated recharge per unit length of the wadi

Fig. 18.13 Simulated and observed behaviour of the December 1973 flood

The long-term mean recharge was estimated initially from the groundwater model steady state simulation. The steady state inputs were permeability, evapotranspiration values for each node and aquifer geometry. Permeability was estimated from a large number of borehole pump-test and shallow dug-well data as 60 m/day. Evapotranspiration was estimated using a vegetation map delineating areas of cultivation and natural vegetation. Each vegetation type was assumed to consume water at a specified proportion of the Penman Eo rate. The total evapotranspiration loss was estimated at 65 Mm^3. The measured piezometric surface within the aquifer at the end of a prolonged dry period was assumed to approximate that corresponding with hydrodynamic equilibrium. The problem of estimating the quantity and distribution of long-term mean recharge on the wadi bed was avoided initially by fixing the piezometric evaluation of the wadi bed nodes at their assumed equilibrium level. Recharge into the nodes was thus head-controlled and the amount necessary to maintain the levels was implicit in the conditions under which the model was operated.

Further steady state simulations were carried out introducing lateral recharge from side wadis and a redistribution of total recharge amount between wadi bed nodes to produce the result shown by Figure 18.14.

A time-controlled simulation was carried out to fix the storage coefficient. This was achieved by retaining transmissivity distribution as proven by steady-state simulation and simulating a historic prolonged groundwater level decline and the response to subsequent floods and the floods which occurred during the whole of the record period. The storage coefficient which produced the best results was 0.25 (Figure 18.15).

The results obtained from the independent groundwater and runoff models were approximately consistent. The groundwater model estimate of recharge was 90 Mm^3 whereas the runoff model estimate of runoff was about 82 Mm^3. The groundwater model estimate of long-term mean recharge was taken as the basis for the recharge-outflow simulation runs described above. The two 30-year synthetic runoff sequences were selected such that the two separate approaches were mutually consistent.

The estimated volume of water in storage in the aquifer system within the study area was 3,000 Mm^3, determined from the aquifer geometry, piezometry and specific yield. Part of an additional estimated 2,000 Mm^3 storage capacity was available for short and medium-term regulation.

Simulation runs were carried out assuming reservoir storage at the upstream limit of the alluvial aquifer and a constant outflow rate to determine the optimum reservoir dimensions for maximum enhancement of the aquifer yield.

18.3.3.5 Case 9. The Khulays Alluvial Basin, Saudi Arabia

1. Introduction

The Quaternary aquifer of the Khulays alluvial basin is an important source of groundwater. In 1967 some 5.9 Mm^3/annum were being transferred from the basin to supply domestic water to the city of Jeddah. Since that time a number of new wells have been opened in the basin to provide additional supplies. In 1967 local agriculture in the basin was extracting about 10 Mm^3/annum (net) for irrigation. By 1973, when the methodology and data requirements for a updating assessments of aquifer potential were examined, the area under irrigation had been expanded.

2. Aquifer Recharge

The saturated alluvium that constitutes the primary aquifer of the Khulays basin is mainly replenished by the infiltration of flood waters which enter the basin on its eastern margin from the outlets of Wadis Merwani and Ghoran. In exceptional years (e.g. 1969) flood waves traverse the basin and leave the system.

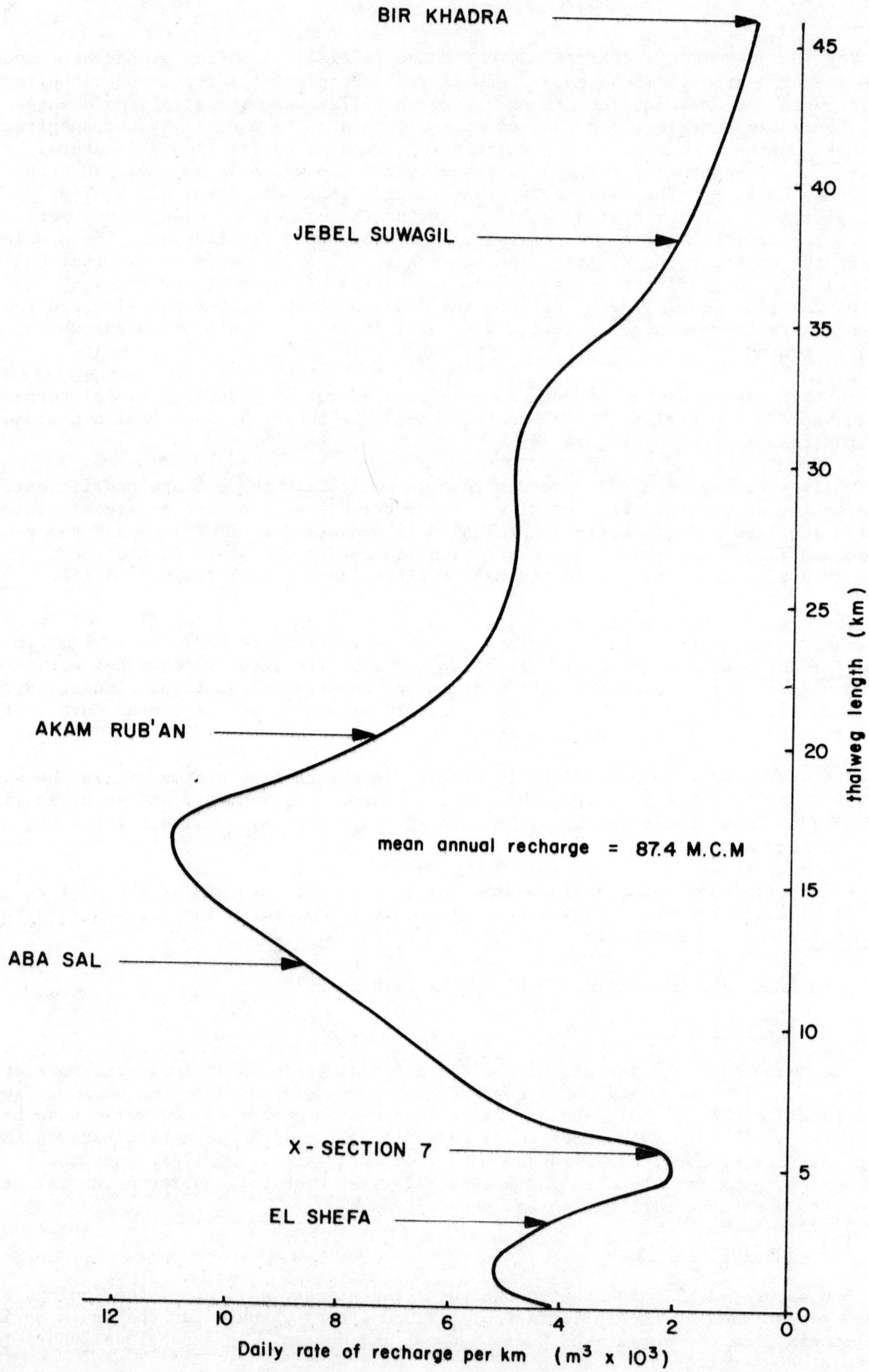

Fig. 18.14 Distribution of alluvial channel recharge (steady state) Wadi Najran, Saudi Arabia

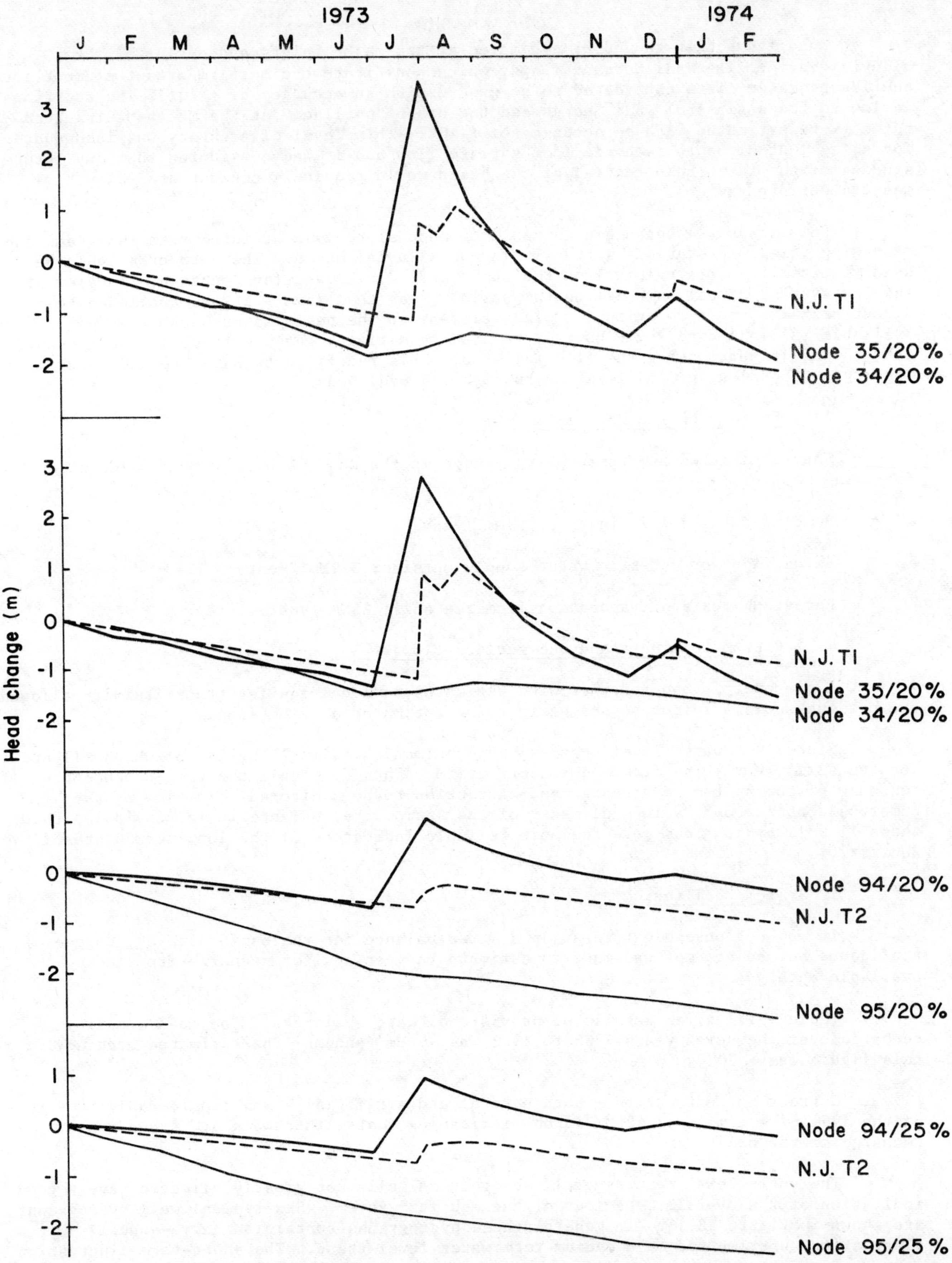

Fig. 18.15 Comparison of simulated and observed water level responses to recharge from 1973 floods with two values of specific yield

Runoff has been gauged or estimated at the major inlets and outlet of the alluvial basin. The Wadi Merwani flows, which constitute the most important element for aquifer recharge, were considered to be gauged with an accuracy of ± 20%. The additional inflow to the basin from Wadi Ghoran and the basin "spillage" at the Mecca-Medina road outlet were estimated with an accuracy of ± 40%. This level of accuracy was inadequate for assessment of basin recharge from surface flow and it was considered that any future studies of the resource potential of the basin would require a concentrated flow measurement programme.

Point rainfall measurements had been made since 1966 at three totalising and one recording raingauge stations sited within the alluvial basin. These records could be used to estimate areal rainfall on the basin and direct aquifer recharge. In view of the low average annual rainfall on the basin - less than 75mm - it was concluded that direct recharge would constitute a minor element in the basin water balance. The available rainfall data could however be used to estimate the magnitude of local runoff from the hills surrounding the alluvial area. This runoff contributes to indirect recharge by ponding and infiltration through the wadi beds.

3. Extraction and Consumptive Use

Water extracted and consumptively used in the Khulays basin was grouped under the following headings:-

a. Pumping for local irrigation: 10Mm^3/year

b. Extraction for domestic water supply outside: 5.9Mm^3/year.

c. Phreatophytes and evaporation from the soil: 26Mm^3/year.

4. Piezometric Data and Sub-surface Flows

A piezometric map together with estimates of local aquifer transmissivity allowed the net sub-surface inflow to the basin to be estimated as 9 Mm^3/annum.

Since 1966 water level changes had been monitored in 21 wells, of which 11 are located sufficiently far from major abstraction points to enable the time-dependent response of the system to recharge and extraction to be monitored. In view of the numerous doubts about various elements of the basin water balance it is considered that these 11 well hydrographs were the most reliable indicators of the long-term state of the aquifer.

5. System Equilibrium

It was not possible to draw up a water balance for the basin with any degree of confidence but an attempt was made to estimate the net aquifer recharge from the available data.

Table 18.11 gives details of provisional water balances. The mean annual recharge over the seven years 1966 to 1972 was 46 Mm^3/annum. The estimated accuracy of this figure was ± 30%.

Current net extraction - both natural and artificial - was considerably in excess of the 1966/67 estimate of 42 Mm^3/annum indicating that withdrawals are exceeding recharge in the basin.

The water level records of 11 observation wells not greatly affected gave a good indication of the overall behaviour of the aquifer system. Six typical well hydrographs are shown in Figure 18.16. In general these hydrographs contain an intra-annual fluctuation superimposed on a medium-term water level trend. The short-term fluctuation

TABLE 18.11

Provisional Water Balance for the Khulays Basin (Mm^3)

Year	Surface Inputs			Gross Surface Input	Outflow at Mecca/Medina Road	Net Surface Inputs	Net Sub-Surface Input	Net Basin Inflow
	Merwani Outlet	G'Arran Outlet	Direct Recharge					
1966	29.7	2.5	0.5	32.7	4.0	28.7	9.0	37.7
1967	17.7	1.3	0.8	19.8	0.9	18.9	9.0	27.9
1968	26.4	1.5	2.5	30.4	5.7	24.7	9.0	33.7
1969	138.2	4.3	1.9	144.4	51.3	93.1	9.0	102.7
1970	18.5	0.6	1.8	20.9	1.0	19.9	9.0	28.9
1971	46.1	0.1	0.4	46.6	0.0	46.6	9.0	55.6
1972	26.8	0.1	1.2	27.1	0.0	27.1	9.0	36.1

EXTRACTION

Estimated extractions based on 1966/67 figures	Natural Evapotranspiration	60(+)
	Net artificial Extractions	16(+)
	TOTAL	42(+)

Notes:

(1) Average net recharge to basin alluvium = 46 Mm^3/annum.

(2) Direct Recharge estimated as 10% of areal rainfall on aquifer.

(3) Estimates of extractions based on data in Water Supply Development Reports - 1967 and 1969.

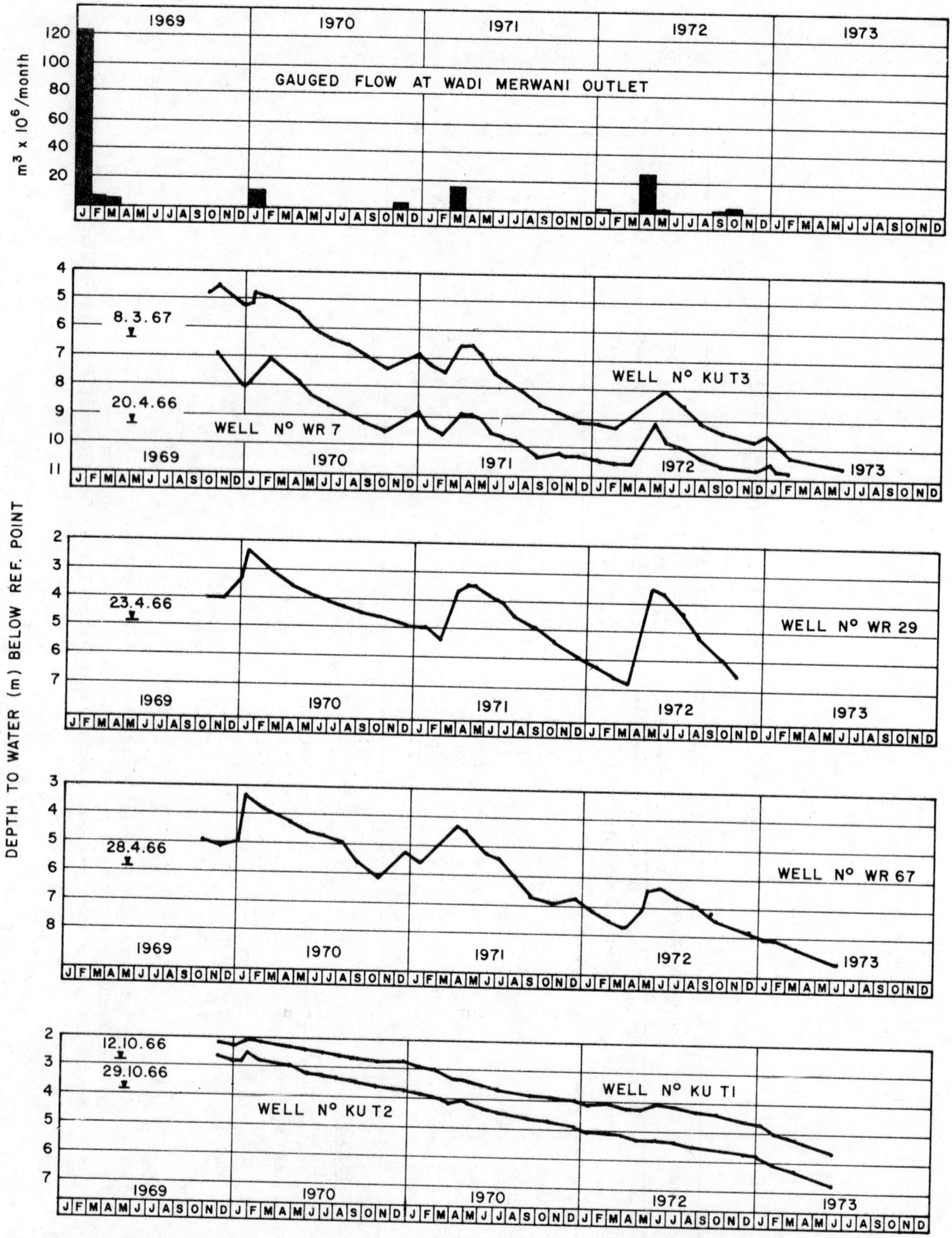

Fig. 18.16 Groundwater levels in the Khulays basin

was a result of the local extraction and recharge and was clearly correlated with flood events. The medium-term trend showed little response to normal flood events although it was clearly dependent on the recharge of exceptionally large floods - e.g. Jan. 1969 - and was considered to indicate the general degree of development of the system.

Well WR-29 is near the aquifer boundary in proximity to the wadi channel and an area of cultivation. The water level trace indicates both components of water level response. Well KU-T1 is located in the lower wadi course in the vicinity of the north western outlet of the system. The water level in this well is only slightly affected by the largest flood events which traverse the full length of the alluvial basin (e.g. 1969) and hence exhibits only small short-term fluctuations. The medium-term trend of the system is clearly indicated by this well.

All well hydrographs indicated that the water table had fallen by between 5 and 6 metres since the peak levels achieved after the high recharge flood event of 1969. Water levels in all wells were between 2 and 3 metres lower than in 1966.

This indicates that since 1966 extraction in the vicinity of the observation wells had exceeded recharge and some local mining of groundwater has been necessary to meet demand. Discussions with villagers confirmed this conclusion and, in 1972-73, a number of wells had become dry for the first time since their construction. The well hydrographs for the down-gradient section of this aquifer indicated that, in this area, the regional water table had been falling at about 1.8 metres per year.

The fact that the available records of well levels indicated that head was being lost over the full areal extent of the aquifer, and the conclusion drawn from the water balance that extraction had possibly exceeded recharge over the previous seven years, gave grounds for concern about the long-term stability of the system.

It was considered that the promotion of an intensive data collection programme would allow a digital simulation model of the Khulays basin to be developed. Once established, such a model could be employed to make forecasts of the long-term stability of the aquifer under the existing extraction rates and the utility of various proposed policies of aquifer management (e.g. removal of phreatophytes) could be assessed in economic terms.

18.3.3.6 Case 10. A Preliminary Model Study of the Wadi Bishah Aquifer.

1. Introduction

The prime objective of the study was to test the overall suitablility of the proposed digital modelling approach to the determination of the potential resources of the wadi aquifer, with due regard for the quality of the available data.

2. Review of the Available Data

2.1 Aquifer Characteristics

2.1.1 Physical Boundaries and Degree of Confinement

The isopach maps of saturated alluvial deposits contained in geophysical reports enable the base and lateral boundaries of the system to be determined.

A typical section of the aquifer was shown to consist of:-

(a) a central channel filled with coarse grained, recent alluvium.

(b) flood plain deposits of silt and muddy sand and

(c) a zone of sands and gravels which underlie the other layers.

Generally, groundwater was extracted from the third zone. The reports indicated that test wells in the vicinity of the silt deposits exhibited "artesian rises" at the time of construction. The conclusion was that the water contained in the regions of the aquifer overlain by silt deposits was, to some extent, in a confined condition.

When reviewing the information concerning the confined conditions, the following points were noted:

- A number of the exploration wells (Figure 18.17) drilled in the wadi course were reported to have exhibited artesian rises at the time of drilling.

  e.g. well Bi T1 - rise 3.2 metres
  Bi T3 - rise 3.4 metres
  Bi T4 - rise 6.5 metres
  Bi P8 - rise 6.0 metres
  Bi P14 - rise 1.4 metres
  Bi P29 - rise 5.0 metres.

  However, large amounts of recharge water pass through the wadi bed to the aquifer and the records of well level observations for wells located more than one kilometre away from the course of the wadi clearly showed the rapid response of the system to recharge. The fact that the aquifer can accept such recharge indicated that it is not confined for any appreciable lateral extent in the vicinity of the main wadi.

- A comparison of the elevations at which water was struck (and above which it rose) in the drilled exploration wells with those of the bottoms of hand-dug wells in the vicinity of the drilled wells indicated that the bottoms of many of the hand-dug wells are higher than that level at which water was reported to have been struck in the drilled wells. If confined conditions existed in the vicinity of these drilled wells then any hand-dug wells excavated there and not penetrating at least to the "confining layer" (i.e. level at which water was struck in the test wells) could not yield much water. In fact the many wells in this condition which yield appreciable quantities of water indicated the lack of significant confinement.

- Examination of the well point inventory indicated that the depth of water standing in the majority of the hand-dug wells (when not being pumped) was of the order of 1.0 to 2.0 metres, although a few wells had up to 9 metres of standing water. If confined conditions existed in the former wells the water level in these hand-dug wells would also have shown artesian rise once the confining layer had been penetrated. Thus the depth of water standing in the wells would be expected to be of the same order as the artesian rise (about 6 metres).

In view of these points it was concluded that the aquifer complex is not confined for any appreciable lateral extent. Hence it was proposed that for the course of the current modelling investigations the system should be considered to be in general unconfined, with probably some small areas of local confinement. The reported rises in the drilled wells are possibly due to an increasing hydraulic potential with depth, probably associated with stratification of zones of high permeability at lower depths which could be attributed to bands and lenses of silts and gravels.

A detailed model study of the Bishah aquifer would require these assumptions to be thoroughly checked against well logs and site visits as well as by discussion with villagers engaged in the construction of hand-dug wells.

2.1.2 Hydraulic Permeability

The available reports gave details of pumping tests conducted in the drilled wells and the analysis of the results of these tests by the classical techniques of well theory (type-curve etc). It was noted that different theoretical approaches in individual wells and the analysis of different data sets (e.g. draw-down and recovery) gave individual values of permeability (Darcy's K) that varied up to one order of magnitude for an individual well and exhibited similar variations throughout the well-field. This variation is to be expected because the field conditions at Bishah depart radically from the theoretical assumption of an isotropic material of semiinfinite lateral extent inherent in the classical well theory. The Bishah alluvial aquifer is highly anisotropic and the proximity of the boundaries necessarily influence the results of long-term heavy pump tests.

In view of the limited usefulness of classical techniques in such conditions, the pumping test data were reworked using a semi-empirical approach found to give satisfactory results in similar conditions elsewhere. This approach is based on a linear relationship between the measured specific capacity of a well and the aquifer transmissibility. When determining the specific capacity of the Bishah wells allowance was made for the effects of the low efficiency of the test wells. These calculations indicated a range of permeability of 100 to 1,000 metres/day with a mean of 300 metres/day. Hydrogeological tests indicate that the permeability range of clean sand with silt to gravels is of the order of 400 to 800 metres/day. When interpreting these calculated values of permeability it should be borne in mind that the value given represents a mean "K" for the total thickness of aquifer cut in the region of the area of influence of the pumped well.

A number of the results quoted in the well-point inventory for shallow, hand-dug wells was analysed. The theoretical treatment of shallow wells was adopted in order to determine representative values of permeability for these shallow wells. A set of realistic assumptions for well diameters yielded a range of Darcy's "K" is typical of coarse poorly-graded sands with some silts. It was therefore concluded that this latter range of "K" is typical of the top few metres of the saturated aquifer.

When the calculated 'K values were plotted on a map of the aquifer no regional groupings of high or low values was apparent. It was therefore concluded that on the regional scale the aquifer could be considered to possess a fairly uniform lateral permeability with a tendency for "K" to increase in the vertical direction.

2.1.3 Specific Yield of the Aquifer

Estimates of the specific yield of the aquifer based on the result of time-drawdown pump tests were available from previous reports. The results of the analysis of these tests gave values of specific yield ranging from 3×10^{-5} (i.e. typical of confined conditions) to 3×10^{-1} (water table conditions).

In order to obtain a first approximation of the overall system specific yield the behaviour of the piezometric surface to the recharge-producing flood of March 1970 was examined. By deducting the gauged flows at Sada near the downstream limit of the study area, from those at Heifa near the upstream limit, and allowing for the intervening catchment, it was estimated that the volume of water going to recharge as a result of this flood event was some 42 Mm^3. An estimate of the increase in volume of saturated alluvium of 210 Mm^3 as a result of this flood was developed using pre and post-flood piezometric surfaces superimposed on the polygonal network developed for the digital model. On the basis of these figures an overall specific yield of 2×10^{-1} is obtained for the system. This latter figure is in agreement with hydrogeological experience in similar aquifers. In the context of a comprehensive study of the aquifer the above calculations could be repeated for all significant recharge events and improved estimates of this important system parameter would be obtained.

2.1.4 Piezometry

Since 1967 records had been kept of the water level fluctuations in a number of wells on approximately a monthly basis. These data were used to construct piezometric maps for a number of the hydrogeologically important stages of the aquifer's recent development. In particular the piezometric surfaces found at the end of lengthy periods of no recharge (e.g. early January 1970 and early March 1971) were examined in some detail. These piezometric surfaces describe the state of the system after more than one year without recharge, and give a good indication of the depletion behaviour of the aquifer. Comparison of these various piezometric surfaces shows that over the last six years the depletion characteristics of the system have not greatly varied. Although a large variation in the magnitude of individual recharge events and associated piezometric response had occurred, the aquifer had fallen to more or less the same piezometric level on each occasion about one year after the cessation of recharge. Hence if mean long term annual conditions were considered, this piezometric surface represents a quasi-steady state picture of the system (that is after all local "moundings" of the piezometric surface have been dissipated). From the physical view point this "steady state" piezometric surface is the distribution of hydraulic head which is in equilibrium with the net mean annual recharge and for the purpose of digital simulation this distribution of head may be considered to be independent of the system storage coefficient.

Piezometric surfaces for immediate post flood (i.e. recharge) situations were prepared. These maps indicated that in general local recharge mounds of between 2 and 6 metres occur along the course of the wadi the higher figure being applicable to the upper reaches of the aquifer. The time-dependent degradation of such a piezometric surface as it approaches the quasi-steady state levels described above is a function of net extractions and aquifer properties. From the point of view of digital modelling of aquifer response, assuming realistic estimates of extraction are available, simulation of this depletion with time would enable the overall validity of the proposed model to be tested.

2.1.5 Recharge and Extraction

Consideration of the seven years (1966-72 inclusive) of surface flow data collected at the gauging stations of Heifa and Sada and making an allowance for the flows of the ungauged wadis contributing recharge to the aquifer, indicated that the mean value of recharge passing to the aquifer from the wadi was about 44 Mm^3/annum. It should be stressed however that the high annual variability of runoff dictates that such a mean value based on a few years of record should be treated with caution. In this context it is noted that if the long, sustained period of runoff, giving optimum recharge conditions, that occurred in 1972 were not used in the determination of this mean, the value would fall to 30 Mm^3/annum.

Details of a water use survey conducted in 1967 indicated that the gross extraction between Heifa and the low reaches of Wadi Bishah in that year was around 83 Mm^3/annum. It was also reported that the irrigation efficiency in the Wadi Bishah is such that net extraction was about 50% of the total gross extraction.

Agricultural expansion had proceeded in the vicinity of the Bishah Oasis since 1967 but no estimates of 1973 extraction have been made. A revised estimate of agricultural use of water in Wadi Bishah could be based on consumptive use figures and a determination of actual crop areas using recent large scale aerial photographs of the region.

After formulation of the groundwater model of the Bishah system and the construction of representative piezometric maps for various states of the system since 1967, attention was devoted to the development of a sub-surface flow balance for the portion of the aquifer considered (i.e. north from approximately the Heifa gauging station to the locality of the village Najadhi in the lower course of the wadi). This analysis, based on values of permeability developed from the considerations discussed in Section 2.1.2 and the provisional calibration runs of the model (see Section 4), suggested that this section of the aquifer had a net sub-surface outflow of approximately 10 Mm^3/annum. If the assumptions of the quasi-steady state conditions outlined in Section 2.1.4 are correct and the system may be considered to be in equilibrium then the above calculation implies that the mean excess of recharge over consumptive use (i.e. net mean annual recharge) over this period was also about 10 Mm^3/annum.

Consideration of the currently available estimates of net recharge based on estimates of mean gross recharge (1966 to 1971) and net recharge of about 2 Mm^3/annum (44-42) over the previous seven years. The apparent discepancy between these two estimates of net mean annual recharge might at first sight appear somewhat disconcerting, but it should be realised that the two estimates are derived from independent data sources (one based on surface flows and extraction estimates, the other being strictly hydrogeological in nature). It was considered that the relative agreement of these figures implies that they are likely to be of the correct magnitude. A number of fairly coarse assumptions are involved in both these estimates of net annual recharge and for this stage of the study these estimates of net recharge should be conservatively interpreted to indicate that the system is currently near equilibrium conditions.

3. The Digital Model

3.1 Model Type and Development

Bearing in mind the quality of the available data and in the light of experience in other studies it was concluded that the most suitable type of situation model for the Bishah system would be one based on the work of Tyson and Weber (1963). In this digital representation of the behaviour of groundwater system, the aquifer is considered to be divided into a number of nodal blocks which in plan constitute a network of Thiessen polygons. The governing equation of groundwater flow in a continuum (Laplace equation) is represented in discrete form in the model by a system of first order finite difference equations which describe the variation of the flows across individual nodal faces with time. This system of equations is solved for discrete intervals of time and specified boundary conditions by a modified form of the well known Gauss-Seidel relaxation procedure.

A suitable nodel network was devised for the section of the Bishah aquifer laying north of the Heifa gauging station and south of the village of Najadhi. The chosen configuration of nodes consisted of 65 internal nodes and 58 external ones. It was considered that the adopted nodal network would enable the principle components of the aquifer to be efficiently modelled for a minimum of computing costs. In particular, the aquifer lying below the course of the main wadi, was represented by a series of small, generally rectangular, nodes that would enable the localised effects of channel recharge to be simulated at the relevant stage of the study. A sketch of the model, indicating the nodal network, is shown on Figure 18.17.

Estimates of the elevation of bedrock at each of the nodal points were made using the geophysical maps given in previous reports. Surface elevations at each of these points were estimated from a coarse topographic map which the consultant prepared from available data on the ground elevation at a number of well points. Recourse to this technique was necessary as no more detailed maps of the region could be obtained.

The boundary conditions of the model were defined from the available geophysical data. Of the 58 external nodes (i.e. boundary nodes) 53 represent "no flow" boundaries which constitute the physical limits of the aquifer. The remaining five represenmt the connection of the section of aquifer being modelled to other areas of saturated alluvium. Examination of the historical piezometric records indicated that the boundary conditions at the main inflow and outflow sections of the model could best be interpreted as boundaries of constant hydraulic gradient and two boundary connections could be considered to be boundaries of constant head.

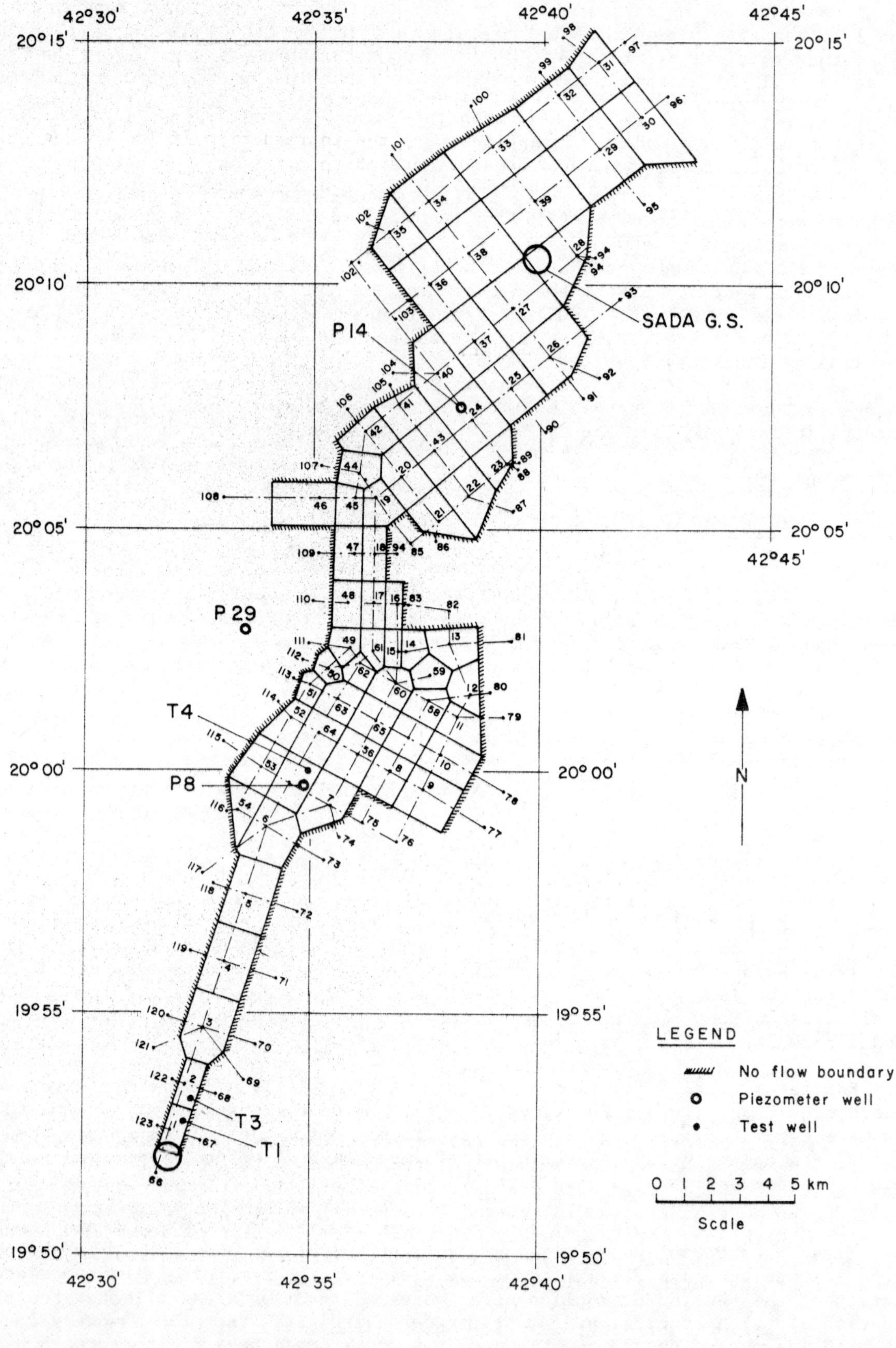

Fig. 18.17 Digital model of Wadi Bishah aquifer - nodal network

In the light of Section 2.1.1 it was decided to treat the system initially as an unconfined aquifer for the purpose of the provisional model study; this condition could be modified at a later date if further evidence from the field or the model necessitated the introduction of local areas of confinement. Examination of the results of well tests indicated that no marked pattern of regional distribution of hydraulic permeability was apparent. Hence a constant value of permeability of 300 metres/day was assigned to most of the nodal faces in the model. Exceptions were the introduction of low values of permeability on a few faces which had been orientated to coincide with the known location of local "inselbergs" of basement rock which cut through the alluvial deposits. The distribution of permeability was considered solely as a provisional "first estimate" of the field situation and was adjusted during the later overall model calibration procedure, in order to achieve the required agreement between field and model piezometry.

All necessary data were prepared in punch card form for input to a Fortran groundwater model programme. After all data checks had been completed the model was set up for running on an I.B.M. 360/50 computer.

The geometrical shape of the aquifer - and hence the adopted nodal network - gave rise to some concern with regard to the speed with which the iterative solution of the finite difference expression of the groundwater flow equations could be expected to converge. It is a general feature of long "shoe-string" type systems that the lack of direct contact between the various regions of the digital model cause normally efficient numerical techniques to be somewhat slow in convergence.

An important aspect of the provisional model study was to consider the implications of any such numerical difficulties for a comprehensive simulation study.

3.2 Test runs of the Provisional Model

In order to test the conceptual basis of the proposed model and to assess the general validity of the various assumptions regarding aquifer properties outlined earlier in this report, it was decided to attempt to simulate the quasi-steady state head distribution described above.

Apart from the definition of the system boundaries implied in the establishment of the model, simulation of the quasi-steady state head distribution requires values of aquifer permeability and estimates of net equilibrium recharge to be entered in the model. In this context the net equilibrium recharge equals the net sub-surface outflow from the aquifer. Examination of the "steady state" piezometry and flow boundary properties indicates that this net equilibrium recharge is about 10 Mm^3/annum. A number of trial runs of the model were made with various reasonable distributions of the equilibrium recharge located along the main wadi course. These ranged from concentrating this recharge in the upper regions of the wadi to a linear distribution of the recharge down the full wadi length. As this "calibration" of the model proceeded, increasingly refined estimates of the distribution and quantity of steady state equilibrium recharge were developed. A similar process of adjustment was employed to refine the provisional estimates of permeability entered in the model.

The results of this provisional calibration of the model were very encouraging and it was possible to obtain a good general agreement between the simulated and measured piezometry. The difference between model and field results was generally less than 1 metre at positions where field data are well controlled. Some further improvements in the quality of the simulation could be readily obtained by further calibration runs, but it was felt that the degree of agreement obtained by the earlier runs served to show that the proposed model could be developed into a valid tool for the simulation of the Bishah aquifer and that any further requirements would not be justifiable at this stage.

The numerical behaviour of the selected relaxation method was as anticipated, somewhat sluggish. This necessarily caused the computer runs of the model to be somewhat lengthy. Possible techniques for speeding up the rate of convergence of the iterative solution of the system equations, to improve the efficiency of the numerical method, were examined.

4. Development and Use of a Comprehensive Model

The successful simulation of the quasi-steady piezometry of the aquifer indicates that a comprehensive model of the Bishah aquifer could readily be developed. Full investigation would require a complete review of all available data and would necessitate some minor fieldwork both to remove the degree of uncertainty surround some aspects of the physical characteristics of the system described earlier and to obtain revised estimates of the current level of consumptive use. The necessary stages in the development of such a model were summarised as:-

(a) Refinement of the provisional model to finalise the distribution of permeability - i.e. completion of calibration on the basis of the quasi-steady state head levels introduced.

(b) Determination of the distribution of specific yield - and introduction of any necessary areas of local confinement - by simulation of the time- variant behaviour of the piezometric surface, i.e. calibration of model response to recorded recharge events.

(c) Sensitivity testing of the model.

The successful completion of the above would result in an aquifer model with the proven ability to simulate the historic behaviour of the Bishah system. Such a model could then be used to determine the long-term implications of various levels of water use.

Once a comprehensive model of the aquifer had been developed and tested it could be used to treat the behaviour of the system under a number of different hypothesised operating policies. This experimentation by simulation would enable a new optimum development plan for the system to be formulated. Utilisation of such a model to investigate alternative aquifer management policies would minimise the economic and social costs of the development of this important local resource. A choice could be made between alternative development policies without the disruptions, delays and risks that would necessarily attach to any similar form of "experimentation" on the actual aquifer.

5. Conclusions and Recommendations

After examination of the available data, a provisional digital model of the Quaternary aquifer of Wadi Bishah was developed, tested and successfully operated. The purpose at this stage of the investigation was not to assess the potential of any proposed development strategies but rather to judge the suitability of the digital simulation approach for this end.

With the exception of restrictions on further development no aquifer management policy was being followed in 1972 with regard to the Bishah aquifer. On the available evidence it appeared that the system was near equilibrium under these conditons. Nevertheless, over the previous six years (1966 to 1972) an average annual surface flow of some 55 Mm^3 was estimated to have been rejected by the aquifer, and to have flowed to the sand areas of the lower reaches of the wadi. In the context of agricultural expansion in the Bishah area, this rejected surface flow must be considered as a lost resource. This 55 Mm^3/annum unused surface runoff is about 30% greater than the estimated total current consumptive use from the aquifer between the Heifa gauging station and the village of Najadhi in the lower reaches of the developed section of the Wadi.

It was considered that any further agricultural expansion in the Bishah area should be based on an aquifer management policy that would ensure the utilisation of this currently rejected surface flow.

Before any such development could be envisaged a comprehensive assessment of the state of aquifer development and an evaluation of the most efficient strategy for the exploitation of the currently rejected wadi flows was required. In the light of the results of the study it was considered that the establishment of a comprehensive digital model of the Bishah system would serve this end.

18.4 ARTIFICIAL GROUNDWATER RECHARGE

18.4.1 Introduction

Groundwater reservoirs have such advantages over surface water storage schemes (for the reasons given in 19.1) that augmenting recharge by artificial means appears an attractive conservation measure. Experience shows, however, that unless conditions are favourable the cost of maintaining the recharge works may be prohibitively high in relation to the benefit achieved.

The main problems are due to sedimentation or the interference with sedimentation processes. For example spate-breakers recharge basins and detention storage dams may rapidly silt up and regulated streamflow, cleansed by sediment deposition in storage reservoirs, may create erosion problems downstream due to its enhanced capacity for sediment transport.

However in suitable locations recharge works may well be worth considering. The following case studies illustrate situations in which artificial recharge may well produce economic benefits.

18.4.2 Case 11. Artificial Recharge Possibilities in Oman

18.4.2.1 Background

Areas of primary interest for groundwater recharge are valley alluvial deposits although areas of younger, apparently highly karstic limestone may be good aquifers. Active alluvial channels 20 to 80 km west of Muscat have vertical walls 0.5 to 4m high and may be completely filled and rescoured during the same storm and flood sequence. The deposits in which these channels are incised in the alluvial fan area contain highly permeable gravels with occasional well-rounded boulders. Maximum percolation rates would be expected to exceed 1m/day. The long-term average in spreading basins would be about 50 cm/day. In the sandy lower end of alluvial fans the larger channels are generally shallower than 1m and the smaller channels 20 cm or less. During runoff, deeper gravels may be exposed by scour of the superficial sands. Maximum percolation rates here would be about 50 cm/day and long-term rates in spreading grounds about 20 cm/day.

Runoff flowing in the alluvial channels is generated on the hard-rock catchment and occasionally reaches the ocean. There is clear evidence that surface runoff also occurs directly from both gravel and sand portions of the alluvium.

18.4.2.2 Locating Artificial Recharge Areas

On the basis of field surveys it was concluded that artifical recharge areas should be located as follows:

1. In or near an active channel so that water will be available.

2. Where coarse surface material of high permeability is found.

3. Where depth to groundwater is more than 5 to 10 metres (certainly not where there are shallow infiltration galleries or other evidence of shallow water level).

4. Where natural recharge does not already occur to the extent that no surface runoff remains.

5. Where it is practical to construct and maintain recharge structures.

6. Where damage to roads or other man-made structures, due to the recharge works, may be avoided.

18.4.2.3 Types of Artificial Recharge Structures

Introduction

Any artificial means to slow down, store and percolate runoff can be considered artificial recharge. Small dikes, ditches, dams, pits, spreading grounds, galleries and wells have all been used. Most references discuss the more spectacular and expensive works while few describe the simpler and in some places more economical methods.

Recharge works in very arid areas may be difficult to justify economically if the economic method does not take into account the cost of obtaining the water from other sources. In the case of Oman the only other source of water with which to compare the cost of recharge is desalination. Furthermore, there is loss to the sea, which could be put to beneficial use.

It is clear in the north cost of Oman that simple artifical recharge possibilities exist.

It should be emphasized that the stream channels are only controlled in a few places and that, therefore, more costly approaches to artificial recharge should be included in channel control works. The simpler techniques discussed next, however, minimise such problems.

18.4.2.4 Simpler Artificial Recharge Methods

Methods requiring rebuilding after each major event.

1. Ploughed ditches - where the main channel can be easily diverted into an adjacent flat gravelly area a series of ploughed ditches with a slight gradient similar to a flood or ditch irrigation system can be effective. The ploughed ditch or furrow is upstream of the ploughed material. The short flood peaks in Oman probably would not result in much water being diverted.

2. Small gravel barriers - these can be considered where a main channel is fairly deep (say, more than one metre) and large dikes are needed to divert water to nearby flat areas, smaller dikes, say, about 70 cm high can be constructed at frequent intervals down the channel. The idea is to slow the flow as much as possible and if the dikes are all or partly washed out they are simply put up again as soon as possible. It is essential not to make the individual dike so high that a lot of storage occurs which could cause a higher flood downstream. In general this method is best for very wide channels which are seldom completely occupied by the flood. The material for the dike is simply bulldozed along the line, excavating upstream of the dike. If the channel is wide enough the dikes may be angled upstream from alternative sides, thus forming deeper pools on one side and allow the water to overflow on the other shallow side of the dike.

In any case it is best to put a small starting spillway at alternating ends of the dikes so that if flows are small the dike will be topped only in a small area and easily repaired.

Where channels are very narrow there is no alternative but to put up small dikes more or less directly across the channel.

It is essential to understand that the very process of washing out of these small dikes is important to maintain recharge rates. In general it is best to have as much surface area as possible with minimum storage. Depth of the stream generally has little influence on recharge rate compared to channel conditions.

18.4.2.5 More Complex Artificial Recharge Methods

All other recharge methods require diversion from the main channel, and some means to keep the recharge works relatively free of clay.

A possibility is the use of gravel pits resulting from excavations for the purpose of artificial recharge. In the lower end of the north coastal alluvial area shallow pits containing water may be observed following rainfall. Deeper pits here may considerably augment recharge.

A. Diversion works - the diversion works are simply to divert as much water as feasible from the channel into the recharge pit. These can be simple, requiring heavy maintenance after each storm or more costly and permanent.

B. Settling pit - even with a deep pit it may be best to have a smaller settling pit between the main recharge channel and the larger recharge pit. The inlet should be such that the walls are not eroded by the inflowing water and the outlet to the nearby recharge pit should be suitably constructed to avoid collapse.

C. Recharge pit - the recharge pit should be fenced and have suitable inlet works to control erosion. If located to the side of a main channel there is usually no problem of overflow of the pit.

The walls of the pit should be as nearly vertical as possible. Silt on the floor of the pit will generally impede vertical recharge; the walls of the pit are the recharge surface through which most water will enter the aquifer. With some care a pit may last for many years without significant maintenance. The size of pits depends on the flood flows expected. In general pits can be 30 or 40 metres deep if properly designed but in the area investigated 10-20m deep pits are appropriate. 20 metre deep pits in areas of deeper water with perhaps a minimum of 10 metres. It is better to go deep than to make the area larger.

Very shallow sand and gravel pits might be made more effective by controlling inflow and removing silt after each flood.

18.4.2.6 Outline of Programme

1. Reconnaissance for potential recharge areas according to the above criteria and types of structures.

2. Construction of a few 70 cm to 1 metre high simple dikes across the channel over a 1/2 km reach to gain practical experience in construction problems. Priority should be given to localities accessible by vehicle. The location of the dikes should be marked by a rock cairn or other marker outside the channel area so that the site can be found and inspected if the dike is washed away.

3. If there are sites for the ditch type recharge, several of these could also be tried.

4. Before a major work is undertaken in any one area a sketch map at least should be prepared, using air photos and whatever else is available. A plan can then be roughed out for perhaps 4 or 5 km of the channel, taking into account the dangers of channel shifting, potential flood damage, etc.

5. Experience should be gained with some of the simpler recharge systems if technical staff are not available or resources for proper designs of more complex work are not possible. Experience in any case will be of value. If 1/25,000 scale topographic maps and more detailed plans of the local area (say, 1/10,000 scale) are available full designs should be made. It would be useful to have a civil engineer or surveyor available in the early days of construction to assure uniformity and level tops of the dikes.

6. More costly rock groynes, etc. are not recommended until much more information is available on the channels and their action during floods.

7. Spreading grounds and pits are also not recommended until more detailed information becomes available. Use of the very shallow sand and gravel pits could perhaps be improved by removal of the silt after each flood.

18.4.3 Case 12. Investigation of Sand River Storage Potential - Mahalatswe River - Botswana

18.4.3.1 Physical Description

The land surface in the upper Mahalatswe catchment lies at slightly more than 1,000m altitude and falls to about 900m at the confluence of the Limpopo. The area is characterised by relatively flat topography except where rocks of the Basement Complex crop out at the surface.

The channel of the river is well entrenched over most of the distance, having relatively steep banks as much as 3m high. Tributary drainage is generally through poorly defined channels. The section of the river included in the investigation extended for a distance of 82km.

The Mahalatswe River normally flows a few days to a few weeks each year.

18.4.3.2 Method of Investigation

The present investigation attempted to derive exploration techniques to define accurately the extent of the sand river aquifer. An electrical resistivity unit and a hydraulic power auger drill were utilized to carry out the study. Surveying teams defined the width and altitude of the river channel and determined depth to water in the sand.

Approximately 100 electrical resistivity traverses were made across the river over a 17km reach of the river. The number of electrical probes per traverse varied from three to seven with a total of approximately 500 probes. More than 100 auger holes were put down on these traverses to correlate electrical resistivity interpretation with auger hole data. The results of the correlation were disappointing as no consistent basis for interpretation could be determined.

Drilling was carried out with a hydraulic power auger to determine the depth to bedrock, the channel configuration and the type of materials composing the river bed. The auger drill was able to penetrate gravel, sand, clay and weathered bedrock to a maximum depth of 15m. Previous exploration by hand probing often was stopped by clay layers or gravel beds.

Auger holes were drilled on electrical resistivity traverses in the upper 17km of the study area but in the downstream reach holes were put down only in the centre of the river bed at 600m intervals. Samples of the auger cuttings were described at 0.5m intervals in each hole. Three hundred and sixty-one holes were bored ranging in depth from less than one metre to more than 12m.

Ground profiles were surveyed at 150m to 200m intervals. Mean sea level altitude and depth to water in the sand were determined at each section. With these data it was possible to establish the river bed and water table gradients.

18.4.3.3 Analysis of Results

Results of auger drilling and cross-section surveys were plotted for each station to calculate the total volume of sediments in the river channel.

Based upon the auger drilling and general field observation, the bedrock profile was drawn assuming side slopes of one vertical to two horizontal. The boundaries of the various types of material were plotted for each section based on an interpretation of the description of material penetrated by the auger holes. Water levels in the river bed determined from August through October 1971 were plotted on each section. These water levels are representative of the conditions which probably exist in most years near the end of the dry season. With the above data it was possible to make a reasonable estimate of the volume of water in storage in the river bed aquifer.

For the purpose of calculating the quantities of water available over different periods of time, the river bed aquifer was divided into three units as defined below. The bank storage which lies above the sand surface level and which drains away shortly after flood has receded was not considered as part of the river bed aquifer.

A. The aquifer below the sand surface level, but above the end of dry season saturated zone, is defined as "shallow-depth storage". The water in excess of that lost by evaporation and transpiration in shallow depth storage is probably lost over the dry season through seepage down the hydraulic gradient.

B. The aquifer below the end of the dry season water level to a further depth of 3.5m is defined as "mid-depth storage". Water in this zone is available for development in a normal year where full recharge occurs each wet season. However, after an exceptional series of dry years where a maximum period of 30 months has elapsed since the last recharge, all the water in the "mid-depth storage" zone would be exhausted by evaporation and seepage and would not, therefore, be available for development. The "mid-depth storage" generally lies 4 to 5.5m below the sand surface level.

C. The aquifer below the mid-depth storage and extending down to bedrock is defined as "deep storage" and is the storage which is available after a period of 30 consecutive months of no flow in the river.

Using the above definition, the volume of each zone was calculated over the 82km distance of the study. These aquifer zones were further subdivided according to their aquifer potential on the basis of clay content. The volume of these zones was estimated as 32.2 Mm^3.

Houston (1972), from laboratory determinations, considered the Mahalatswe River aquifer to have permeabilities ranging from 20 m/d to 250 m/d with a median of 80 m/d. He concluded that the permeability of the sand with clay (which typically had only a small clay fraction) was likely to be at least as permeable as the coarse sand. Houston reported the specific yield would range between 10 and 30 percent From laboratory permeability tests on both disturbed and undisturbed soil sanples from the Mahalatswe River the permeability was estimated to be about 110 m/d.

Pumping tests in the Motloutse indicated that one or two days of surface flow would be sufficient to completely recharge the aquifer and that probably no surface flow could occur unless the aquifer were fully saturated. From these pump tests the storage potential of the aquifer was estimated to be 17.3 percent which was the average result of all the tests.

From the above results, it was assumed that 15 percent of the total volume of the "shallow", "mid" and "deep storage" zones was extractable water.

18.4.3.4 Water Balance

A tentative water balance was made for a 10km section of river. A study of rainfall records from 1911 indicate that seasonal rainfall totals varied up to nearly 40 percent of the long-term average above or below the average. A progressive ten year average indicated that there has been a sharp decrease in rainfall over the past 10-15 years at Mahalapye. However, the decline had reached a minimum and the next decade was expected to produce an overall upward trend. Reliable runoff records for the Mahalatswe River are given in Table 18.12.

TABLE 18.12

RUNOFF RECORDS FOR THE MAHALATSWE RIVER

Year	Number of floods	Total discharge (Mm^3)	Peak discharge (m^3/s)	Annual rainfall at Mahalapye (mm)	Runoff as % of rainfall
1967/68	0	0	0	332	0
1968/69	4	3.19	108.0	391	0.96
1969/70	8	1.61	32.0	330	0.50
1970/71	6	48.99	752.0	399	16.5

Table 18.12 indicates that there is little relationship between mean annual rainfall and runoff.

The individual flood data for 1968/69 and 1969/70 were analysed, the potential evaporation loss calculated and the total recharge to the 10km section of stream estimated as 8 percent of inflow volume.

It was assumed that subsurface inflow and outflow were equal and amounted to approximately 100 m^3/d. Based on the above, the provisional water balance for two years for the 10km section of river is shown in Table 18.13.

The water balance figures indicate that the 10km section of the Mahalatswe River was in approximate equilibrium; however there should be sufficient water in storage to sustain draft over a period of two complete dry seasons without recharge.

From the auger hole survey it was calculated (assuming an average porosity of 25 percent) that the maximum amount of water held in storage (as distinct from water extractable) in the 10km stretch of the river is 0.63 Mm^3. This consists of 0.09 Mm^3 in shallow, 0.45 Mm^3 in mid-depth and 0.09 Mm^3 in deep storage.

TABLE 18.13

WATER BALANCE FOR 10KM SECTION OF MAHALATSWE RIVER

Year	Input Mm^3		Output Mm^3	
1968/69	Surface inflow	3.19	Surface outflow	2.75
	Subsurface inflow	0.04	Subsurface outflow	0.04
	Rainfall	0.32	Extraction	0.10
			Evaporation and transpiration	0.50
			Gain to storage	0.16
	Total	3.55	Total	3.55
1969/70	Surface inflow	1.61	Surface outflow	1.31
	Subsurface inflow	0.04	Subsurface outflow	0.04
	Rainfall	0.28	Extraction	0.10
			Evaporation and transpiration	0.48
			Gain to storage	0
	Total	1.93	Total	1.93

18.4.3.5 Present Water Development and Estimated Future Demand

In 1971 the Rhodesia Railways were extracting approximately 360 m^3/day of water from the sands of the Mahalatswe River and the town of Mahalapye was extracting approximately 320 m^3/day.

In order to meet expected future increased water demands, two alternative schemes for the construction of a dam to provide a surface water supply were proposed but found not to be economically viable. One was a small scheme to impound 3.7 x Mm^3 and the other a larger scheme to impound 10.7 Mm^3 of water. Both schemes were based on a mean annual design runoff of 14.8 Mm^3.

18.4.3.6 Available Water Resources

At full saturation, the sand aquifer over the 82km length of the river contains 4.790Mm^3 of extractable water. Of this, 1.570Mm^3 are in the shallow storage zone which depletes very rapidly in the first one or two months after recharge. When recharge occurs approximately every six months as is normal, the balance of 3.220 Mm^3 in mid and deep storage would be available for development.

During an exceptional period where there was no recharge for 30 consecutive months, 0.910Mm^3 in deep storage would be available. The bulk of these deep resources lie in pools formed behind subsurface rock barriers downstream of Mahalapye.

Surveys carried out by the project have indicated that there is potential for small sand river extraction schemes down the entire length of the river. From estimates of total extractable water per km it appears that below the section of river used to supply Mahalapye town, it should be possible to extract 0.045Mm^3 or more of water per km per year.

18.4.3.7 Possible Use of Surface Reservoir and Sand Dams for Increasing Recharge

From the auger hole survey it has been found (assuming an average porosity of 25 percent) that the maximum amount of water held in storage (as distinct from water extractable) in the length from Madiba to the Limpopo is as follows.

Shallow storage	=	2.65 Mm^3
Mid-depth	=	3.88 Mm^3
Deep storage	=	1.53Mm^3
Total storage	=	8.05 Mm^3 (25% of total of all materials)

Records (Table 18.12) indicate that in 1967/68 there was no flow and in 1969/70 the total discharge was less than shallow storage. On the other hand, discharge exceeds shallow storage by about 20 percent in 1968/69 and by 18 times in 1970/71. The 1970/71 flood exceeded total storage by more than six times; therefore most of the runoff from the 1970/71 floods was lost to drainage into the Limpopo River.

The proposed large dam would have stored 10.7 Mm^3. Of this 4.5 Mm^3/year would have been lost through evaporation. The feasibility of conserving water by recharging the sand rivers should be considered as a part of any surface water development scheme.

An alternative to increased recharge from surface reservoirs is the construction of "sand dams". These dams are constructed for the purpose of increasing the sand bed thickness and thereby enlarging the potential for subsurface storage of water. Such dams have been used very economically and successfully in parts of Africa for many years to supply farms, villages, mining operations and other purposes.

18.5 WATER BALANCES

18.5.1 General

Although the size of the elements of an arid zone water balance are usually imprecisely known, setting them out as inputs and outputs to a water resource system is useful for assessing their orders of magnitude and revealing any false assumptions. Water balances may theoretically be estimated for any area from entire basins to small plots. However, the magnitude of rainfall P and evapotranspiration ETa in relation to runoff R and recharge REC makes it impractical to attempt basin-wide balance assessment in arid zones. We cannot estimate runoff from rainfall and potential evapotranspiration ETp because:

$$ETp \gg P \gg ETa \gg R$$

We can estimate direct recharge DREC realistically only where the size of the rooting depth constant C and maximum soil moisture deficit D can be estimated with reasonable precision (see Section 18.3.2) and only at a point where P is measured and R is negligible.

Usually however, it is only necessary to estimate the balance for relatively small areas, for example alluvial aquifers already developed for agriculture. In such cases the elements of the water balance may be as shown in Tables 18.14 and 18.15. The return flow from irrigation may play an important role in such balances; in one area of Iran this represented 60% of recharge.

Loss of water by evaporation from playas may represent a high proportion (or all) of the output from a resource system. For example in the closed Azraq basin of East Jordan, an estimate of the water balance may be summarised as:

TABLE 18.14

Mean Annual Water Balance for the Azraq Basin

Inputs	
Underflow from recharge outside the basin	31 Mm^3
Internal recharge	5 Mm^3
Total	36 Mm^3
Outputs	
Springs and seeps	19 Mm^3
Evaporation from playas	17 Mm^3
Total	36 Mm^3

It is sometimes necessary to break balance elements down into smaller components, for example the water balance for the Geropotamos Plain was estimated as shown in Table 18.15.

18.5.2 Errors in Water Balances

In preparing water balances as an aid to resources assessment it is clearly necessary to have some idea of the error of estimate for each element of the balance. This is especially important where storage change S is a major factor and is affected by short or medium term trends, for example increased pumped extraction from alluvial aquifers for irrigation over the past twenty years.

An illustration of a short-term water balance estimate for an alluvial aquifer is provided by Example 18.3. This example shows the approximate ranges within which true values of balance components are thought likely to lie.

The change in groundwater storage between maximum "equilibrium" piezometric surface of mid - 1969 and minimum levels at the end of 1973 was estimated from changes of level in 7 observation boreholes. The range of groundwater level decline in these boreholes was from 0m to 7m.

The 4 1/2 year storage change was calculated as the product of the weighted mean water level decline of 3m applied to the upper 200 km^2 of the aquifer surface and the estimated mean specific yield of 0.25.

The difference between estimated input and output for the 4 1/2 year period is 91 Mm^3. The estimated tolerances, at 67% confidence, on the balance components show quite clearly that an "equally likely" balance could be produced in which input would equal output. It is probable that surface water input is underestimated and possible that evapotranspiration losses are overestimated. The discussion on runoff gauging problems in Section 12 suggests that systematic underestimation occurs because bed scour during runoff cannot be taken into account. Evapotranspiration is often overestimated because the ratio of cropping density to total cultivated area over the period of study may be well below what is assumed.

TABLE 18.15

Water Balance for Geropotamos Plain

Mean annual flows in Mm^3

Sub-Area 2

A.	SURFACE WATER	
	Input	
1.	Rainfall	13
2.	Streamflow	17
3.	Springs	0
	Total input	+30
	Output	
4.	Infiltration (from rainfall)	3.5
5.	Infiltration (from streamflow)	5.8
6.	Water use (net)	1
7.	Evaporation	10
8.	Streamflow	10
	Total Output	-30.3
B.	GROUNDWATER	
	Input	
9.	Groundwater inflow	8.2
10.	Direct recharge (from rainfall)	3.5
11.	Indirect recharge (from streamflow)	5.8
	Total input	+17.5
	Output	
12.	Springs	0
13.	Water use (net)	1.3
14.	Groundwater outflow	16.2
	Total output	-17.5

The balance components discussed above are estimated long-term mean values under assumed conditions of hydrodynamic equilibrium, so storage changes do not enter the balance.

Example 18.3

Water Balance for Wadi Najran Alluvial Aquifer
July 1969 to December 1973

INPUT	Mm^3	Standard Error % (1)	67% (2) Confidence Interval-Mm^3
1. Surface water inflow, main wadi	160	- 20 to + 50	128 - 240
2. Surface water inflow, tributaries	20	- 50 to + 100	10 - 40
3. Groundwater inflow	18	± 50	9 - 27
4. Groundwater storage change (3)	150	± 50	120 - 360
Total	348		
OUTPUT			
5. Surface water outflow	27	- 50 to + 200	13 - 81
6. Groundwater outflow	126	± 20	101 - 151
7. Evapotranspiration from irrigated area	216	± 50	108 - 324
8. Evapotranspiration from natural vegetation	68	± 50	34 - 102
9. Evaporation from channels	2	± 50	1 - 3
Total	439		

Note:
(1) A positive error is an underestimate of the true value expressed in terms of the estimated value. A negative error is an overestimate of the true value expressed in terms of the estimated value.
(2) Range within which the true value is estimated to lie at 67% confidence.
(3) Estimated from the weighted mean water level change in seven observation boreholes.

REFERENCES

Agricultural Research Service. Groundwater Recharge Hydrology, U.S. Dept. of Agriculture.
1970 ARS 41-161.

Alam, A.M.Z., Kennedy, J.F. "Friction Factors to Flow in Sand Bed Channels" ASCE
1969 Vol. 95 No. HY6 Proc. Paper 6400 Nov. 1969.

Allison, L.J. et al. "Tropical Cyclone Rainfall as Measured by the Nimbus 5 ESMR".
1974 Bull. Amer. Met. Soc. 55.

Amiran, D.H.K. "Arid Zone Development : A Case of Limited Choices" in Arid Zone
1963 Development: Potentialities and Problems. Ballinger Publishing Co., Cambridge, Mass.

Amorocho, J. An Application of Satellite Imagery to Hydrologic Modelling; the
1975 Upper Sino Basin, Columbia. MMH Symposium, Bratislave

Askew, A.J. "Deviation of Formulae for Variable Lag Time". Journal of Hydrology
1970 Vol. 10, No. 3, 1970.

Astier, J.L. Geophysique Appliquee a Hydrogeologie. Masson et Cie, Paris.
1971

Attour, I.S., Ibbitt, M.E. Flood Probabilities of the Yarmouk and Zergua Rivers.
1966 Natural Resources Authority, Hashemite Kingdom of Jordan, Amman, Sept. 1966

Barrett, E.C. "The Estimation of Monthly Rainfall from Satellite Data", Mon. Wea. Rev.
1970 98.

Barrett, E.C. "The Tropical Far East: ESSA Satellite Evaluations of High Season
1971 Climatic Patterns". Geog. J., 137.

Barrett E.C. Curtis, L.F. Introduction to Environmental Remote Sensing. Chapman and
1976 Hall.

Barrett, E.C. "Monitoring Precipitation: a Global Strategy for the 1980's". in:
1977a Monitoring Environmental Change by Remote Sensing. Remote Sensing Society (U.K.)

Barrett, E.C. The Assessment of Rainfall in North-Eastern Oman through the Integration
1977b of Observations from Raingauges and Meteorological Satellites. FAO Consultants Report, W/KT629.

Barrett, E.C. Mapping Rainfall across Algena, Libya, Morocco and Tunisia. FAO
1977c Consultant's Report, AGP/DL/RS/1

Barrett, E.C. Rainfall Monitoring in the Region of the North-West African Desert
1977d Locust Commission in 1976-77. FAO Consultants Report, AGP. LCC/77/9.

Barrett, E.C. (ed), Satellite Remote Sensing Application in Agroclimatology and
1979a Agrometeorology. Proceedings of an International Training Course, Rome, Italy, 2-13 Oct., 1978, ESA/FAO/WMO.

Barrett E.C. Space Research 19.
1979b

Barrett, E.C., Lounis, M. Progress Towards the Establishment of an Operational
1979 Remote Sending Unit for Desert Locust Monitoring and Control in North-West Africa FAO. (Report in Press 1979)

Beard, L.R. Statistical Methods in Hydrology. Corps of Engineers, Sacramento.
1962

Bell, F.C. "Generalised Rainfall-Duration-Frequency Relationships" ASCE. Vol. 95,
1969 No. HY1, January 1969

Benson, M.A. Measurement of Peak Discharge by Indirect Methods. WMO Technical Note No.
1968 90.

Besbes, M., Delhomme, J.P., De Maisily, G. "Estimating Recharge from Ephemeral
1978 Streams in Arid Regions: A Case Study at Kairouan, Tunisia", Water Resources Research Vol. 14, No. 2 April 1978.

Beverage, J.P., Culbertson, J.K. "Hyperconcentrations of Suspended Sediment" ASCE Vol.
1964 90, No. HY6, Proc. Paper 4136, Nov. 1964

Bize, J., Bourguet, L., Lemoine. Alimentation Artificielle des Nappes Souterraines.
1972 Masson et Cie, Paris

Bottomley, A., Simpson, J., Nudds, D., Stephenson, N.W. "Some Economic Aspects of
1974 an Arabian Irrigation Scheme Design." ICID Bulletin, July 1974.

Campbell, M.D., Lehr, J.H. Water Well Technology, McGraw-Hill.
1973

Central Water Authority. Hashemite Kingdom of Jordan, Technical Paper No. 34,
1964 Part I: "30-year Average Rainfall in Jordan 1931-1960" Part II: "Rainfall in Jerusalem 1846-1964" Dec. 1964.

Chow, V.T. Handbook of Applied Hydrology. McGraw-Hill
1964

Chow, V.T. Open Channel Hydraulics. McGraw-Hill
1959

Clarke, R.T. Mathematical Models in Hydrology. FAO Irrigation and Drainage Paper
1973 No. 19

Colby, B.R. Hembree, C.H. Computation of Total Sediment Discharge, Niobrara River, Near
1955 Coby, Nebraska. USGS Water Supply Paper No. 1357, Washington D.C.

Colby, B.R. "Relationship of Unmeasured Discharge to Mean Velocity" Transactions
1957 of the American Geophysical Union, Vol. 38, No. 5 Washington D.C.

Colby, B.R., Hubbell, D.W. Simplified Methods for Computing Total Sediment Discharge
1961 with the Modified Einstein Procedure. USGS Water Supply Paper No. 1593, Washington D.C.

Cooper, R.H., Peterson, A.W. Discussion of Paper 6884 - T. Blench. Coordination
1970 in Mobile-Bed Hydraulics. Nov. 1969 ASCE Vol. 96, No. HY9, Proc. Paper 9747 Sept. 1970.

Corbett, D.M. et al. Stream-gaging Procedure. USGS Water Supply Paper 888
1945

Dalrymple, T.D. Flood Frequency Analysis: Manual of Hydrology, Flood Flow Techniques.
1960 Part III, 1960. USGS Water Supply Paper No. 1543-A.

Davy, E.G., Mattei, F., Solomon, S.I. An Evaluation of Climate and Water Resources for
1976 Development of Agriculture in the Sudano-Sahelian Zone of West Africa. WMO Special Environmental Report No. 9.

Dawdy, D.R. Depth-Discharge Relations of Alluvial Streams - Discontinuous Rating
1961 Curves. USGS Water Supply Paper No. 1498-C.

Dincer, T., Mugrin, A.L., Zimmerman, U. "Study of the Infiltration and Recharge
1974 through the Sand Dunes in Arid Zones with Special Reference to the Stable Isotapes and Thermonuclear Tritium" Journal of Hydrology, Vol.23, 1974

Diskin, M.H., Lane, L.J. "A Basinwide Stochastic Model of Ephemeral Stream Runoff in
1972 South-Eastern Arizona" IASH Bulletin XVII, No. 1, 4/1972.

Doorenboos, J. et al. Crop Water Requirements. FAO Irrigation and Drainage Paper No.
1977 24.

Duckstein, L., Fogel, M.M., Thames, J.L. "Evaluation Effects on Rainfall: A Stochastic
1973 Model" Journal of Hydrology, Vol. 18, 1973

Einstein, H.A. Handbook of Applied Hydrology. Section 17.2, Editor V.T. Chow,
1964 McGraw-Hill

Ellis J.C., Green, M.J. River Gauging by Ultrasonics. Technical Memorandum 114
1975,1977 (1975), Technical Reports 46(1977) 95(1978), W.R.C. Medmenham
1978

Evenari, M., Shannan, L., Tadmore, N. The Negev. Harvard University Press
1971

Evenari, M. "Ancient Desert Agriculture and Civilisations: Do they point the way
1977 to the future? in Arid Zone Development: Potentialities and Problems. Ballinger Publishing Co., Cambridge, Mass.

Fiering, M.B., Jackson, B.B. Synthetic Streamflow. AGU Water Resources Monograph
1971 No. 1.

Fleming, G. "Design Curves for Suspended Load Estimation" Proc. Institution of
1969 Civil Engineers, May 1969.

Follansbee, W.A. Estimation of Average Daily Rainfall from Satellite Cloud Photograph.
1969 NOAA Technical Memorandum, MESS 44, Washington D.C.

Follansbee, W.A., Oliver, V.J. A Comparison of Infrared Imagery and Video Pictures in
1975 the Estimation of Daily Rainfall from Satellite Data. NOAA Technical Memorandum, NESS 62, Washington D.C.

Follansbee, W.A. Estimation of Daily Precipitation over China and the USSR using
1976 Satellite Imagery. NOAA Technical Memorandum, NESS, 81 Washington D.C.

FAO Agro-meteorological Field Stations. FAO Irrigation and Drainage Paper No. 27
1976

FAO Water Law in Muslem Countries. FAO Irrigation and Drainage Paper No. 2.
1973

Fowler, M.G. et al. The Estimation of Rain Rate over Land from Microwave Sensors,
1979 AWRA Symposium, Sioux Falls, S.D.

Gibb, Sir Alexander & Partners. Water Resource Survey, Northern Oman, June 1976
1976

Giddings, L.E. Extension of Surface Data by use of Meteorological Satellites NASA
1976 Technical Memorandum, LEC-8377, Houston, Texas

Griffiths, C.G., et al. Rainfall Estimation from Geosynchronous Satellite Imagery
1976 during Daylight Hours. NOAA Technical Report, ERL 356 - WMPO 7, Boulden, Colo.

Griffith, C.G., et al. "Rain Estimation from Geosynchronous Satellite Imagery -
1978 Visible and Infrared Studies". Monthly Weather Review, 106.

Grosh, R.C. et al. "Cloud Photographs from Satellites as a Hydrological Tool in
1973 Remote Equatorial Regions". Journal of Hydrology 20.

Halcrow, Sir William & Partners. Report on the Water Resources of the Trucial
1969 States

Hellwig, D.H.R. "The Loss of Water into the Atmosphere from a Sandy River Bed
1973 under Arid Climatic Conditions" Journal of Hydrology, Vol. 18, No. 3/4 March 1973

Henderson, F.M. Open Channel Flow. Macmillan
1966

Herbst, P.H., Shaw, E.M. "Determining Raingauge Densities in England from Limited Data
1969 to give a Required Precision for Monthly Rainfall Estimates" Journal of the Institution of Water Engineers, Vol. 23 No. 4 1969

Herndon, A., Woodley, W.L., Miller A.H., Samet, A., Senn, H. Comparison of Gauges and
1973 Radar Methods of Convective Precipitation Measurement, National Oceanic and Atmosphere Admin. Technical Memorandum ERL OD-18, Boulder, Colorado, USA.

Herschy, R.W. The Magnitude of Errors at Flow Measurement Stations. Water Resources
1969 Board TN No. 11, January 1969.

Herschy, R.W., Loosmore, W.R. The Ultrasonic Method of River Flow Measurement
1974 WRC/WDU Symposium on River Gauging by Ultrasonic and Electomagnetic methods, Reading.

Huysmans, L. Groundwater Recovery, MacMillan.
1972

International Atomic Energy Agency. Guidebook on Nuclear Techniques in Hydrology. IAEA
1968 Technical Report No. 91, Vienna.

International Organisation for Standardisation. Accuracy of Velocity Area Measurements.
1978 ISO Data No. 2.

Jackson, I.J. "Tropical Rainfall Variations over a small area" Journal of Hydrology,
1969 Vol. 8 No.1, May 1969.

Jones, J.E. Calculation of evapotranspiration using colour-infrared photography.
Undated USGS Reston, Virginia.

Kohler, M.A., Nordenson, T.J., Fox, W.E. Evaporation from pans and lakes. U.S.
1955 Weather Bureau Research Paper No. 38.

Kruseman, G.P., De Ridder, N.A. "Analysis and Evaluation of Pumping Test Data",
1970 Int. Inst. for Land Reclamation and Improvement, Bulletin 11, Wageningen, The Netherlands.

Kunkel, F. Data Requirements for Modelling a Groundwater System in an Arid Region.
1973 USGS Water Resources Investigations 4-73.

Laurenson, E.M. "A Catchment Storage Model for Runoff Routing" Journal of Hydrology,
1964 Vol.2, 1964

Lloyd, J.W. The Hydrogeology of the Southern Desert of Jordan UNDP/FAO.
1969

Maddock, T. Jr. "Drought is the Dilemma: The Philosophy of Water Management" in
1977 Arid Zone Development: Potentialities and Problems. Ballinger Publishing Co. Cambridge, Mass.

McCulloch, J.S.G. "Tables for the Rapid Computation of the Penman Estimate of
1965 Evaporation" East African Agricultural and Forestry Journal, Vol. 30, No. 3, 1965.

MacMahon, T.A. "A Model for the Simulation of Stream Flow Data from Climatic Records"
1971 Journal of Hydrology, Vol. 13, No. 4, 1971.

Maxey, G.B., Eakin, T.E. Groundwater in White River Valley, White Pine, Nye and
1949 Lincoln Counties, Nevada. Nev. Stage Eng. Water Resources Bulletin No. 8.

Martin, D.W., Scherer, W.D., "Review of Satellite Rainfall Estimation Methods". Bull.
1973 Amer. Met. Soc., 54.

Merritt, E.S (ed) EarthSat Spring Wheat Yield System Test, 1975. NASA Contract
1976 Report, NAS 9-14655, Houston, Texas.

Midgley, D.C. et al. Design Flood Determination in South Africa. Hydrological
1972 Research Unit, Univ. of Witwatersrand, Report 1/72, Dec. 1972.

National Oceanic and Atmospheric Administration Rainfall Estimation from Geosynchronous
1976 Satellite Imagery During Daylight Hours. NOAA Technical Report ERL 356-WMP07, Boulder, Colorado, USA.

Nilsson, B. Development of a Depth Integrating Water Sampler. Dept. of Physical
1969 Geography, Uppsala University, Report No. 2.

Oltman, R.E. Measurement of Sediment Concentration and Computation of Sediment
1962 Transport. WMO Flood Control Series No. 22.

Osborn, H.B., Lane, L.J. "Depth-area Relationships for Thunderstorm Rainfall in
1972 Southeastern Arizona" Trans. ASCE Vol. 15, No.4, 1972.

Osborn, H.B., Lane, L.J. "Optimum Gauging of Thunderstorm Rainfall in Southeastern
1972 Arizona" Water Resources Research, Vol. 8, No. 1, Feb. 1972.

Park, A. Programme Plan for Developing the Capability of Forecasting Crop Production.
1975 FAO Consultant's Report, AGD (RS) 2/75.

Penman, H.L. Vegetation and Hydrology. Commonwealth Bureau of Soils, Technical
1963 Communication No. 53.

Perrin De Brichambaut, G., Wallen, C.C. A Study of Agroclimatology in Semi-Arid
1963 Zones of the Near East. WMO Technical Note No. 56.

Pike, J.G. "Evaporation of Groundwater from Coastal Playas (Sabkhah) in the Arabian
1970 Gulf" Journal of Hydrology, Vol. 11, No. 1, 1970.

Pike, J.G., Chidley, T.R.E. "A Generalised Computer Programme for the Solution
1970 of the Penman Equation for Evapotranspiration" Journal of Hydrology, Vol.10, 1970.

Pike, J.G., Harhash, I., Gemmel, B.A.P. Rainfall and Groundwater Recharge over
1975 Qatar. FAO/UNDP Integrated Water and Land Use Project; Ministry of Industry and Agriculture, State of Qatar, Technical Note No. 24.

Pike, J.G., Parker, D.H. Integrated Water and Land Use Project, Qatar, FAO.
1977

Prickett, T.A. "Modelling Techniques for Groundwater Evaluation". in Advances
1975 in Hydro Science, ed. V.T. Schav. Academic Press.

Raikes & Partners. Hydrological Services Project, Final Report, January 1971.
1971

Rao, M.S.V., et al. Satellite-Derived Global Oceanic Rainfall Atlas (1973 & 1974), NASA
1976 SP-410, NASA Washington D.C.

Rapp, Anders; Berry, Len; Temple, Paul. Studies of Soil Erosion and Sedimentation
1972 in Tanzania. Bureau of Resource Assessment and Land Use Planning, University of Dar-es-Salaam, Dept. of Physical Geography, University of Uppsala.

Reeves, R.G., Anson, A., Landen, D. editors. Manual of Remote Sensing, American
1975 Society of Photogrammetry, Virginia, USA.

Robinson, T.W. Introduction, Spread and Areal Extent of Salt Cedar (Tamarix) in
1965 the Western States. USGS Prof. Paper No. 491-A, 1965.

Robinson, T.W. Evapotranspiration by Woody Phreatophytes in the Humboldt River
1970 Valley, Near Winnemucca, Nevada. USGS Prof Paper No. 491-D, 1970.

Schick, A.P. "Suspended Sampler" Ref. Geormorphol Dynamique, Vol. 17 No. 4, 1967
1967

Schofield, R.A., Oliver, V.J. A Scheme for Estimating Convective Rainfall from
1977 Satellite Imagery, NOAA Technical Memorandum, NESS 86.

Searcy, J.K., Hardison, C.H. Double Mass Curves: Manual of Hydrology Part 1.
1960 USGS Water Supply Paper No. 1541-B, Washington.

Sellers, W.D. Physical Climatology. University of Chicago Press
1965

Sharon, D. "The Spottiness of Rainfall in a Desert Area" Journal of Hydrology,
1972 Vol. 17, 1972

Shulits, S. et al. "Bedload Formulas". Dept. of Civil Engineering, Hydraulics
1968 Laboratory, Pennsylvania State University, Bulletin, Dec. 1968

Sikdar, D.N. "AT3-3 Observed Cloud Brightness Field Related to a Meso-to-subsynoptic
1972 scale rainfall pattern". Tellus, 24.

Simon, D.B., Richardson, E.V. Resistance to Flow in Alluvial Channels. USGS Prof.
1966 Paper No. 422J.

Simpson, J. On the Possibilities of Rain Augmentation by Cloud Seeding in the
1976 Persian Gulf Area (Qatar and Oman) FAO Consultant's Reort, WS/H7493.

Slayter, R.O., Mabbut, J.A. Hydrology of Arid and Semi-Arid Regions, Section 24
1964 of Handbook of Applied Hydrology. Ven Te Chow, Editor in Chief, McGraw-Hill

Smith, R.E., Schreiber, H.A. "Point Processes of Seasonal Thunderstorm Rainfall"
1973 Water Resources Research Vol. 9, No. 4, August 1973

Sokolov, A.A., Rantz, S.E., Roche, M. Flood Flow Computation. UNESCO Press
1976

Sutcliffe, J.V., "The Assessment of Random Errors in Areal Rainfall Distribution"
1966 Int. Assoc. Sci. Hydrol. Bulletin Vol. 11, No. 3, 1966.

Tanner, C.B., Fuchs, M. "Evaporation from Unsaturated Surfaces: A Generalised Combination
1968 Model" Journal of Geophysical Research, Vol 73, No. 4, 1968.

Thom, H.S.C. Some Methods of Climatological Analysis. WMO Technical Note No.81
1966

Todd, D.K. Ground Water Hydrology. John Wiley and Sons.
1959

Tyson, H.N., Weber, E.M. Use of Electronic Computers in the Simulation of the
1963 Dynamic Behaviour of Groundwater Bearing. Paper Presented at ASCE Water
Resources Conference, May 1963.

UNDP/FAO Integrated Planning of Irrigated Agriculture in the Varamin and Garmsar
1971 Plains. UNDP/FAO Technical Report No. 3.

UNESCO Arid Zone Research - XV: Plant - Water Relationships in Arid and Semi-Arid
1960 Conditions

UNESCO Groundwater Studies, an International Guide for Research and Practice
1977

U.S. Bureau of Reclamation. Drainage Manual, Washington D.C.
1978

Van Bavel, C.H.M. "Potential Evaporation: The Combination Concept and its Experimental
1966 Verification" Water Resources Research Vol. 2, No. 3, 1966

Van Hylckama, T.E.A. "Water Use by Salt Cedar" Water Resources Research Vol. 6,
1970 No. 3, 1970

Vanoni, V.A. Sedimentation Engineering Manual, ASCE
1975

Walton, K. The Arid Zones. Hutchinson University Library
1969

Wan, P. "Point Rainfall Characteristics of Saudi Arabia" Proc. of the Institution of
1976 Civil Engineers, Vol. 61, Part 2, March 1976

Water Resources Council. A Uniform Technique for Determining Flood Flow Frequencies.
1967 Water Res. Council. Bulletin 15, Washington D.C.

Watson, P., Sinclair, P., Waggoner, R. "Quantitive Evaluation of a Method for
1976 Estimating Recharge to the Desert Basins of Nevada" Journal of Hydrology
Vol.31

Weiss, G. "Shot Noise Models for the Generation of Synthetic Streamflow Data (AGU)"
1977 Water Resources Research, Vol. 13, No. 1, Feb. 1977.

White, W.N. A Method of Estimating Groundwater Supplies Based on Discharge by
1932 Plants and Evaporation from Soil. USGS Water Supply Paper No. 659-A

Williams, J.R., Berndt, H.D. "Sediment Yield Computed with Universal Equation"
1972 ASCE Vol. 98, No. HY12, Dec. 1972

Winghart, J. Connaissance Actuelle du Mechanisme du Transporte Solide, Annales
1965 des Ponts et Chaussee, Paris

Wischmeier, W.H., Smith, D.D. Predicting Rainfall - Erosion Lossess from Cropland East
1965 of the Rocky Mountains. U.S. Department of Agriculture, ARS Agricultural
Handbook 282

Woodley, W.L. et al. Rainfall Estimation from Satellite Cloud Photographs NOAA
1972 Technical Memorandum, ERL OD-11, Boulder, Colo.

World Meteorological Organisation. Guide to Agricultural Meteorological Practices WMO
1963 TP61 No. 134

World Meteorological Organisation. Measurement and Estimation of Evaporation and
1966 Evapotranspiration WMO. Technical Note No. 83

World Meteorological Organisation Guide to Meteorological Instrument and Observing
1969 Practices. WMO TP3, No.8.

World Meteorological Organisation. Working Group of the Commission for Hydrometeorology:
1969 Estimation of Maximum Floods. WMO Technical Note No. 98

World Meteorological Organisation. Guide to Hydrological Practices, 3rd Edition
1974 WMO No. 168

Yang, C.T. "Incipient Motion and Sediment Transport" ASCE Vol.99, No. HY10 October 1973
1973

Zohdy, A.A.R., Eaton, G.P., Mabey, D.R. Application of Surface Geophysics to Groundwater
1974 Investigations, USGS Technique of Water Resources Investigations, Book 2
Chapter DI, Washington D.C.

ANNEX 1

RAINFALL INTENSITY ANALYSIS

A.1.1 Method 1. Depth-Duration Analysis (Example A.1.1)

A.1.1.1 Criteria and Assumptions

For the purpose of applying this method:

(a) Rainfall during a storm period is assumed to be continuous if the apparent break between storm segments, indicated by a horizontal line on the recorder chart, is less than half an hour.

(b) Storm periods separated by breaks of more than half an hour are ascribed to the date on which they began and are numbered sequentially for each day.

(c) Storm periods during which rainfall is less than 4mm are not analysed for intensity.

(d) The duration for which mean intensities are calculated are:- 10, 20, 30 minutes and 1, 2 and 3 hours.

(e) The maximum duration for analysis is the period specified in (d) less than the storm duration.

(f) The maximum 10-minute intensity for each storm is the maximum for <u>any</u> 10-minute period, regardless of chart time-lines.

A.1.1.2 Rainfall Intensity Analysis Procedure

A typical procedure, is shown by Example A.1.1 and Figure A.1.1.

- Enter the first storm equal to or greater than 4mm as "Period No." 1.
- Read the rainfall depth corresponding with the 10-minute time-line immediately prior to commencement of the storm and enter the observed value in column 4. (5.4mm at 1940 hours).
- Enter the rainfall depth corresponding with the next 10-minute time-line in column 4. (5.6mm at 1950 hours).
- Repeat the procedure for all 10-minute time-lines to the end of the storm period.
- Calculate the rainfall increment for each 10-minute time interval and enter the value in column 5.
- Calculate the intensity (in mm/h) within each 10-minute interval by multiplying the depth increment by 6.
- For each storm period enter, in the box headed "Actual Fall", the maximum depths for the specified durations. In the case of the 10-minute value this should be determined by inspection of the chart as it is probable that the maximum increment for any duration will not coincide with a discrete time-line interval. In the example, the maximum ten-minute increment for Period No. 1 is 4.0mm between 2025 and 2035 hours. For all other durations the maximum "actual fall" is calculated by trial and error, by inspection of the tabulated increments in column 5.

EXAMPLE A.1.1

SHEET No. 1

Rainfall Intensity Analysis Sheet

Gauge Type SIAP RECORDER

Station No. Ta - 204

Storm Date 15.11.1967

Sheet No. 1 OF 3

Period No.	Hrs.	Mins.	Observed rainfall mms	Increment	Intensity mm/hr.
1	19	00			
		10			
		20			
		30			
		40	5.4		
		50	5.6	0.2	1.2
	20	00	6.0	0.4	2.4
		10	6.6	0.6	3.6
		20	8.4	1.8	10.8
		30	10.6	2.2	13.2
		40	13.6	3.0	18.0
		50	14.0	0.4	2.4
	21	00	14.2	0.2	1.2
		10	14.2	–	–
		20	14.4	0.2	1.2
		30			
		40			
		50			
2	22	00			
		10			
		20			
		30			
		40	14.6		
		50	20.6	6.0	36.0
	23	00	28.6	8.0	48.0

MAXIMUM PERIOD INTENSITY

MINUTES 10	MINUTES 20	MINUTES 30	HOURS 1	HOURS 2	HOURS 3	
ACTUAL FALL						
4.0	5.2	7.0	8.4	–	–	–
8.4	14.0	17.8	22.6	29.2	35.6	42.0
RATE OF FALL						
24.0	15.6	14.0	8.4	–	–	–
50.4	42.0	35.6	22.6	14.6	11.9	10.5

NOTES

Total daily rainfall = 63.6 mm

Computed By Ahmed Salem

Date 2/12/67

Checked By

Date

SHEET No. 2

EXAMPLE A.1.1.

Rainfall Intensity Analysis Sheet

Gauge Type SIAP RECORDER **Storm Date** 15.11.1967

Station No. Ta - 204 **Sheet No.** 2 OF 3

Period No.	Hrs.	Mins.	Observed rainfall mms	Incre-ment	Intensity mm/hr.
2	23	00	28.6		
		10	32.4	3.8	22.8
		20	34.2	1.8	10.8
		30	36.0	1.8	10.8
		40	36.6	0.6	3.6
		50	36.8	0.2	1.2
	24	00	37.0	0.2	1.2
		10	37.4	0.4	2.4
		20	39.2	1.8	10.8
		30	41.6	2.4	14.4
		40	43.2	1.6	9.6
		50	44.6	1.4	8.4
	01	00	45.4	0.8	4.8
		10	46.6	1.2	7.2
		20	48.2	1.6	9.6
		30	49.0	0.8	4.8
		40	49.6	0.6	3.6
		50	50.4	0.8	4.8
	02	00	51.4	1.0	6.0
		10	52.6	1.2	7.2
		20	53.6	1.0	6.0
		30	54.6	1.0	6.0
		40	56.0	1.4	8.4
		50	57.4	1.4	8.4
	03	00	58.8	1.4	8.4

MAXIMUM PERIOD INTENSITY

MINUTES			HOURS		
10	20	30	1	2	3

ACTUAL FALL

RATE OF FALL

NOTES

Computed By ....

Date ....

Checked By ....

Date ....

SHEET No. 3

EXAMPLE A.1.1

Rainfall Intensity Analysis Sheet

Gauge Type SIAP RECORDER

Storm Date 15.11.1967

Station No. Ta - 204 KM5

Sheet No. 3 OF 3

Period No.	Hrs.	Mins.	Observed rainfall mms	Incre-ment	Intensity mm/hr.
2	03	00	58.8		
		10	60.0	1.2	7.2
		20	60.8	0.8	4.8
		30	61.8	1.0	6.0
		40	62.4	0.6	3.6
		50	63.4	1.0	6.0
		00	65.2	1.8	10.8
		10	66.4	1.2	7.2
		20	67.2	0.8	4.8
		30	67.8	0.6	3.6
		40			
		50			
		00			
		10			
		20			
		30			
		40			
		50			
		00			
		10			
		20			
		30			
		40			
		50			
		00			

MAXIMUM PERIOD INTENSITY					
MINUTES			HOURS		
10	20	30	1	2	3
ACTUAL FALL					
RATE OF FALL					

NOTES

Computed By ....

Date ....

Checked By ....

Date ....

SUMMARY OF RAINFALL INTENSITIES

Area VIII

Period : 1971-1972

	Date	Maximum fall in mm. during period in min						Total	Occ.	Remarks
		10	20	30	60	120	180			
Ta-231 -										
MAHANI	10.3.71	8.0	9.0	10.0	12.0	13.6				
	25.3.71	5.2	5.4	6.6	8.0	8.8				
	"	5.8	8.8	11.4	17.0					
	"	8.0	9.6	10.2	13.4	14.2				
	12.4.71	1.6	2.8	3.0	4.0	5.2				
	2.5.71	5.2	5.6	5.6	7.0					
	"	1.2	2.2	2.8	4.8					
	26.4.72	2.6	3.8	4.6	6.2	9.4	11.4			
	27.4.72	8.4	13.2	17.8	26.6	29.0				
	28.4.72	3.6	4.2	5.2	7.2					
	8.6.72	6.0	7.4							
	29.10.72	5.4	7.8	11.6	14.6	16.0				
	"	3.2	3.8	4.6	8.8	9.6				

- Enter the maximum falls for subsequent storm periods on the same day on Sheet 1 of the analysis sheet.

- Calculate mean intensities corresponding with maximum falls for each period and enter the values under "Rate of Fall". Mean intensity values are calculated in mm/h:

- Enter the total daily rainfall for the period(s) covered by the analysis, in the space provided under "Notes".

In the example shown on Figure A.1.1 it will be noted that the rainfall trace is displaced by about 0.4mm with respect to the ordinate scale of the chart.

A.1.1.3 Summary of Rainfall Intensities

The data in the "Actual Fall" box are transferred to a summary of depth-duration maxima for each analysed storm at the station. The summary is a concise numerical presentation of the data for each station and the basis of graphical presentation of the data and analysis of storm rainfall depth-duration maxima (e.g. Figure A.1.3).

A.1.2 Method 2. Intensity Segment Analysis (Example A.1.2)

A.1.2.1 Criteria for Application of the Analysis

(i) Individual rainstorms equal to or greater than 1mm - in the case of computer analysis - and 5mm in the case of manual analysis.

(ii) Suitable for "open-scale" chart records, on punched tape and magnetic tape recording instruments to ensure discrimination of hydrologically significant variations in rainfall intensity.

A.1.2.2 Rainfall Intensity Abstract

The following procedure applies to manual analysis of autographic recorder charts only.

- For rainfall events equal to or greater than 5mm, mark all distinct intensity (slope) changes of the recorder trace as shown on Figure 6.2. The criterion for decision as to the degree of refinement required in this process is determined, in theory, by the intensity exceedance classes selected. In practice, however, it is impossible to detect differences between closely spaced intensity intervals by eye, especially on charts with a curvilinear ordinate scale, and it is necessary, to abstract the coordinates of all discernible intensity changes.

- Tabulate the co-ordinates of the selected intensity change points on the intensity abstract form which is self-explanatory. This assumes zero initial loss since all storm rainfall has been accounted for.

- For each of the selected initial loss classes (5, 10, 15, 20 and 50mm) tabulate the coordinates for the partial storm i.e. after deducting the depth corresponding with each initial loss class from the first part of the storm.

- Calculate the depths D and durations T for each intensity segment.

- Calculate the intensities of each segment

- Enter check marks in the appropriate duration-intensity class column.

EXAMPLE A.1.2

RAINFALL INTENSITY ABSTRACT Gauge No. Ta 204

Legend: t_1 = time of beginning of rain
t_2 = time of end of rain
d_1 = depth of rain at beginning of event
d_2 = depth of rain at end of event

$t_2 - t_1 = T$ = Duration of rainfall event in mins.
$d_2 - d_1 = D$ = Depth of rain in rainfall event in mins.

I = intensity of rainfall = $\frac{D}{T}$ in mms./hour.

Time Classes for Analysis:

I: 0 - 10 mins; II:10 - 20 mins; III: 20 - 30 mins
IV:30 - 60 mins; V:60 -120 mins; VI: > 120 mins.

Note:- All figures in brackets - class T,D,I refer to partial periods of longer storms.

Initial Loss Class: (a) 0mm (b) 5mm (c) 10mm (d) 15mm (e) 20mm (f) 50mm.

Date	t_1	t_2	d_1	d_2	D	T	I	Class (Check)					
								I	II	III	IV	V	VI
15.11.67	2247	2305	4.8	21.6	16.8	18	56		✓				
	2305	2312	21.6	22.7	, 1.1	7	9	✓					
	2312	2317	22.7	23.4	0.7	5	8	✓					
	2317	2325	23.4	25.7	2.3	8	17	✓					
	2325	2332	25.7	26.5	0.8	7	7	✓					
	2332	0010	26.5	27.7	1.2	38	2				✓		
	0010	0025	27.7	30.0	2.3	15	9		✓				
	0025	0032	30.0	32.1	2.1	7	18	✓					
	0032	0040	32.1	33.5	1.4	8	10	✓					
	0040	0050	33.5	34.9	1.4	10	8		✓				
	0050	0100	34.9	35.6	0.7	10	4		✓				
	0100	0110	35.6	36.8	1.2	10	7		✓				
	0110	0118	36.8	38.4	1.6	8	12	✓					
	0118	0133	38.4	40.0	1.6	15	6		✓				
	0133	0210	40.0	42.9	2.9	37	5				✓		
	0210	0248	42.9	47.6	4.7	38	7				✓		
	0248	0312	47.6	50.0	2.4	24	6			✓			
	0312	0330	50.0	52.0	2.0	18	7		✓				
	0330	0340	52.0	52.6	0.6	10	4		✓				
	0340	0350	52.6	53.6	1.0	10	6		✓				
	0350	0400	53.6	55.3	1.7	10	10		✓				
	0400	0410	55.3	56.6	1.3	10	8		✓				
	0410	0420	56.6	57.4	0.8	10	5		✓				
	0420	0430	57.4	57.9	0.5	10	3		✓				
					53.1								

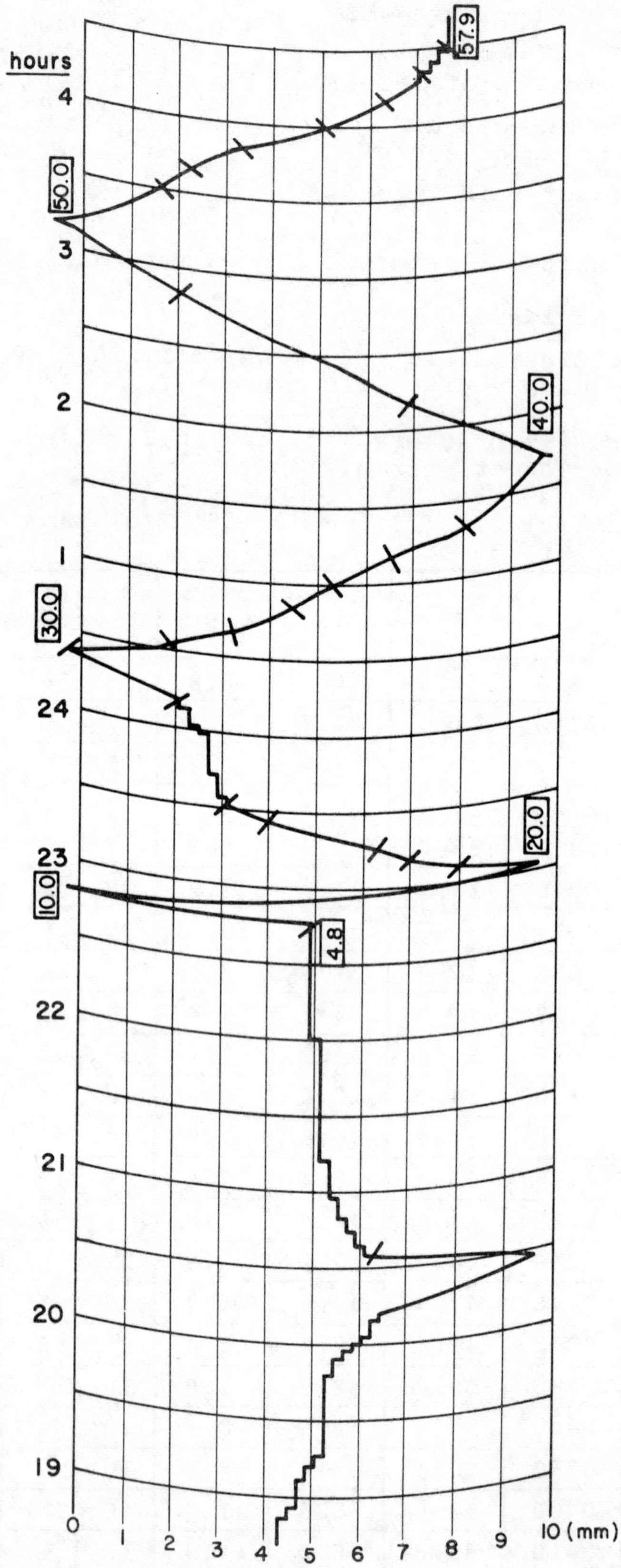

Fig. A.1.1 Rainfall intensity analysis - intensity segment method

FIGURE A.1.1

Rainfall Intensity Analysis - Intensity Segment Method

The figure shows an autographic raingauge chart on which intensity changes have been demarcated by eye. Intensity differences in the range up to 1 mm/hour are disregarded as being hydrologically unimportant. Rainfall depths are classified according to intensities and the total amount above specified intensity thresholds is calculated as shown by Example A.2.

Rainfall depths within discrete time intervals, rather than intensity classes, are sometimes abstracted for analysis as shown by Example A.1.

The hyetograph shown on the chart illustrates the intensity variations within such discrete time intervals. The maximum intensity calculated by the intensity-segment method is 56 mm/hour for a duration of 18 minutes whereas the discrete time interval method reveals an intensity of 50 mm/hour for 10 minutes and 42 mm/hour for 20 minutes.

The former method produces results which are much more likely to be related to runoff response than are those produced by the latter.

EXAMPLE A.1.2

GAUGE No.: Ta-204

RAINFALL EXCESS INTENSITY ANALYSIS BY INITIAL LOSS CLASSES.

CLASS.:- (a)

DATE	WHOLE STORM INTENSITY ≥ mm/hr DEPTH mm.						PARTIAL STORMS INTENSITY ≥ mm/hr DEPTH mm.					
	5	10	15	20	25	50	5	10	15	20	25	50
15.11.67	50.1	25.9	21.2	16.8	16.8	16.8						
TOTALS mm.												

TOTAL DEPTH B/F.

" " THIS PAGE ______________

" " C/F.

A.1.2.3 Rainfall Intensity Analysis Sheet

This form is used to group storm elements into intensity-exceedance and initial-loss classes and for calculating the total depth of rainfall in each class.

The procedure is as follows:

- Enter the date of the storm on the left hand side of the form according to the intensity classes indicated. The depth is shown in each column up to the column appropriate to the particular storm segment intensity.
- Enter the depths of each segment of the partial storm in the columns provided on the right-hand side of the form for each initial loss class.
- Repeat the procedure for subsequent storms.
- Enter the totals at the foot of each column, bringing forward from and carrying forward to other pages as necessary for individual storms.
- Enter the total depths of rainfall for all analysed storms in the space provided at the foot of the page.

A.1.3. Interpretation of Analysis Results

The procedures described for depth-duration and intensity-segment analysis of storm rainfall hyetograms have reduced storms to the constituents of their temporal distribution. It is now necessary to present the results in a form suitable for hydrological interpretation, either graphically or numerically.

A.1.3.1 Depth-Duration-Frequency Relationships

A plot of mean intensity maxima for individual storms against duration from the rainfall intensity analysis sheet (Example A.1.1) produces a series of vertical columns of data points converging asymptotically on the physical maximum intensity for the given duration. The asymptote cannot be drawn but, if the sample of storms is adequate, the percentage occurrence of mean intensity maxima for given durations can be regionalised and plotted as shown on Figure A.1.2. Frequency curves cannot yet be drawn for single stations and the graphical plot is merely an illustration of the available data.

Frequency estimates are possible by invoking the "station year" concept to combine several stations records to form a single synthetic "station" record and determining the frequency distribution of the required intensity parameter. For this technique to be applied it must be established that each station is within a homogeneous rainfall type and depth zone and that all items of data are independent.

A.1.3.2 Depth-Intensity Relationships

The intensity segment analysis of the hyetograph (Examples A.1.2) shows the total analysed depth of rainfall by intensity exceedence classes. For single stations with an adequate length of good quality record, or for groups of stations in homogeneous rainfall zones, graphs may be prepared showing the percentage of rainfall for any intensity exceedance level by initial loss classes. The relationships indicated by these curves, or derived analytically may be used together with estimates of infiltration rates to produce estimates of run-off or recharge.

The procedure for preparation of curves is as follows:

- From the analysis sheet, calculate the total depth of rainfall for the period covered by the analyses at each station, including unanalysed rainstorms of less than 5mm. Any missing periods of record should be ignored.

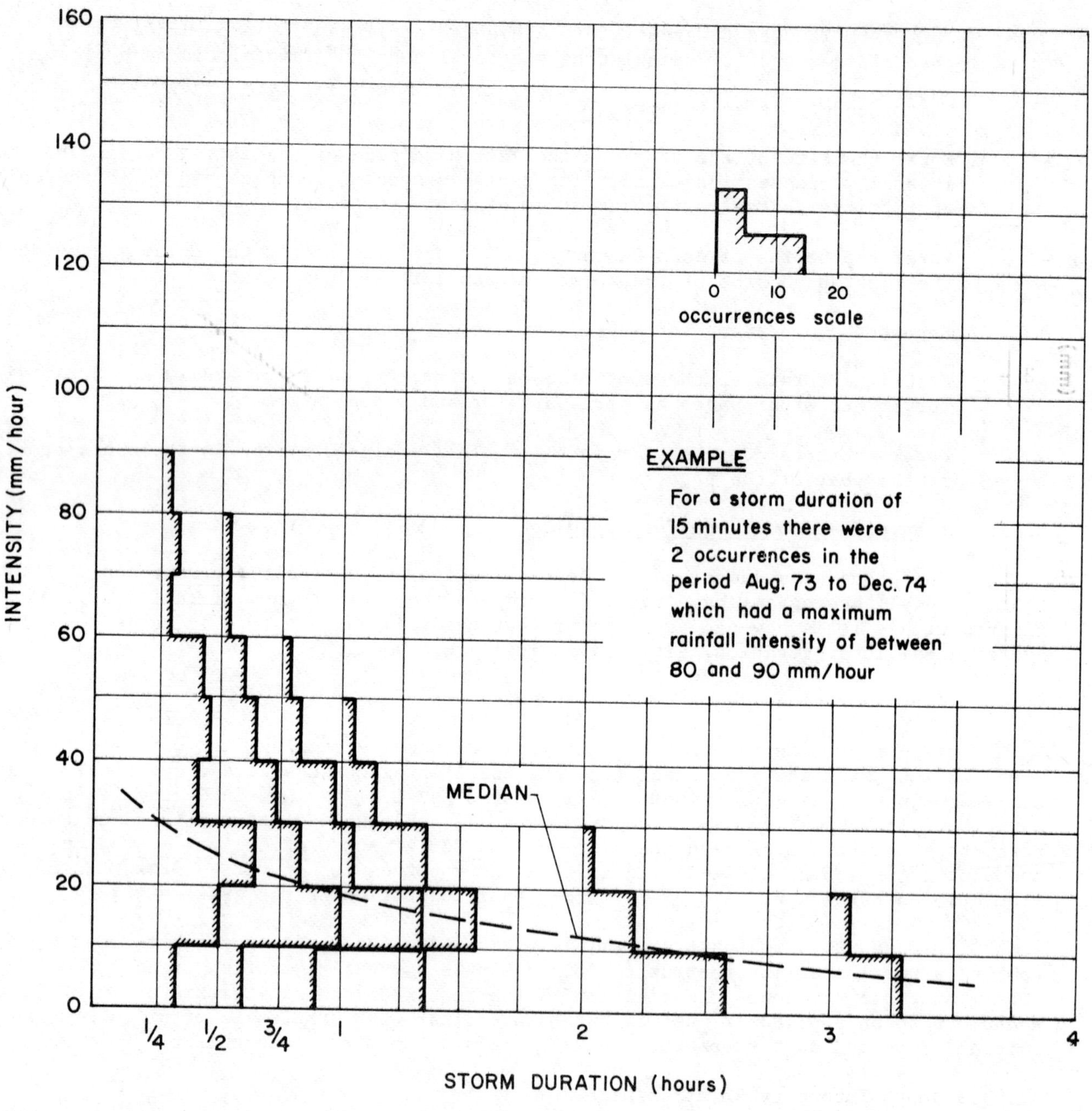

Fig. A.1.2 Distribution of occurrences of maximum storm intensities for specific durations

The figure shows the frequency distribution of maximum storm intensities for specified durations. Histograms are prepared from summaries such as that illustrated in Example A.1.1. As the record length increases the shape of the frequency distribution will be better defined and greater intensities will be experienced.

The median line shows typical storm intensity maxima for each duration. Thus, in the case illustrated, 50% of storms would be expected to have maximum one-hour intensities of about 20 mm/hour.

- Express the depths of rainfall in each intensity exceedence class and for each initial loss class as percentages of the total analysed rainfall depth.

- Plot these percentages as abscissae on semi-logarithmic paper against the intensity exceedance level as ordinate for each initial loss class.

- Draw curves of relation by eye through the plotted points. (A typical family of curves as shown by Figure A.1.3).

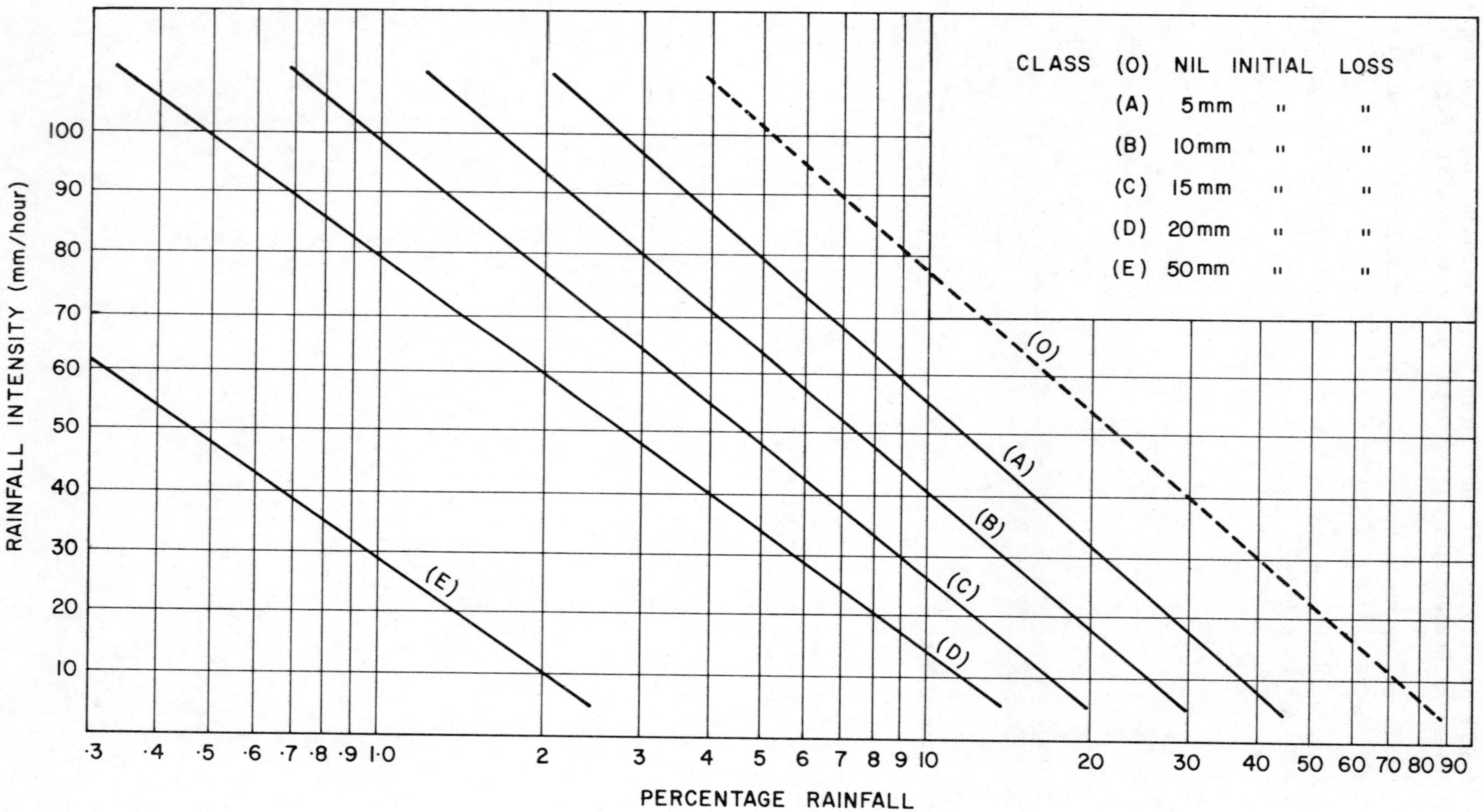

Fig. A.1.3 Rainfall intensity - percentage of rainfall equal to or greater than specified for different initial loss classes

FIGURE A.1.3

Rainfall Intensity - Percentage of Rainfall Equal to or Greater than Specified for Initial Loss Classes

The figure illustrates the results of the intensity segment method of analysis.

Rainfall is classified by intensity and initial loss classes as shown by Example A.1.2. The total rainfall depth in each class for all analysed storms is expressed as a percentage of total analysed rainfall. Percentages are cumulated from high to low classes and plotted against the intensity threshold for each initial loss class.

Thus, in the example illustrated, after the first 10 mm of each storm, 10% of rainfall occurs at intensities in excess of 40 mm/hour.

The range of percentage corresponding with each intensity class is considerable as between storms. The graphs represent the medians for all analysed storms.

ANNEX 2

ERROR ASSESSMENTS

A.2.1 Double Mass Analysis.

An illustration of double-mass analysis is given below, based upon the data for three stations in the vicinity of Khartoum. The plot suggested that Khartoum (Stack Laboratories)(1) and Khartoum (Irrigation)(2) are mutually consistent but that the relationship between Khartoum (North)(3) and (1) changed, in 1927. If the three stations weré used as a base group covering the period 1913-37 the result would be as shown on Figure A.2.1.

The ratios of the double-mass segment slopes for the single station and base group plots are 1.25 and 1.13 respectively. This is due to the fact that the presence of the suspect station within the small base group has a compensatory effect. A suspect record within a much larger group would have correspondingly less effect and the slope ratios for the single station and base group plots would be about the same.

It should be noted that, in the case illustrated, the slope of the double-mass plot of stations (1) and (2) for the period 1908 to 1945 approximates unity and gives confidence in both stations which, for the vicinity of Khartoum, would be expected to have similar mean rainfall. It may be assumed that in this locality, the early part of the record for station (3) (1913 to 1928) may be suspect.

The F significance test based on the analysis of variance and covariance given as Example A.2.1 suggests that the break in slope of the station (3) vs base group double-mass curve is not significant. A similar test based on station (3) vs a base group composed of stations (1) and (2) only shows the break in slope to be significant.

The analysis shown here is not according to usual practice, which requires large base groups. It is suggested, however, that single station double-mass plots (or numerical analyses) are capable of revealing anomalies in records which would otherwise go undetected. It is a rapid way of testing the homogeneity of record segments with respect to other stations in the vicinity. The graphical method is easier than numerical methods which would require programmable desk-top computers to achieve the same objective.

For the example given it would normally be concluded that station (3) under-recorded during the period 1913-1928 and that stations (1) and (2) should be used to assess the annual rainfall characteristics of the area.

A word of warning is appropriate in relation to double-mass analyses in particular and generally in the use of hydrological data for which the quality and history are uncertain. Very often, records are completed by correlation procedures by means of which the records of one station are estimated from that of another. The fact that a record (or part of it) has been estimated in this way may not appear in the station history sheet. It is evident that double-mass analysis of data which is partially synthesised by correlation with others in the same base group is useless. It may be the purpose of the analysis to produce effectively complete records. The data used for this purpose must not themselves be synthetic.

EXAMPLE A.2.1.

Khartoum Annual Rainfall Series (mm)

	North Stn. 3	Stack Laboratories Stn. 1	Irrigation Stn. 2	Mean
1913	102	101	92	98
14	107	101	140	116
15	(45)	176	87	103
16	122	146	124	131
17	68	76	68	71
18	46	93	79	73
19	74	75	105	85
20	304	341	287	311
21	66	246	190	167
22	306	366	414	362
23	248	338	265	284
24	81	158	85	108
25	59	102	101	87
26	(88)	92	101	94
27	157	216	229	201
28	(57)	121	96	91
29	280	254	266	267
30	248	223	231	234
31	143	205	195	181
32	215	192	175	194
33	66	126	106	99
34	185	228	235	216
35	154	225	221	200
36	124	122	91	112
37	(160)	118	119	132

Example A.2.1 Significance Test of Change of Double Mass Slope - Station (3) vs Group (Stations (1), (2) and (3) by Analysis of Variance and F-Test (Searcy and Harrison, 1960)

Symbols

N = 25 years - total period (1913-1937)

N_1 = 16 years - period (1) (1913-1928)

N_2 = 9 years - period (2) (1929-1937)

X_1 = Annual rainfall for the base group in period (1)

X_2 = Annual rainfall for the base group in period (2)

Y_1 = Annual rainfall for station 3 in period (1)

Y_2 = Annual rainfall for station 3 in period (2)

x = Difference between an X value and the period mean.

y = Difference between a Y value and the period mean.

$\frac{\Sigma x^2}{N-1}$ or $\frac{\Sigma y^2}{N-1}$ = Variance of X or Y.

$\frac{\Sigma x.y}{d.f.}$ = Convariance of X and Y.

d = Residual - the vertical (Y) difference between a line through the plotted points and a plotted point.

d.f. = Degrees of freedom.

Calculations are made using the data given in Table A.2.1.

$\Sigma X_1 = 2384$ $\quad \Sigma Y_1 = 1936$

$\Sigma X_2 = 1638$ $\quad \Sigma Y_2 = 1575$

$\Sigma X = 4022$ $\quad \Sigma Y = 3511$

$\Sigma X_1^2 = 481488$ $\quad \Sigma Y_1^2 = 348574$

$\Sigma X_2^2 = 322866$ $\quad \Sigma Y_2^2 = 309851$

$\Sigma X^2 = 804354$ $\quad \Sigma Y^2 = 658425$

$(\Sigma X_1)^2/N_1 = 355216$ $\quad (\Sigma Y_1)^2/N_1 = 234256$

$(\Sigma X_2)^2/N_2 = 298116$ $\quad (\Sigma Y_2)^2/N_2 = 275625$

$(\Sigma X)^2/N = 647059$ $\quad (\Sigma Y)^2/N = 493085$

$\Sigma X_1 Y_1 \ = \ 403168$

$\Sigma X_2 Y_2 \ = \ 312687$

$\Sigma X_1 Y \ = \ 715855$

$\Sigma X_1 \ \Sigma Y_1 / N_1 \ = \ 288464$

$\Sigma X_2 \ \Sigma Y_2 / N_2 \ = \ 286650$

$\Sigma X \ \Sigma Y / N \ = \ 564850$

Analysis of Variance and Covariance

	(1) Degrees of Freedom	(2) Σx^2	(3) Σy^2	(4) Σxy	(5) Degrees of Freedom	(6) Σd^2	(7) $\Sigma \bar{d}^2$
a) Among periods	1	6273	16796	10264	1	2989	2989
b) Within periods	23	151022	148544	140741	22	17384	790
c) Total	24	157295	165340	151005	23	20373	886

a) $$\Sigma x^2 = \frac{(\Sigma X_1)^2}{N_1} + \frac{(\Sigma X_2)^2}{N_2} - \frac{(\Sigma X)^2}{N}$$

b) $$\Sigma x^2 = X^2 - \frac{(\Sigma X_1)^2}{N_1} + \frac{(\Sigma X_2)^2}{N_2}$$

c) $$\Sigma x^2 = \Sigma X^2 - \frac{(\Sigma X)^2}{N}$$

a) $$\Sigma y^2 = \frac{(\Sigma Y_1)^2}{N_1} + \frac{(\Sigma Y_2)^2}{N_2} - \frac{(\Sigma Y)^2}{N}$$

b) $$\Sigma y^2 = \Sigma Y^2 - \frac{(\Sigma Y_1)^2}{N_1} + \frac{(\Sigma Y_2)^2}{N_2}$$

c) $$\Sigma y^2 = \Sigma Y^2 - \frac{(\Sigma Y)^2}{N}$$

a) $$\Sigma xy = \frac{\Sigma X_1 \ \Sigma Y_1}{N_1} + \frac{\Sigma X_2 \Sigma Y_2}{N_2} - \frac{\Sigma X \ \Sigma Y}{N}$$

b) $$\Sigma xy = \Sigma XY - \frac{\Sigma X_1 \ \Sigma Y_1}{N_1} + \frac{\Sigma X_2 \ \Sigma Y_2}{N_2}$$

c) $$\Sigma xy = \Sigma XY - \frac{\Sigma X \Sigma Y}{N}$$

$$\Sigma d^2 = \Sigma y^2 - \frac{(\Sigma x y)^2}{x^2}$$

c) $\Sigma d_t^2 = 20373$ (t denotes total)

b) $\Sigma d_w^2 = 17384$ (w denotes within)

a) $\Sigma d_a^2 = \Sigma d_t^2 - \Sigma d_w^2 = 20373 - 17384 = 2989$ (a denotes among)

$$F = \frac{2989}{790} = 3.78$$

The tabulated value of F for 1 and 22 degrees of freedom is 5.79 at 5% level. Therefore the break in the double mass curve is not significant on the basis of this test.

The above calculations can be performed rapidly by means of desk-top computers, some of which have standard programmes for analysis of variance and covariance.

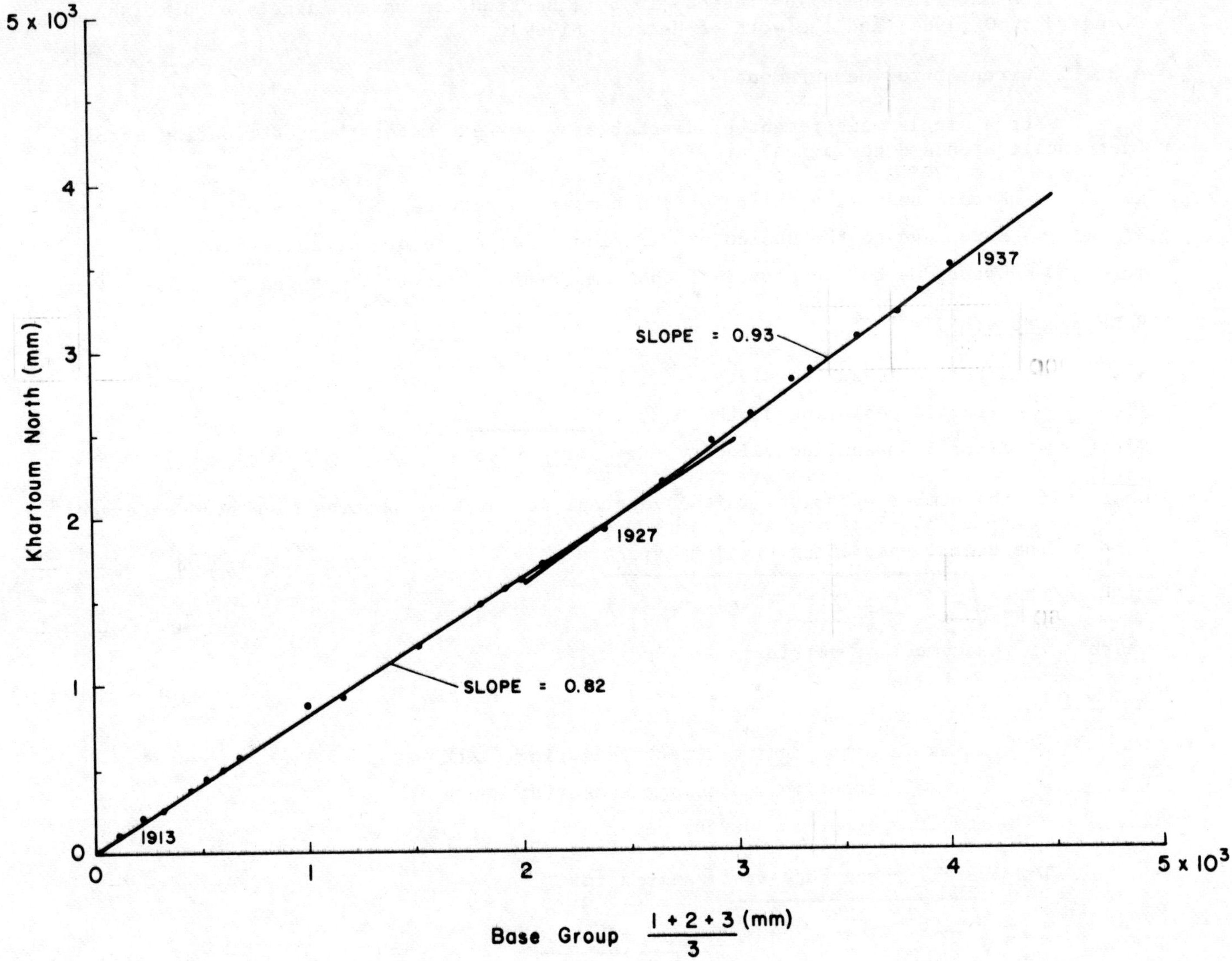

Fig. A.2.1 Double mass analysis - Khartoum North vs. Mean of three stations

The figure shows a typical double-mass plot in which cumulated annual rainfall for Khartoum North is plotted against the cumulated annual mean rainfall for the three stations Khartoum North, Irrigation and Stack Laboratories. The change of slope at year 1927 was found, by analysis of variance and the F-test, not to be significant at the 5% level, as shown.

The effect of including the suspect station in the base group is to reduce the ratio of the double-mass slopes before and after the change. With a larger base group the effect of the suspect station would be diminished. In the absence of such a group the suspect station should be plotted against the other two. In this case the F-test would have indicated that the break of slope <u>is</u> significant.

A.2.2 Errors in Flow Measurement Estimation

The material and illustrations in this Section are based largely on British Standard 3680 (1964) and the work of Herschy (1969).

A.2.2.1 Current Meter Measurement

For a single measurement of discharge by current meter the contributing errors (percentage standard deviation) are:

Xm = Error due to the choice of the number of verticals

Xf = Error due to the choice of duration of exposure of current meter

Xo = Error due to the choice of the number of points in a vertical

Random errors:-

X'b = Error in measuring width

X'd = Error in measuring depth

X'v = Error in measuring velocity = $\pm\sqrt{\frac{Xf^2}{p} + Xo^2}$

where p is the number of points in the vertical at which velocities have been observed.

The overall percentage random error

$$X'q = \pm\sqrt{X\,m^2 + \frac{1}{m}(X\,b^2 + X\,d^2 + X\,v^2)}$$

where m is the number of verticals

Systematic errors:-

X"b = systematic error in instrument measuring width

X"b = systematic error in instrument measuring depth

X"v = systematic error in current meter

The overall percentage systematic error

$$X''q = \pm\sqrt{(X''b^2 + X''d^2 + X''v^2)}$$

The overall error is then $XQ = \pm\sqrt{(X'q^2 + X''q^2)}$

The overall systematic error is normally small and for present purposes is set at 2%.

Suggested values for random errors are as follows:

Xm Error due to the number of verticals m

m	Xm Normal Conditions	Xm Poor Conditions
8	± 5%	± 10%
15	± 3%	± 6%
50	± 1%	± 2%

Xf Error due to choice of duration of exposure of current meter.

For an exposure time of 60 seconds the suggested value is Xf = ± 6%

Xo Error due to the choice of the number of points in a vertical.

Method	Xo Normal Conditions	Xo Poor Conditions
Velocity Distribution	0.5%	2%
2 point	3.0%	5%
1 point (0.6 depth)	3.5%	10%
1 point (0.5 depth)(1)	5.0%	10%

(1) This position is recommended where the numeracy of observers is limited.

X'b Error in measuring width.

Errors in measuring width, either between verticals or in total width, by line or winch are usually small and a value of ± 1% is suggested.

X'd Error in measuring depth

These are best determined on site but normally a tolerance range of ± 1% to 5% should be possible to attain at large values of depth and ± 3% to 10% at shallow depths. The higher values relate to mobile sand-bed dune formation, high velocities causing velocity-head pile-up on the measuring rod, sinking of the rod, suspension cable angles etc.

Example A.2.2: Estimate of Error in Current Meter Measurement

It is required to calculate the overall tolerances XQ on two discharge measurements from the following particulars.

Number of verticals used	15
Exposure of meter	60s (1 exposure made)
Method of measurement	(a) 2 point (b) 1 point (0.5d)
Average velocity	above 0.3m/s
Standard rating	
Alluvial channel	(a) plane bed regime (b) dune regime

From the above information the partial (random) errors (at 68% confidence level) can be assessed:-

	(a)	(b)
X m	= ± 3%	± 3%
X f	= ± 6%	± 6%
X o	= ± 3%	± 10%
X'b	= ± 1%	± 1%
X'd	= ± 3%	± 10%

(a)

$$X'v = \pm \left(6^2 + 3^2\right)$$

$$= \pm \ 45$$

(b)

$$X'v = \pm \left(6^2 + 10^2\right)$$

$$= \pm \ 136$$

The overall random error in discharge is then

(a) $X'q = \pm \sqrt{(3^2 + \frac{1}{15}(1^2 + 3^2 + 45))} = \pm\ 3.6\%$

(b) $X'q = \pm \sqrt{(6^2 + \frac{1}{15}(1^2 + 10^2 + 136)} = \pm\ 7.2\%$

The overall systematic error in discharge is $\approx 2\%$.

The overall standard error is then

$$XQ = \pm \sqrt{(X'q^2 + X''q^2)}$$

(a) $XQ = \pm \sqrt{(3.6^2 + 2^2)}$

$= \pm\ 4.1$ approx.

(b) $XQ = \pm \sqrt{7.2^2 + 2^2}$

$= \pm\ 7.5\%$ approx.

XQ at 95% confidence level

(a) $= 2 \times 4.1$

$= \pm\ 8\%$

(b) $= 2 \times 7.5$

$= \pm\ 15\%$

A.2.2.2 Errors in Slope-Area Estimates

1. Manning Equation

Extensive use has been made of the Manning equation for extrapolation of rating curves or to provide approximate checks on linear extrapolations of logarithmic plots of current meter measurements at low-medium discharges. In the absence of discharge measurements, the slope-area method is appropriate for estimating peak discharge and has to be relied upon to produce daily runoff volume estimates. The Manning equation is:

$$Q = \frac{Ar^{2/3}\ S^{1/2}}{n}$$

The accuracy of individual slope-area estimates is highly dependent on the quality of the survey, the confidence with which flood marks are identified, the stability of the alluvial channel and the correctness of the selected roughness value (n).

The overall error in the discharge estimates may be estimated from the equation:

$$XQ = \pm \sqrt{\left(\frac{25}{9} . XA^2\right) + \left(\frac{4}{9}\ XP^2\right) + \left(\frac{1}{4}\ XS^2\right) + X_n^{\ 2}}$$

Where: XQ = Error in discharge Q

XA = Error in area A

XP = Error in wetted perimeter P

XS = Error in slope S

Xn = Error in roughness factor n

For arid region conditions the percentage standard errors may be estimated as:

XA ≃ ± 10% for stable beds

≃ - 33% for unstable beds

XP ≃ ± 5% for wide streams

≃ ± 10% for deep, narrow streams

XS ≃ ± 5% for good conditions, little turbulence

≃ ± 10% for poor conditions, much turbulence

Xn ≃ ± 20% for all conditions

These values assume a high standard of survey in a uniform reach of adequate length.

Combining the partial errors to produce an estimate of the overall error.

For good conditions:

$$XQ = \pm\sqrt{(\tfrac{25}{9} \times 10^2) + (\tfrac{4}{9} \times 5^2) + (\tfrac{1}{4} \times 5^2) + 20^2}$$

$$= \pm 26\%$$

For poor conditions:

$$XQ = -\sqrt{(\tfrac{25}{9} \times 33^2) + (\tfrac{4}{9} \times 10^2) + (\tfrac{1}{4} \times 5^2) + 20^2}$$

$$= - 59\%$$

The result of the second estimate is expressed as a negative value because under conditions of scour such as that indicated (the true area 50% greater than the surveyed area) the true discharge would be greater than the calculated value.

The systematic nature of XA (and thus XQ) is not fully accounted for in this estimate but the order of magnitude of the overall error is probably about right.

2. Modified Manning Equation

The accuracy attainable by application of the modified Manning formula (Section 12.2.2) is lower than for the conventional application in cases where bed and banks are stable and the cross section area of flow is known.

In the normal application Q varies directly as A and n^{-1}, as $r^{2/3}$ and $S^{1/2}$.

In the modified application Q varies directly as P, as $n^{3/2}$, as $S^{-3/4}$ and as $V^{5/2}$. The discharge estimate is thus highly sensitive to errors in V and increasingly so as the value of

$\frac{V + XV}{V}$ (where XV is the error in estimating V)

departs from unity and is more sensitive to errors in n than with the normal application.

With a stable section A and r may be determined with high accuracy from field surveys but, with a mobile bed, A is unknown and must be estimated from the other parameters whose value is known or can be estimated.

In the modified application of the formula described in 12.2.2:

$$Q = VP\left[\frac{Vn}{S^{1/2}}\right]^{3/2}$$

$$= V^{5/2}\ P\ n^{3/2}\ S^{-3/4}$$

For good conditions

$$XQ = \pm\sqrt{\frac{25}{4}XV^2 + XP^2 + \frac{9}{4}X\ n^2 + \frac{9}{16}XS^2}$$

$$= \pm\sqrt{625 + 25 + 900 + 14}$$

$$= \pm\ 40\%$$

For poor conditions (assuming $XV \simeq \pm 20\%$)

$$XQ = \pm\sqrt{(\frac{25}{4} \times 20^2) + 10^2 + (\frac{9}{4} \times 20^2) + (\frac{9}{16} \times 10^2)}$$

$$= \pm\ 60\%$$

Thus in poor conditions the probable error is about the same as for the conventional application of the method but is less likely to be systematic if n is well chosen. The precision of the discharge estimate is highly dependent on the quality of velocity measurement, but the repeated measurements of S and V made during a flood have the effect of reducing the error in the estimate of total runoff.

Absolute errors in gauge readings cause percentage errors in slope estimates which increase with decreasing slope. Slope errors should not, however, be the source of large discharge errors if the slope-reach is uniform, the slope distance adequate for the water surface gradient and water levels (as opposed to flood marks) are observed.

The value of n is critical and especially so in the modified application of Manning's formula. It should be selected from actual field discharge and slope measurements, as mentioned earlier, or by reference to WMO Technical Note No. 90, pages 15-17, or standard text books on open channel hydraulics.

A.2.2.3 <u>Stage - Discharge Curves for Stable Controls</u>

For a stable control the stage-discharge curve is a curve of best fit and should, unless some of the individual gaugings actually lie on the curve, be more accurate than any of the individual gaugings. One method of assessing this accuracy statistically is as follows:

1. Correct stage readings for cease-to-flow level c.

2. Plot discharge vs corrected stage on logarithmic paper.

3. Fit straight-line segments to the data points.

4. Calculate the regression constants a and b in the rating equation $Q = a(h - c)^b$, (where c is the cease to flow stage) from least squares regression of the logarithms of the data points for each segment, and the coefficient of determination r^2.

5. Calculate the discharges Q_2 predicted from the rating equations and the proportional differences Δ from these values and measured discharge Q_1 from

$$\Delta = \frac{Q_2 - Q_1}{Q_1}$$

6. Calculate the standard deviation $\hat{\sigma}$ of Δ to produce the estimated standard error SE of the rating segment expressed as a percentage at, 95% confidence from:

$$SE = t\,\hat{\sigma}_\Delta \qquad \text{Where } t \simeq 2 \text{ for } n > 20$$

7. Calculate the percentage standard error of the mean relationship SMR% at 95% confidence from:

$$SMR = \frac{SE}{\sqrt{n}}$$

Figure A.2.2 shows the stage discharge relationship for a station with a relatively stable control. The outer limit curves in Figure A.2.3 represent the tolerance on a single gauging at the 95% confidence level and should include about 19 out of 20 of the observations used in the determination of the stage discharge equation. If no change occurs in the rating (due to control shift), then 19 out of 20 check gaugings may be expected to fall between these limits. The 2 SMR limits (Figure A.2.3) represent the tolerance at 95% confidence on discharges determined on the basis of the rating equation; that is, in 19 cases out of 20 the true value of Q in the population of Q lies between the limits of $\pm$ 2 SMR.

It should be noted that, if the rating segment equation is based on only a small number of measurements additional check gaugings may fall systematically outside the outer limits, without a control shift having occurred, as discussed in the following section.

A.2.2.4 Student's t-Test to Detect Control Shift

New gaugings falling within the 2SE tolerance band without significant bias indicate that no control shift has taken place. A small number of check gaugings may plot to one side of the curve within the 2SE band without necessarily indicating bias since a larger number may have restored the balance.

Student's t statistic provides a test of the homogeneity of two samples which may be applied to rating curve analysis.

$$t = \frac{|\bar{x}_1 - \bar{x}_2|}{\hat{\sigma}_\Delta / \sqrt{n}}$$

where $\bar{x}_1$ and $\bar{x}_2$ are the means of the two samples and $\hat{\sigma}$ is the standard deviation.

The t value calculated from the sample data is compared with t values for various probability levels and degrees of freedom, given in statistical tables. If calculated t is less than the tabulated value it may be accepted that both sets of data are from the same parent population. Example A.2.3 shows the application of this method to the rating curve shift illustrated on Figure A.2.3.

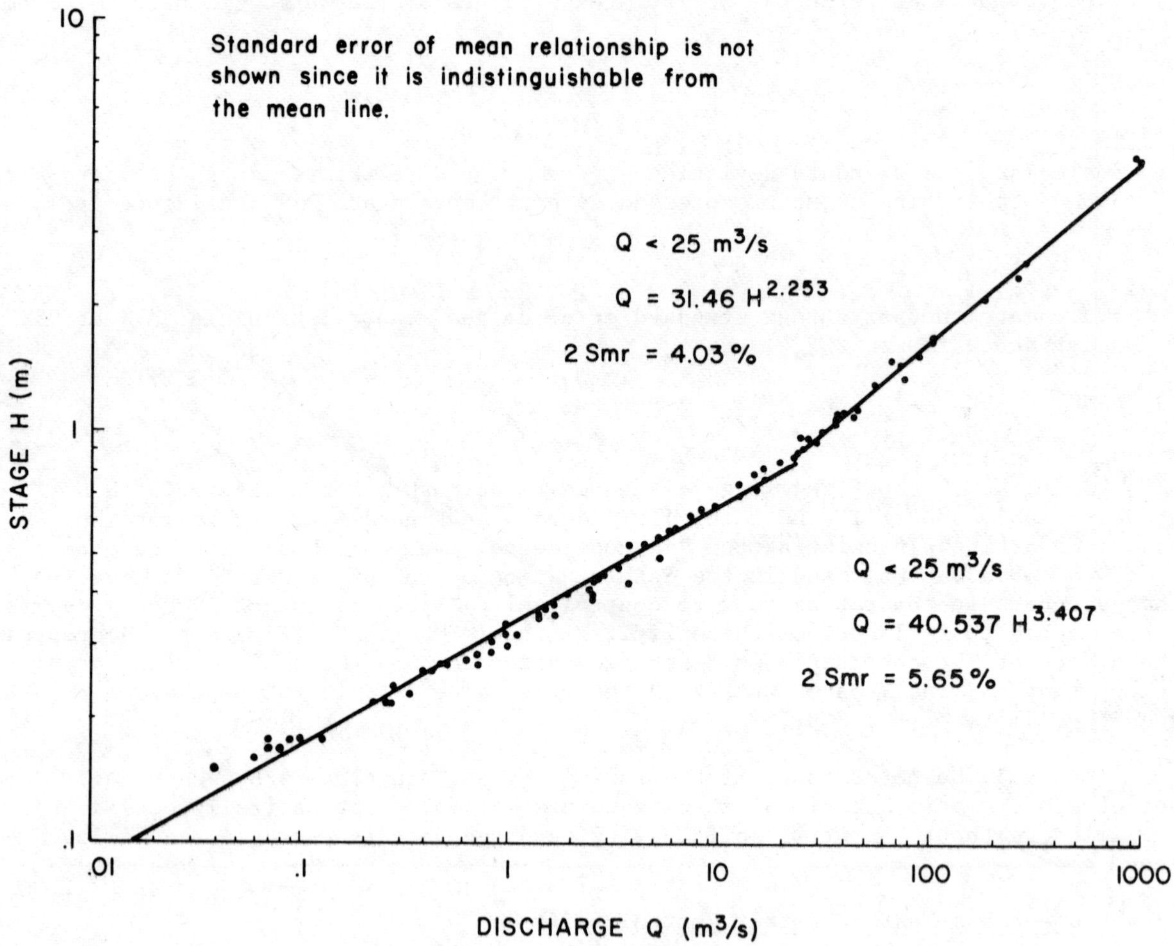

Fig. A.2.2 Stage discharge curve for Wadi Najran at Mudiq (27 December 1973 - 9 May 1974)

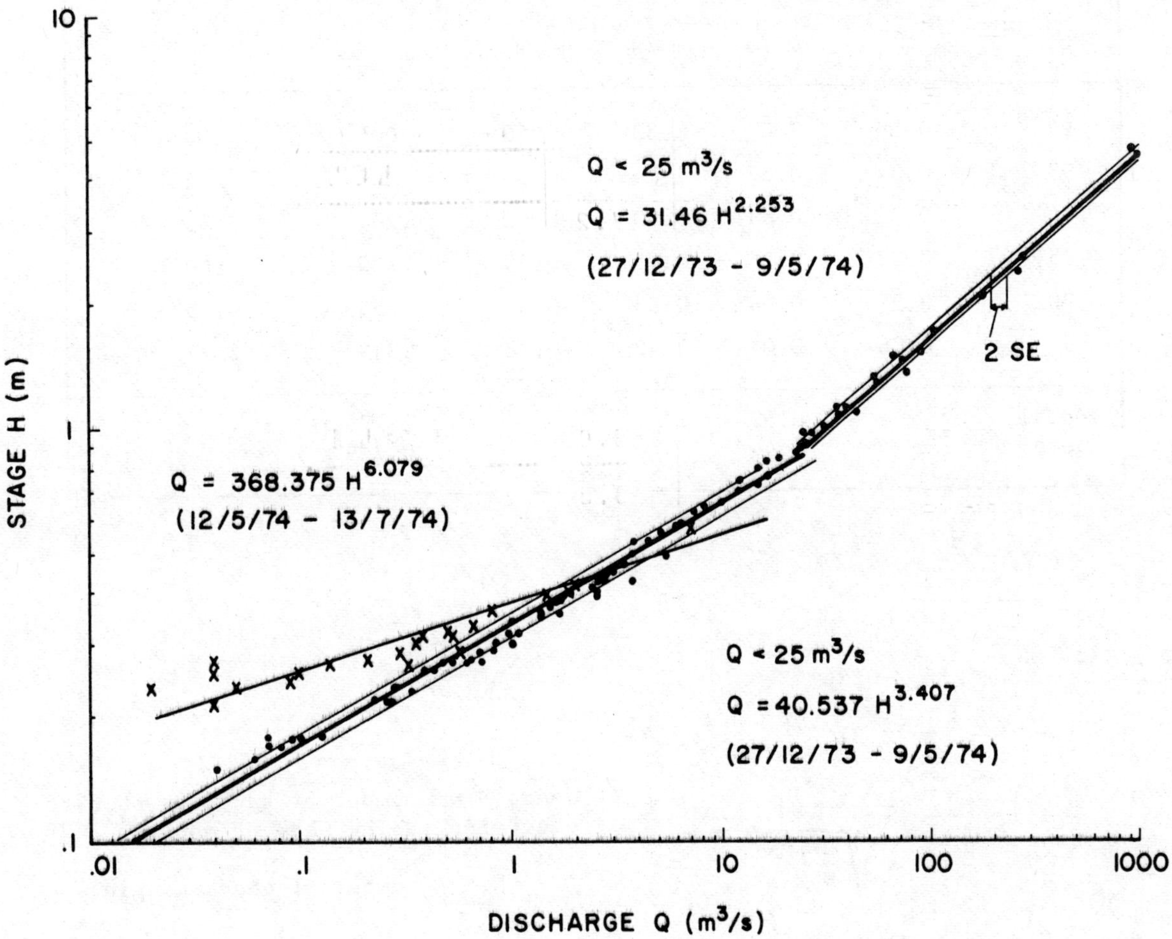

Fig. A.2.3 Stage discharhe curve for Wadi Najran at Mudiq (showing 2 x standard error and control shift)

Example A.2.3

Student's t Statistic Test for Control Shift

Gauging No.	Stage m	Q_2 m^3/s	Q_1 m^3/s	Δ m^3/s	$(\Delta/Q_1)^2$	$(\Delta/Q_1)^2$
92	0.59	7.0	6.50	+0.50	.006	.006
93	0.34	0.67	1.03	-0.36	.122	.128
94	0.33	0.51	0.93	-0.42	.204	.332
95	0.31	0.36	0.75	-0.39	.270	.602
96	0.28	0.04	0.53	-0.49	.855	1.457
97	0.26	0.04	0.42	-0.38	.819	2.276
Total Mean		8.62 1.44			2.276 0.379	

$\Sigma(\Delta/Q_1)^2 = 2.276$

$N = 6$

$\bar{\Delta} = -0.26$

$\hat{\sigma}_\Delta = 0.37$

$$t = \frac{|\bar{\Delta}|}{\hat{\sigma}\sqrt{\frac{N + N1}{N \times N1}}}$$

Where N = number of gaugings used to determine rating segment equation.

N_1 = number of check gaugings.

$$= \frac{0.26}{0.37\sqrt{\frac{69 + 6}{69 \times 6}}} = \frac{0.26}{0.16} = 1.63$$

For $N + N_1 - 2 = 73$ degrees of freedom

The probability level of the tabulated t_t value with which calculated t_c should be compared depends on the nature and precision of the data used to calculate the rating equation. For rating curve analysis, a level of 90% is normally appropriate.

In the present case $t_{t90,73} = 1.67$ which is a statistical indication that a rating shift may have occurred (1.63 is sufficiently close to 1.67 for doubts to exist). In the event, further measurements confirmed the shift which was applied retrospectively.

A.2.2.5 Error in Discharge Due to Stage Error

The tolerance on the stage value used to calculate the discharge at a given time is the square root of the squares of the separate errors due to:

a. reading of the primary and stilling-well gauges

b. relationship between primary and stilling-well gauge readings

c. recording instrument error

The separate errors are difficult to estimate but, typically, the total error Xh in the value used to calculate discharge from the rating curve or table will be in the range 1cm to 5cm with the higher value being more appropriate to high stages.

The effect of this error is to increase the overall discharge error which may now be calculated from:

$$XT = \pm\sqrt{(2SMR)^2 + b^2Xh^2}$$

Where SMR is the standard error of the mean relationship of the rating curve, b is the exponent in the rating equation and Xh is the estimated total stage error.

For the station rating shown in Figure A.2.1 the overall precision at selected stages are as follows:

a) $$XT = \pm\sqrt{5.9^2 + (3.4^2 \times 4^2)}$$

$$= \pm 14.8\%$$

b) $$XT = \pm\sqrt{4.0^2 + (2.25^2 \times 2.5^2)}$$

$$= \pm 6.9\%$$

A.2.2.6 Error in Daily Runoff Volume

If the errors XT_i in individual discharge values calculated from stages selected to depict the hydrograph may be regarded as random, the unweighted estimate of error in daily flow volume Xv may be estimated roughly from:

$$Xv^2 = \frac{\Sigma(XT_i)^2}{N^2}$$

The precision of the daily volume estimate may be calculated from the discharge-weighted percentage errors Xv_i of the hydrograph segments defined in the analysis. If the Xv_i are roughly equal, Xv depends only on the number of points N selected to define the hydrograph.

The calculation of Xv may be tabulated as shown in Example A.2.4.

The actual number of discharges computed for each day depends upon the complexity of the hydrograph. During flood periods a greater number of points are abstracted from the recorder chart than during recession and base-flow periods and these points are randomly distributed in time. In these circumstances the XT_i values should be volume-weighted as shown in Example A.2.4.

Example A.2.4.

Estimate of the Error in Calculation of Daily Runoff Volume

Time	Stage m	Q_i m^3/s	XT_i %	V Mm^3	Xv_i Mm^3
0000	0.25	0.2	15	0.15	0.02
0300	0.95	27.6	12	1.07	0.09
0400	3.60	567	5	1.47	0.08
0500	2.50	248	6	0.71	0.05
0600	2.00	149	7	1.36	0.10
0900	1.70	103	8	0.98	0.08
1200	1.50	78	9	1.45	0.14
1800	1.30	56	10	1.11	0.12
2400	1.20	47	11		
			$(XT_i)^2$ = 845	V=8.30	Xv=0.680

a) Xvw = 0.68/8.3 = 8.2% where Xvw is the weighted Xv estimate

Xv = $845/9^2$ = 10.4% where Xv is the unweighted Xv estimate

ANNEX 3

ESTIMATION OF FLOOD MAGNITUDE

Example A.3.1

SINGLE STATION FLOOD FREQUENCY ANALYSIS

Gumbel Extreme Value Distribution (See WMO Guide No. 168)

Determination of the 100-year peak flood.

Annual Qp Series (Wadi Najran, Saudi Arabia)

Year	1965	1966	1967	1968	1969	1970	1971	1972	1973	1974	$\bar{Q}p$	$\hat{\sigma}_{Qp}$
Qp-m^3/s	1800	100	700	300	2500	250	200	250	250	900	725	807

From the general equation:-

$$Qp_{100} = \bar{Q}p = K_{100}\,\hat{\sigma}_{Qp}$$

for the Gumbel distribution with T = 100 and n = 10 the value of K is 4.3227(Table 5.3 WMO Guide 168).

$$Q_{p100} = 725 + (4.3227 \times 807)$$

$$= 4213, \text{ say } 4200\ m^3/s$$

Determination of the confidence interval of the above estimate.

The 95% confidence limits are $Qp_{100} - t\alpha\,Se.$ and $Qp_{100} + t\alpha\,Se.$

Where $t\alpha =$ 1.96 for 95% confidence limits (Page 5.2, WMO Guide 168).

The standard error is given generally by:-

$$Se = \beta_T\,\hat{\sigma}_{Qp}/\sqrt{n}$$

and for the Gumbel distribution $\beta_T = \sqrt{1 + 1.14K_T + 1.1K_T^2}$

where K_T is the frequency factor, hence $\beta_T = 5.15$

and Se = 5.15 x 807/ 10 = 1314 m^3/s

$$CL_{0.95} = \pm (1.96 \times 1314) = 2575, \text{ say } 2600\ m^3/s$$

At the 95% confidence level the 100 year peak flood lies on confidence interval

$$4200 \pm 2600\ m^3/s = 1600 \text{ to } 6800\ m^3/s$$

Example A.3.2

REGIONAL FLOOD FREQUENCY ANALYSIS - USGS METHOD

ANNUAL PEAK DISCHARGE Qp - m^3/s								
YEAR	J-401	Ta-402	Ta-401	Ta-406	A-401	A-402	B-402	N-401
1965	x	x	x	x	x	x	x	1800
1966	240	20	60	x	0	x	10	100
1967	170	20	110	x	20	x	260	660
1968	130	240	430	2300	90	130	1000	140
1969	940	100	70	400	10	10	160	2500
1970	340	20	10	270	10	35	2100	210
1971	580	1	240	450	40	15	50	250
1972	390	40	90	430	140	60	1400	200
1973	x	x	x	x	x	x	x	280

Example A.3.2 contd.

HOMOGENEITY TEST

Station	n Years	Drainage Area km^2	$Q_{p2.33}$ m^3/s	Q_{p10} m^3/s	$\frac{Q_{p10}}{Q_{p2.33}}$	$Q_{p2.33}$ X6.06 m^3/s	T_{10} Years
J-401	7	2840	363	1150	3.17	2200	8
Ta-402	7	178	36	339	9.41	218	7
Ta-401	7	236	112	575	5.13	680	12
Ta-406	5	3600	617	2140	3.47	3740	20
A-401	7	59	31	275	8.88	188	8
A-402	5	80	43	355	8.25	261	8
B-402	7	11230	708	2820	3.99	4300	17
N-401	9	4800	436	2690	6.16	2640	10
MEAN RATIO $Q_{10}/Q_{2.33}$					= 6.06		

Notes:

(1) $Q_{p2.33}$ and Q_{p10} are taken from visually fitted Gumbel plots for each station.

(2) T_{10} is the return period corresponding with $Q_{p2.33}$ multiplied by the mean $Q_{p10}/Q_{p2.33}$ ratio.

(3) Values of T_{10} plotted against n are within homogeneity test limits for 95% confidence limits. (see USGS Water Supply Paper 1543A).

Example A.3.2. contd.

ESTIMATION OF MEDIAN R ($Qp_T/Qp_{2.33}$ RATIOS)

LOG-GUMBEL DISTRIBUTION

Station	Recurrence Interval T in Years							
	1.1	R	1.5	R	5	R	10	R
J-401	132	.36	224	.62	692	1.91	1148	3.16
Ta-402	5	.14	14	.39	126	3.50	339	9.42
Ta-401	28	.25	58	.52	282	2.52	575	5.13
Ta-406	214	.35	389	.63	1230	1.99	2140	3.46
A-401	4	.13	13	.42	105	3.39	275	8.88
A-402	7	.16	19	.44	141	3.28	355	8.25
B-402	7	.01	81	.11	1514	2.14	2820	3.98
N-401	89	.20	234	.54	1202	2.76	2690	6.16
Median R.		.18		.48		2.64		5.64

Notes:

(1) Q_{PT} values are taken from visually fitted Log-Gumbel plots for each station.
(2) Median R is the average of the 4th and 5th largest ratios.

Example A.3.2 contd.

ESTIMATION OF MEDIAN $Qp_P/Qp_{.50}$ RATIOS

LOG-NORMAL DISTRIBUTION

P-PROBABILITY

STATION	.90			.75			.50			.25			.10		
No.	LogQp	Qp	R	LogQp	Qp	R	LogQp	Qp	R	LogQp	Qp	R	LogQp	Qp	R
J-401	2.06	115	.36	2.27	190	.59	2.50	320	1.0	2.74	550	1.72	2.95	890	2.78
Ta-402	0.58	4	.13	1.02	10	.33	1.50	30	1.0	1.98	95	3.17	2.42	260	8.67
Ta-401	1.25	18	.18	1.60	40	.40	2.00	100	1.0	2.38	240	2.40	2.73	540	5.40
Ta-406	2.28	190	.35	2.50	320	.58	2.74	550	1.0	2.99	980	1.78	3.20	1590	2.89
A-401	0.42	3	.14	0.87	7	.32	1.34	22	1.0	1.84	70	3.19	2.28	190	8.64
A-402	0.72	5	.12	1.14	14	.85	1.60	40	1.0	2.07	120	3.00	2.49	310	7.75
B-402	1.06	11	.04	1.70	50	.20	2.40	250	1.0	3.10	1260	5.04	3.74	5500	22.0
N-401	1.82	66	.18	2.18	150	.41	2.57	370	1.0	2.96	910	2.46	3.32	2100	5.68
R_{50}			.16			.40			1.0			2.73			6.72
LOG R_{50}			.796			.398			0			.436			.828

Notes:

(1) Qp_P values are taken from visually fitted Log Qp plots for each station.
(2) Median R is the average of the 4th and 5th largest ratios.

(a) Log-Gumbel Probability Distribution of $Q_T/Q_{2.33}$

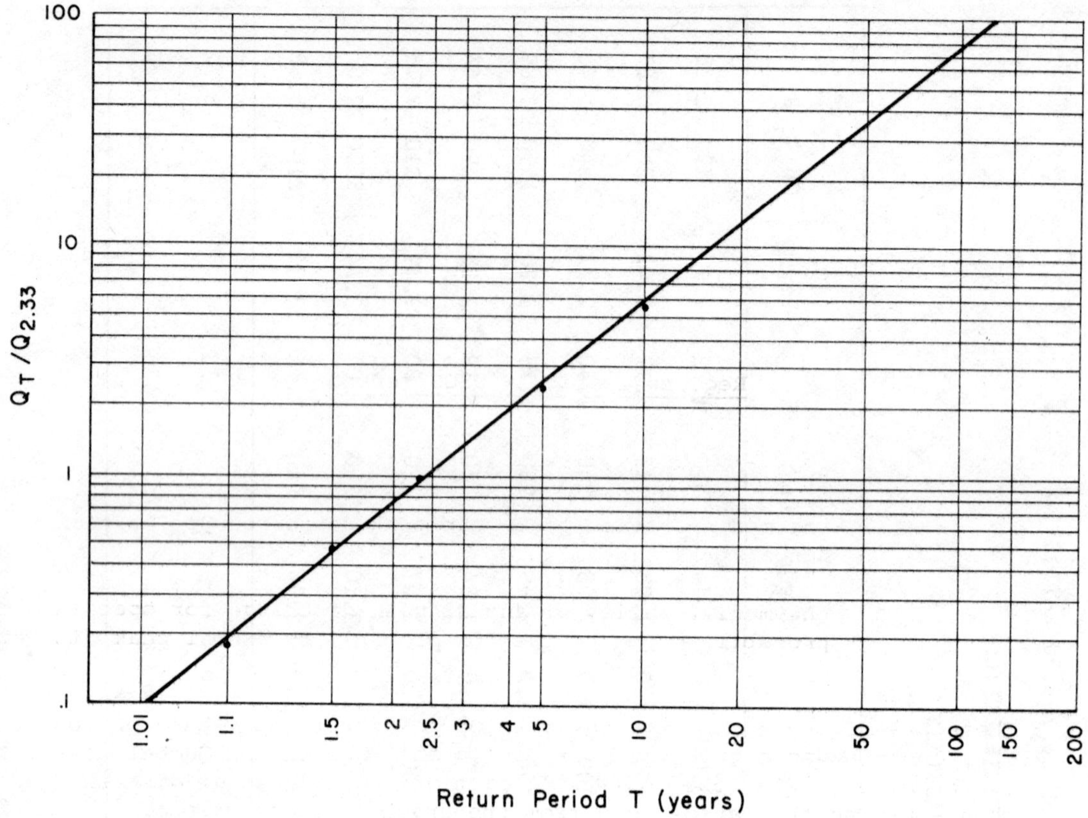

(b) Log-Normal Probability Distribution of $Q_p/Q_{0.5}$

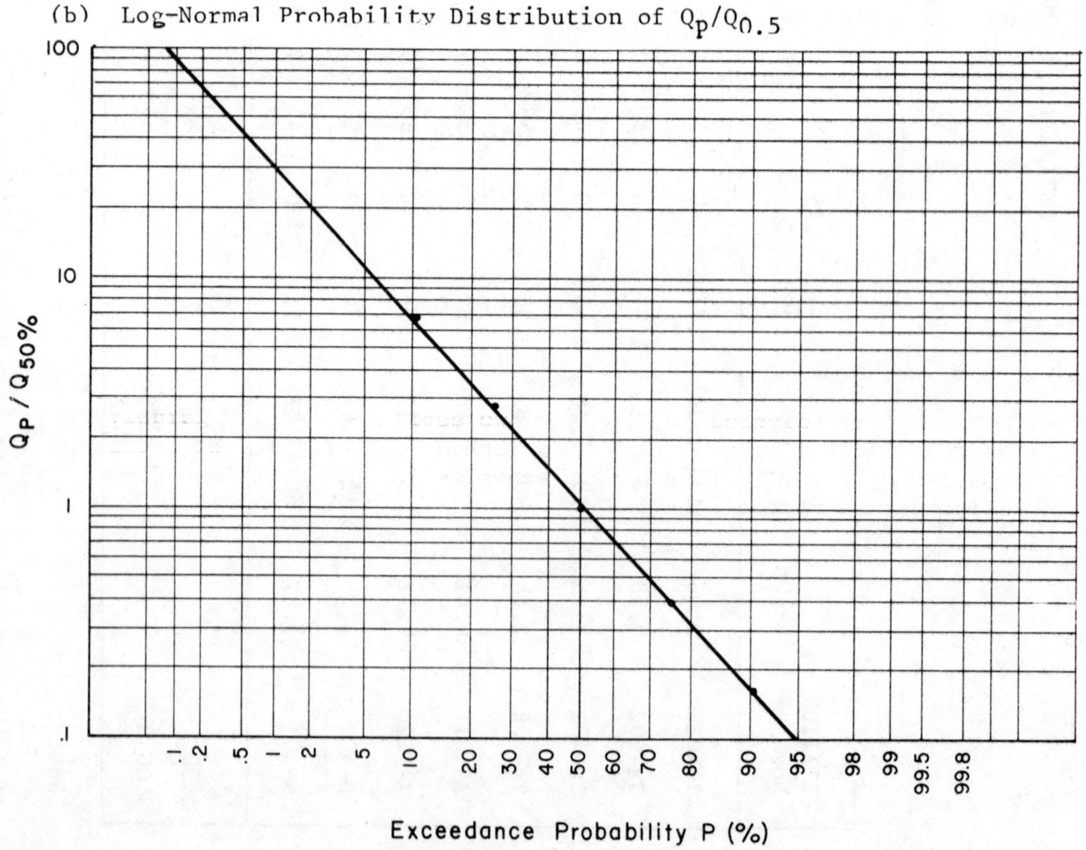

Fig. A.3.1 Regionalisation - USGS method

FIGURE A.3.1

Regionalisation - USGS Method

The figures show, for a number of stations in a region:

- (a) the median ratios of T-year peak discharge Qp_T to mean annual peak discharge $Qp_{2.33}$.

- (b) the median ratios of annual peak discharge for specified probability Qp_p to the 50% probability annual peak discharge $Qp_{0.5}$.

These graphs may be used to estimate Qp for specified return period by first estimating either the mean Qp (in the case of the Gumbel distribution) or the midean (P = 0.5) Qp (in the case of the Log-Normal distribution)and then applying the ratios read from the graph to these values.

For example:

$Qp_{2.33}$ for Najran = 725 m^3/s (Example A.3.1)

Regionalised $Qp_T/Qp_{2.33}$ ratio for T = 100 ≃ 75

Qp_{100} = 725 x 75 = 54 375 - say 54 000 m^3/s

$Qp_{0.5}$ for Najran = 370

Regionalised $Qp_{100}/Qp_{0.5}$ ratio for P = 1% ≃ 30

$Qp_{0.01}$ = 370 x 30 = 11 100 - say 11 000 m^3/s

Note that the divergence of these two estimates results largely from the short period of record. The USGS method is therefore not recommended for small samples. The figures and examples in this Paper are for illustrative purposes only.

Example A.3.3

REGIONAL FLOOD RATIOS
(Station-Year Method)
Annual $Qp/\overline{Q}p$ Series

YEAR	Station No. J-401	Ta-402	Ta-401	Ta-406	A-401	A-402	B-402	N-401
1964	x	x	x	x	x	x	x	3.20
1965	x	x	x	x	x	x	x	1.92
1966	.60	.34	.42	x	0	x	.01	.11
1967	.43	.36	.74	x	.49	x	.37	.75
1968	.33	4.07*	2.98*	2.99*	1.97	1.86	1.41	.32
1969	2.36	1.70	.47	.52	.27	.17	.22	2.67
1970	.85	.31	.08	.35	.22	.50	2.95	.27
1971	.98	.02	1.67	.58	.22	.19	.08	.21
1972	1.45	.68	.63	.56	.93	2.26	1.97	.27
1973	x	x	x	x	3.04	x	x	.27

Mean of $Qp/\overline{Q}p = 0.906$, $\hat{\sigma} = 0.930$

* Floods occurred on the same date and resulted from the same meteorological situation. For the purposes of the analysis the three values are average to produce a single ratio. For the logarithmic transformation the zero value must be ignored.

Correlation coefficients calculated to show independence of stations:

	J-401	Ta-402	Ta-401	Ta-406	A-401	A-402	B-402	N-401
J-401		.082	.401	.609	.296	.345	.087	.781*
Ta-402			.738*	.923*	.856*	.430	.104	.030
Ta-401				.893*	.794	.344	.118	.227
Ta-406					.927*	.502	.022	.214
A-401						.787	.306	.227
A-402							.418	.444
N-401								.313

Statistical tests show the correlation coefficients marked * to be significant at the 5% level.

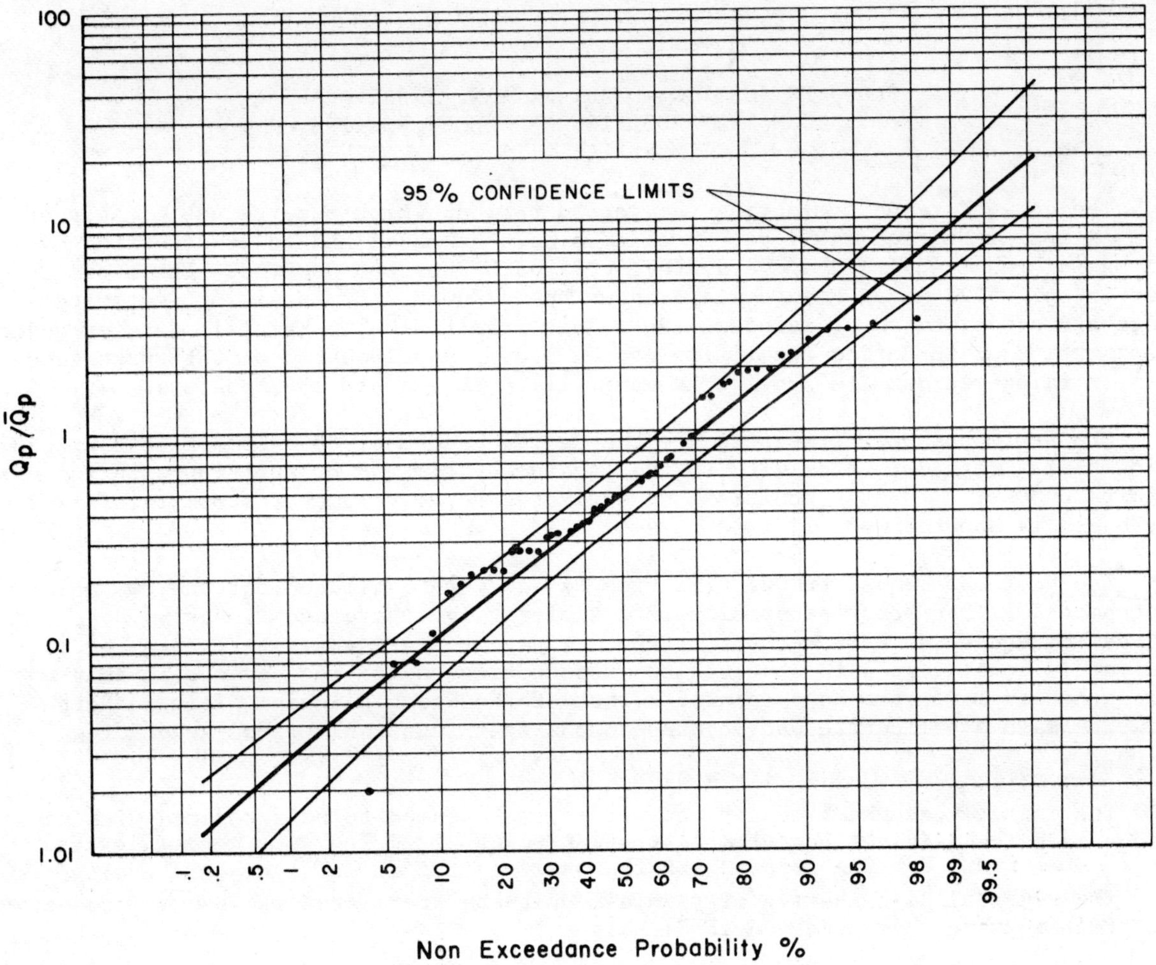

Fig. A.3.2 Regional ratios - station year method - log-normal probability distribution of $Qp/\overline{Qp}$

The figure shows the probability distribution of the ratios of annual peak discharge Qp to mean annual peak discharge $\overline{Qp}$, for 8 runoff stations in Saudi Arabia with records ranging from 5 to 10 years in length. A number of ratios have been omitted as they relate to the same rainfall event; the remainder are assumed to be independent and the station-year approach therefore to be valid.

For the region represented, the T-year flood may be estimated by applying the $Qp/\overline{Qp}$ ratios to the estimated value of $\overline{Qp}$. For example, the Qp_{100} value for Wadi Najran at El Shefa may be estimated from:

$\overline{Qp} \simeq 725\ m^3/s$ (Example A.3.1)

$Qp_{100}/Qp \simeq 9.2$

$Qp_{100} = 725 \times 9.2 = 6\ 670$ - say $7\ 000\ m^3/s$

Example A.3.4

FLOOD ESTIMATION - WADI NAJRAN, SAUDI ARABIA
STORM TRANSPOSITION AND FLOOD ROUTING

The largest rainstorm during the record period, which occurred at El Abis on 2nd March 1970 and might reasonably be expected to have an equal chance of occurring over the Najran basin, was transposed and a synthetic flood hydrograph derived. The storm covered an area of some 5000 km^2. Figure A.3.3 shows the assumed storm position, estimated isohyets and a mass curve of rainfall at El Abis itself. Daily rainfall data at adjacent stations were used to define the limits of the storm, supplemented by interpretation of totalising raingauge data to provide an upper limit to the storm dimensions.

The storm has been transposed to the Najran catchment to assume a most probable position having regard to proximity to the escarpment rim and orientation with respect to moist air movement. In this position the estimated total volume of storm rainfall input to the basin is about 250 Mm^3.

Isochrones of equal travel time from Najran were estimated for the two main sub-catchments. The isochrone spacing was, arbitrarily, increased linearly in a downstream direction to take account of the drainage network characteristics and is approximately consistent with observations and experience for rare events in this area. An allowance was made for the detaining effect of the Sadah alluvial plain in North Yemen. The main storm centre was located at the most favourable flood-generating position.

The initial catchment conditions were first assumed to be saturated with no allowance made for initial loss or infiltration rate.

The temporal distribution of rainfall over the storm area was assumed to occur in direct proportion to that observed in El Abis.

A simple computation was used to route the run-off in hourly volumes to Najran. The resultant discharge is shown in Figure A.3.4 as block diagram of mean hourly discharge and as an inferred instantaneous discharge hydrograph.

The resulting flood peak of 7000 m^3/s is a credible response under the stated conditions.

However, in view of the severity of the assumption of zero initial loss and infiltration, a second estimate allowed for a regional infiltration rate of 5mm/hr as being more probable under virtually any circumstances. The effect of this is to reduce the total run-off volume by approximately 60%. The result is a total run-off volume of about 100 Mm^3 and a flood peak of about 3800 m^3/s.

Figure A.3.3 shows that the mean peak discharge results mainly from the 2-hour fall of 60mm after some 50mm had already fallen. This depth-duration has an estimated return period of 25 years. The estimated 100-year return period 2-hour rainfall for a 200-300mm rainfall zone is 80mm. By incorporating a 2-hour depth of 80mm and retaining the overall storm depth of 154mm by disregarding subsequent storm segments, the resultant flood peak is increased to 5000 m^3/s. This is also shown by Figure A.3.4.

There is scope for large error in the assumptions made above but the result is considered to be a reasonable prediction of the response of a catchment under the stated conditions. It should be pointed out that the figure of 80mm for a 2-hourly rainfall does not necessarily indicate a 100-year flood for the basin. A higher intensity, shorter duration rainfall of the same frequency or lower intensity rainfall covering a larger area, could result in a larger flood. Conversely the ideal run-off producing situation of 50mm rainfall followed by a high intensity storm segment of nearly 80mm must have a much lower joint probability than for the high-intensity segment alone.

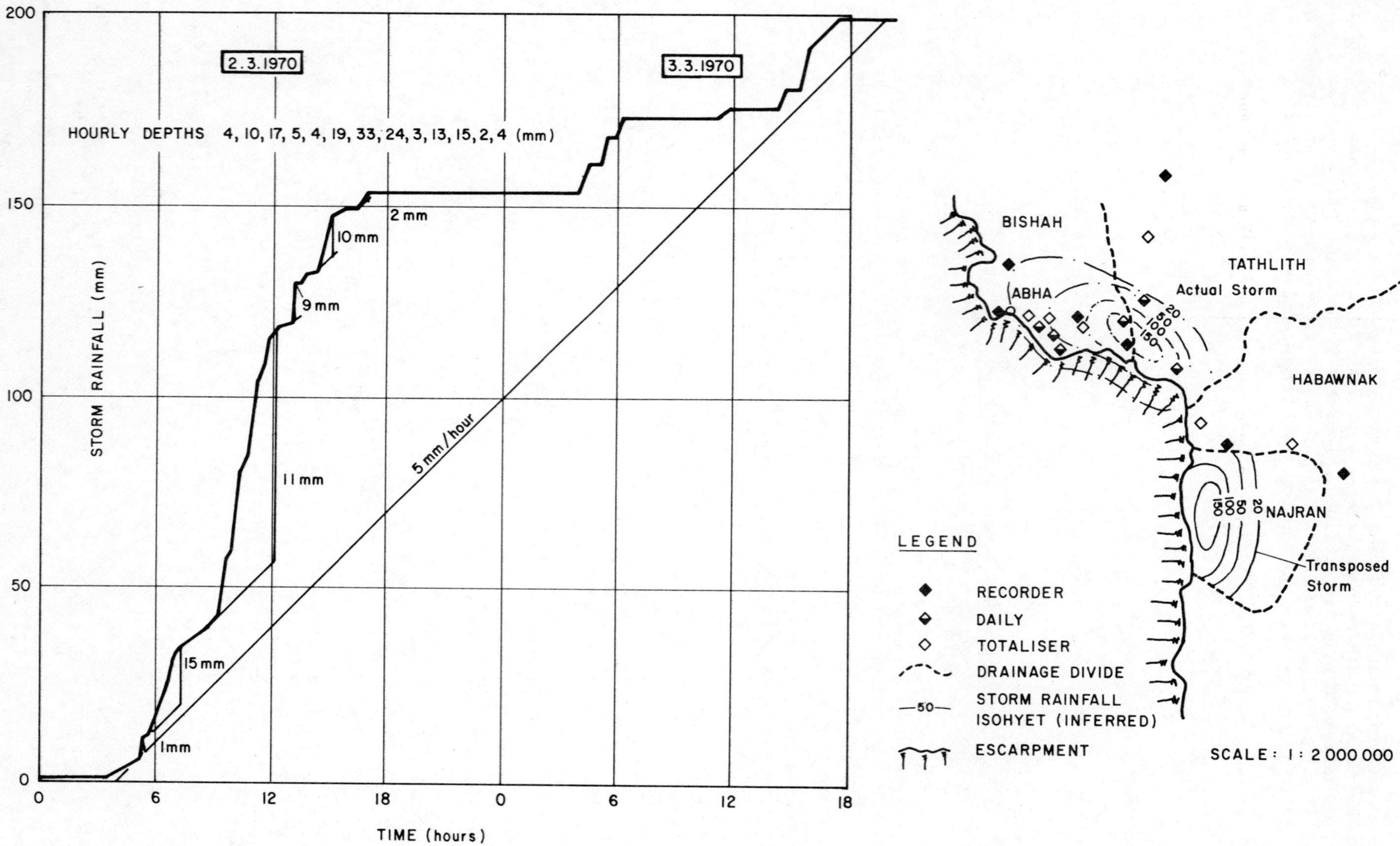

Fig. A.3.3 Storm of 2/3 March 1970 at El Abis

FIGURE A.3.3

Storm of 2/3 March 1970 at El Abis, Saudi Arabia

The figure shows the mass-curve (hyetograph) for a major storm event, the isohyetal pattern of storm rainfall and the location of the storm when transposed to its optimum flood producing location with respect to the study basin.

The hyetograph shows rainfall depth occurring at intensity greater than 5 mm/hour for estimation of the flood hydrograph resulting from an assumption of uniform temporal rainfall distribution and a loss rate of 5 mm/hour.

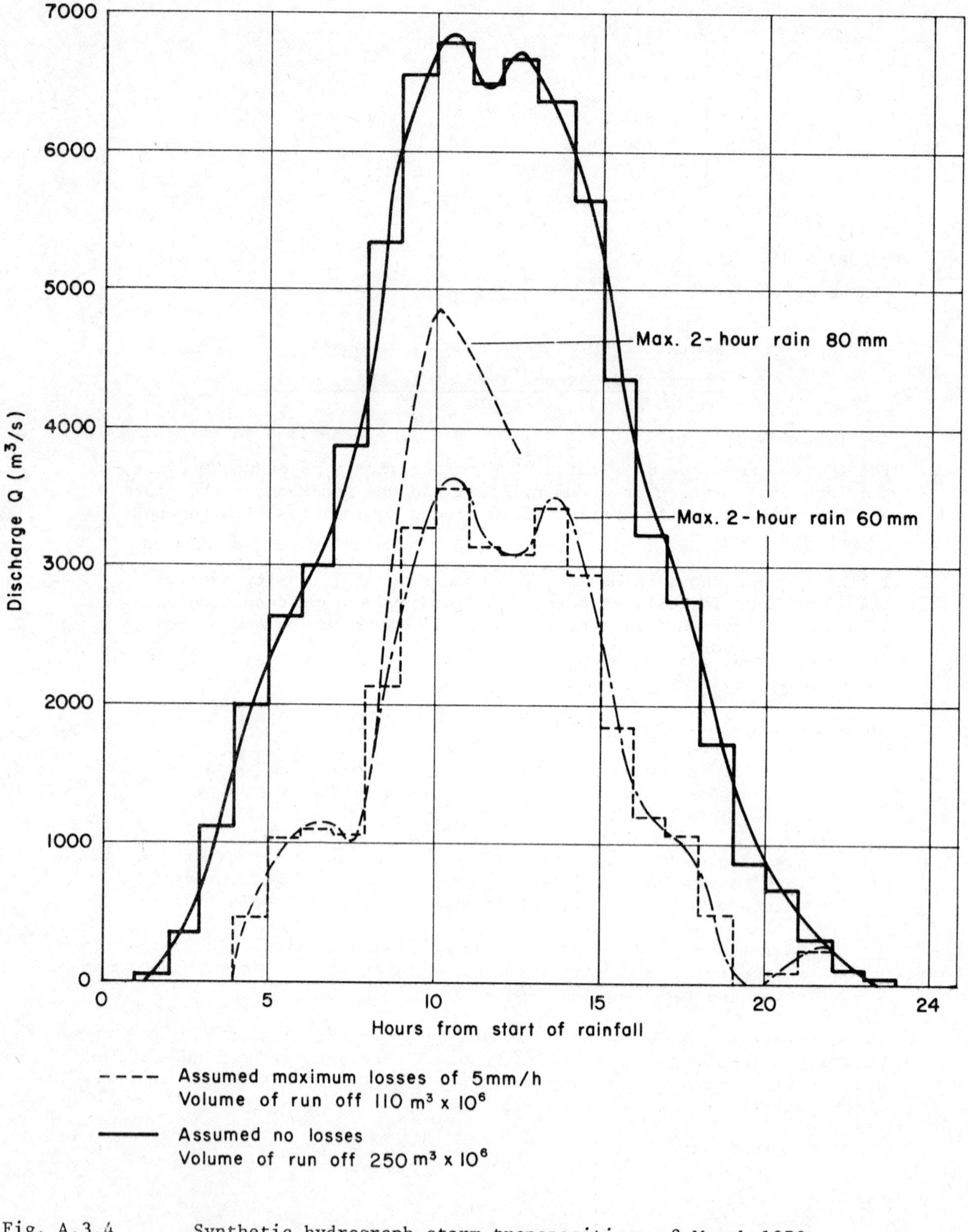

Fig. A.3.4 Synthetic hydrograph storm transposition - 2 March 1970

FIGURE A.3.4

Synthetic Hydrograph Storm Transposition
2 March 1970

The transposed storm rainfall (Figure A.3.3) was assumed to be distributed uniformly in time over the catchment - i.e. the total storm duration was the same areally, with the total storm depth varying according to the isohyetal pattern assumed.

Isochrones of runoff travel time to the basin outlet were estimated. A time-area distribution was calculated and the rainfall depths minus any losses convoluted with the distribution produced the estimated hydrograph response.

The double peak results from the difference in response time of the two major tributaries forming the basin.

Three hydrographs are illustrated:

- Assuming no losses; all rainfall reaches the basin outlet. This is the upper limit of discharge peak and volume under the assumptions made.

- Assuming no <u>initial</u> losses but a continuous loss rate of 5 mm/hour. This possibility produces a lower limit of hydrograph peak and volume since only some 40% of the rainfall runs off.

- Assuming a 2-hour, 100-year return period rainfall depth of uniform intensity to occur after the first 50 mm of the historic storm and retaining the 5 mm/hour loss rate.

It is important to note that no return period can be assigned to the flood peak and volume estimates derived from this analysis. The figure serves only to illustrate a method by which the response to a known or conceivable storm pattern can be estimated.

ANNEX 4

CALCULATION OF SEDIMENT DISCHARGE

Example Computations for Selected Methods
List of Symbols for Variables Used in Bed-load Formulas

q_s	:	sediment discharge per unit width (N/s/m = 0.102 kg/s/m)
Q_s	:	sediment discharge (N/s = 0.102 kg/s)
γ	:	specific weight of water (= 1000 kg/m^3 = 9807 N/m^3)
γs	:	specific weight of sediments (2650 kg/m^3 = 25989 N/m^3)
g	:	acceleration of gravity m/s^2 = 9.81 m/s^2
d_s	:	median diameter of the bed material - d_{50} (m)(= 0.00338m in the examples)
ν	:	kinematic viscosity of water (m^2/s)(= 9.2×10^{-7} m^2/s in the examples)
r_b	:	hydraulic radius representative of the bed (m) (see Section A.4.1.2.2)
S	:	slope of the energy line (= 0.004 in the examples)
y	:	water depth (m)
V	:	mean flow velocity (m/s)
Q	:	flow discharge (m^3/s)
w	:	fall velocity of sediment particles of diameter d_s (m/s)
W	:	river bed width m (= 55m in the examples).

Values of r_b (≃ r'_b) are found in Table 1. of Example A.4.1.
Value of V are found in Table 3, of Example A.4.1.
Values of W are found in Figure A.4.8a.

Example A.4.1 The Einstein Method

The following presentation of the basic principles of the theory tries to avoid undue mathematical detail. At each step, the numerical calculations are applied in the example. The following basic data are required:

i) Topographic survey of the test reach which must be approximated by a trapezoidal, constant-slope channel for the assumption of uniform, steady flow conditions implicit in the Einstein method.

ii) Grain-size analysis of the bed material.

iii) The range of possible water level and the water level - discharge relationship. If the latter is not available an estimate of this relationship forms part of the procedure, as in the example.

iv) Water temperature.

A.4.1.1 Basic Data

1. The Test Reach

The Einstein method is strictly applicable only to uniform and steady-state hydraulic conditions. In practice such conditions do not exist but the "test reach" must be carefully chosen to meet the requirement of uniformity as closely as possible.

The test reach should ideally consist of a straight, regular-shaped channel some hundreds of metres in length with no marked changes in the longitudinal slope. The topographical survey is required to produce a longitudinal profile of the bed (Figure A.4.1) and 5 to 10 cross-sections (Figure A.4.2). An idealised "axis" (mean bed level of constant slope) is drawn to approximate the longitudinal profile. All cross-sections are superimposed on tracing paper and a single symmetrical, trapezoidal cross-section is inferred from the group of cross-sections. In the example the inferred cross-section has first been chosen with a transverse sloping bed and with equal side slopes. The bed has then been flattened to produce a symmetrical trapezoidal shape.

The idealised test reach of the example has the following characteristics:

Base width: $b = 55\text{m}$

Side slopes: $s = 0.65$

Longitudinal slope: $S = 0.004$

Manning - Strickler roughness coefficients for estimating the water level - discharge relationship, are estimated as:

i) Sandy bed: $K_b = 36\ \text{m/s}^{1/3}$ (or $n_b = \dfrac{1}{K_b} = 0.028$)

ii) Rocky walls: $K_w = 20\ \text{m/s}^{1/3}$ (or $n_w = \dfrac{1}{K_w} = 0.050$)

2. Bed Grain-Size Analysis

Einstein considered that each grain-size fraction behaves independently, following its own rules of erosion, transportation and deposition and has no influence on (nor is it influenced by) the dynamic behaviour of other grain-size fractions in suspension.

From bed-material samples and sieve analysis, a diagram is derived of grain-size distribution represented by equal size samples taken from the entire bed from the surface to the maximum scour depth (Figure A.4.3). When the granulometry is uneven across the bed or with depth, more samples should be taken in the centre of the bed than on the sides in order that the sample should reflect the fact that sediment concentration is denser in the centre of flow where velocities and turbulence are greater. The resulting grain-size curve is the average of all samples. Figure A.4.3 illustrates the result of this procedure. Four to ten grain-size classes are then chosen. For Example A.4.1, five classes (I to V) have been selected and the class limits, representative diameters d_i and weight percentage p_i are given on Figure A.4.3.

The granulometry of the example test reach ranges from sand to gravel. Silt and clay are practically non-existent in the wadi bed. This preponderance of relatively coarse material is generally typical of arid zone channel beds although flood plains flanking the main channel may consist almost entirely of Recent alluvial silts.

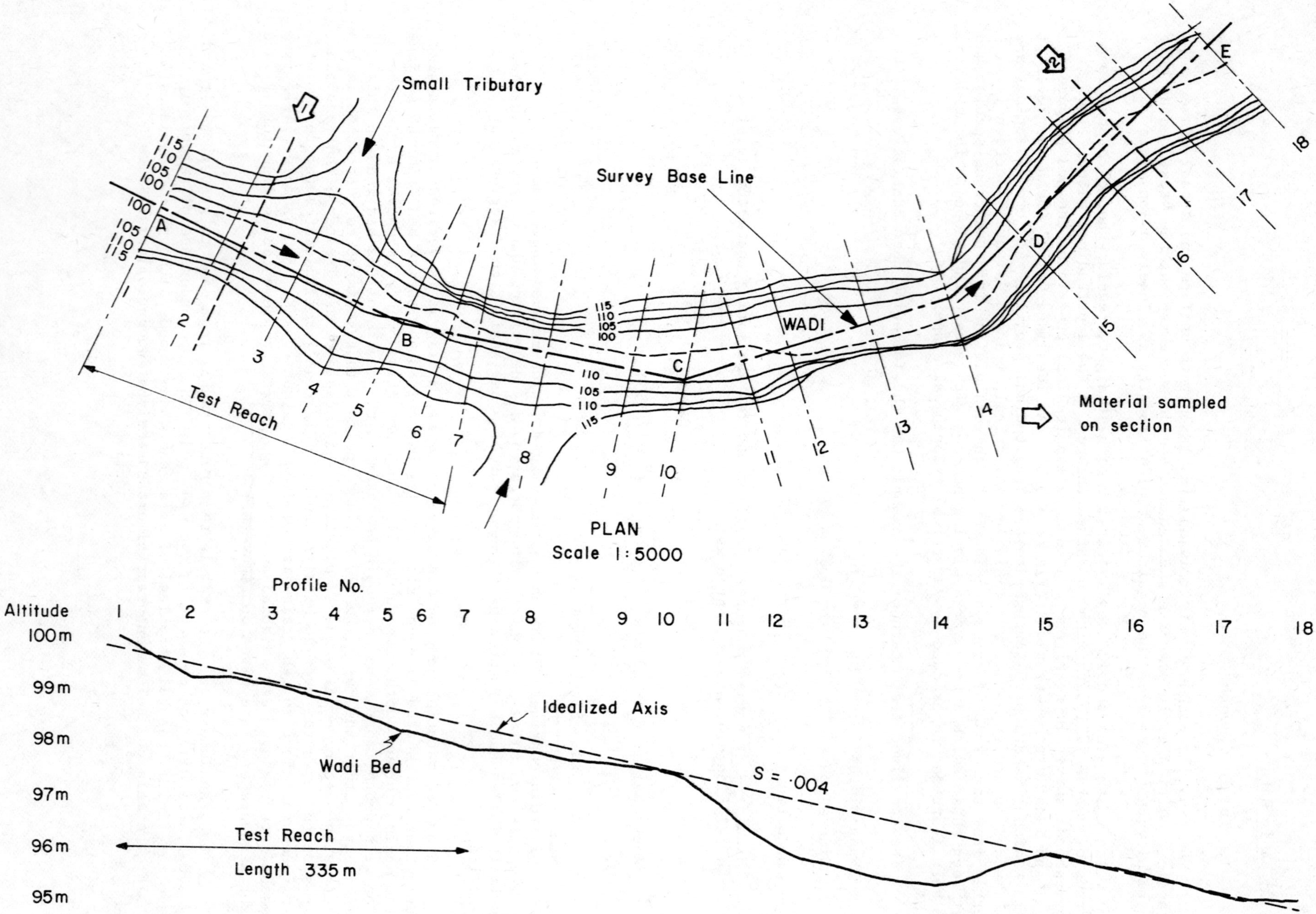

Fig. A.4.1 The test reach

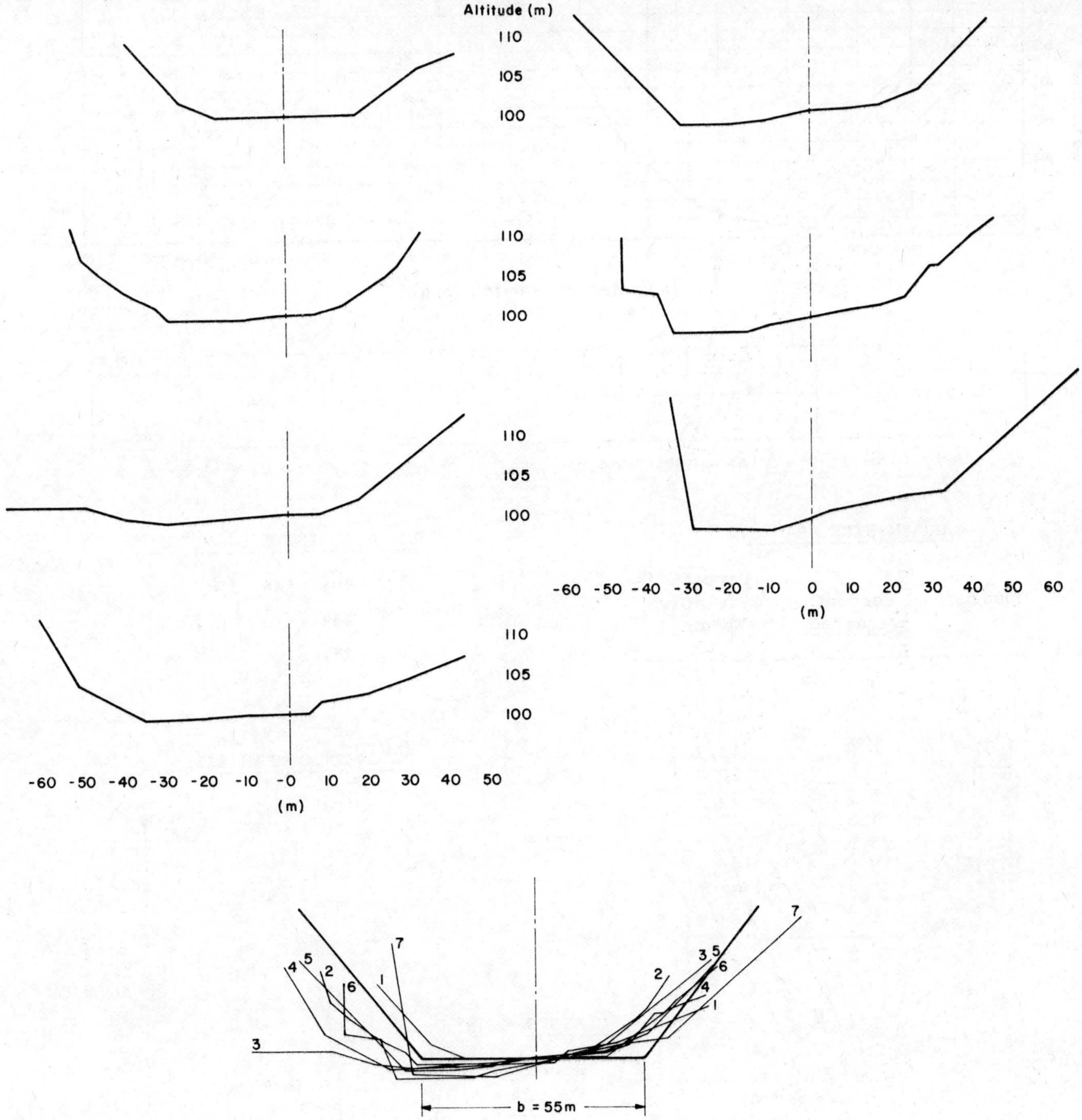

Fig. A.4.2 The test reach - cross sections and equivalent section

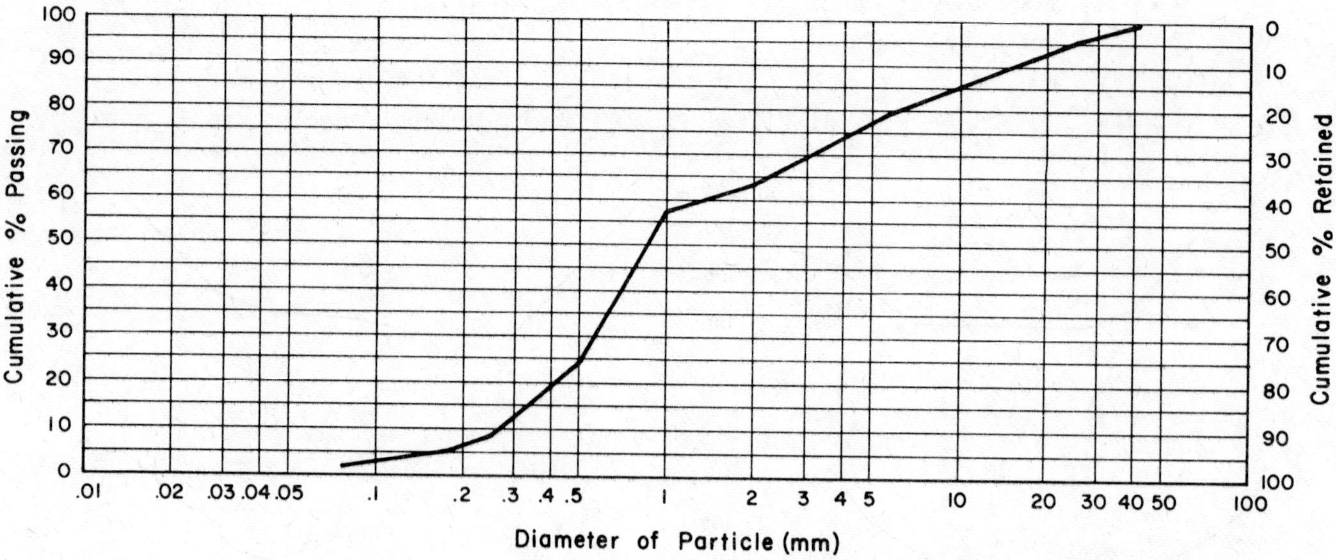

<table>
<tr><td rowspan="2">Clay to Silt</td><td>VF</td><td>F</td><td>M</td><td>C</td><td>VC</td><td rowspan="2">Gravel</td></tr>
<tr><td colspan="5">Sand</td></tr>
</table>

VF - Very Fine F - Fine M - Medium C - Coarse VC - Very Coarse

GRAIN-SIZE ANALYSIS

Opening mm	Percent retained	Percent Cumulative Weight %
38.1	1.4	1.4
25.4	2.5	3.9
19.0	3.6	7.5
9.51	6.8	14.3
5.00	7.7	22.0
2.00	8.7	30.7
1.00	11.5	42.2
0.50	33.5	75.7
0.246	15.5	91.2
0.074	2.3	98.1
<0.074	1.9	100.0
	100.0	

CHARACTERISTIC DIAMETERS

90% size	15.0 mm
65% size	2.2 mm
35% size	0.62mm

GEOMETRIC MEAN SIZE

$$d_s = \frac{\Sigma d_i P_i}{\Sigma P_i} = 3.38 \text{ mm} \simeq d_{67}$$

CLASSES

No.	Limits of Classes (ø mm)	Representative diameter d_i (mm)	Weight percentage Pi
I	0.00 - 0.40	0.25	19
II	0.40 - 1.20	0.65	36
III	1.20 - 3.10	2.00	16
IV	3.10 - 9.00	5.00	14
V	9.00	15.00	15
		Total	100 %

Fig. A.4.3 Diagram of grain size analysis

3. Water Levels

Sediment discharges are calculated for the range of water levels, the interval increasing with depth, in the standardized test-reach. At least five different water levels are recommended.

In the example case, high flood marks were detected up to almost 9m above the wadi bed. The levels chosen were: 0.20, 1.00, 3.00, 6.00 and 10.00m above the bed.

4. Water Temperature

Water temperature $T^{o}C$ is required for determination of kinematic viscosity for calculation of the sub-laminar layer thickness δ . Values of ν for a range of T are given in Figure A.4.4 for fresh water. Water temperatures of the wadi flows in the example case were about $24^{o}C$. From Figure A.4.4 the kinematic viscosity is about = $9.2 \cdot 10^{-7}\ m^2/s$.

A.4.1.2 Preliminary Geometric and Hydraulic Calculations

1. The "Side-Wall" Correction

In the Einstein method the idealised channel section is represented by an equivalent width of a theoretical, infinitely wide rectangular channel of similar bed material and hydraulic characteristics. The method is then applied to a unit width (1m) of this "equivalent" channel. The basic assumptions for this transformation are that:

i) One part of the flow cross-section is attributed to the walls of the channel (Q_w) and another part to the bed (Q_b), the total of $Q_b + Q_w$ being equal to the total discharge Q.

ii) The average velocity $\bar{V}$ is equal in each part of the sub-divided cross-sectional area.

It is observed in practice that the sediment concentration is distributed unevenly across the flow cross-section. The sediments are concentrated mainly in the centroid of flow. The edges, where velocities are low, are almost sediment-free. Hence, the concept that the sediment erosion - transport - deposition process takes place only in the partial cross-section A_b (Figure A.4.5) under the forces exerted by Q_b. In the same manner the hydraulic radius is split into r_w and r_b. The latter value is the representative "depth" of flow in the equivalent channel.

A number of analytical methods exist to calculate the geometric and hydraulic variables associated with the bed, but the recommended method is a graphical construction which approximates the separation between the two components of the flow (Qb and Qw) based on orthogonal lines across the isovels (curves of equal velocities in the flow cross-section). The method of construction of these lines is shown on Figure A.4.5 together with its numerical application to the example.

In arid zones, alluvial sand channels may increase in width and decrease in depth with distance downstream. In such cases, the side-wall correction may be omitted from the calculation, r_b being calculated as A/W without significant loss of accuracy.

2. Decomposition of Hydraulic Radius and Cross-sectional Area

Einstein and Barbarossa proposed that the cross-sectional area A_b and the hydraulic radius r_b of the side-wall-corrected channel should each be regarded as consisting of two independent additive elements:

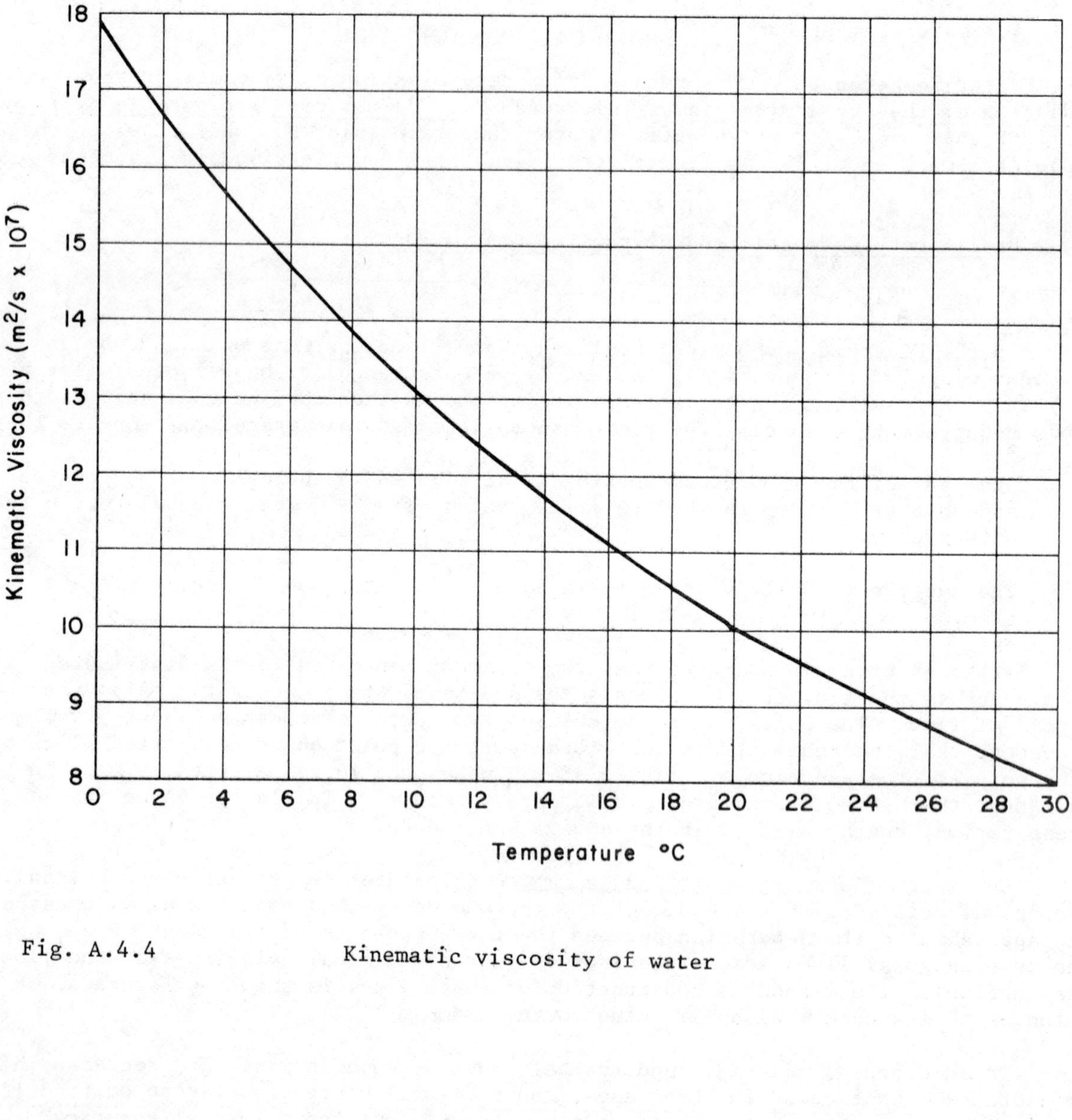

Fig. A.4.4 Kinematic viscosity of water

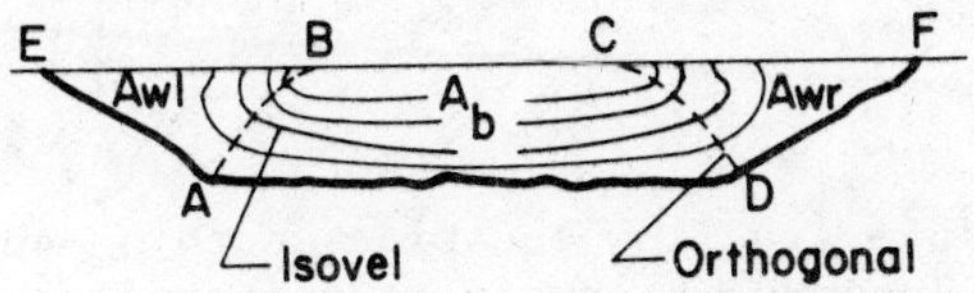

Isovel Diagram with Orthogonals

GRAPHICAL CONSTRUCTION

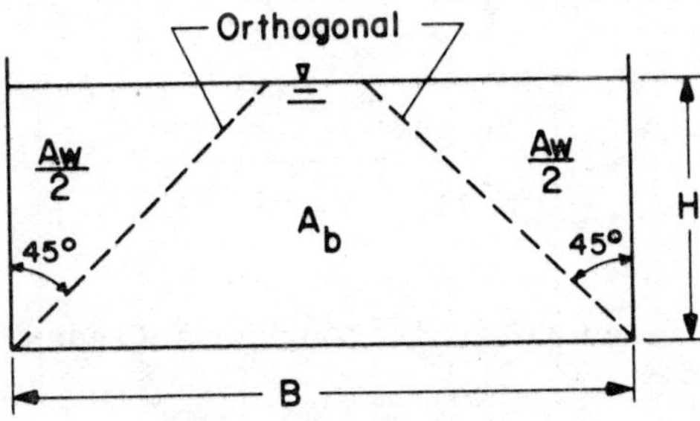

Rectangular Channel

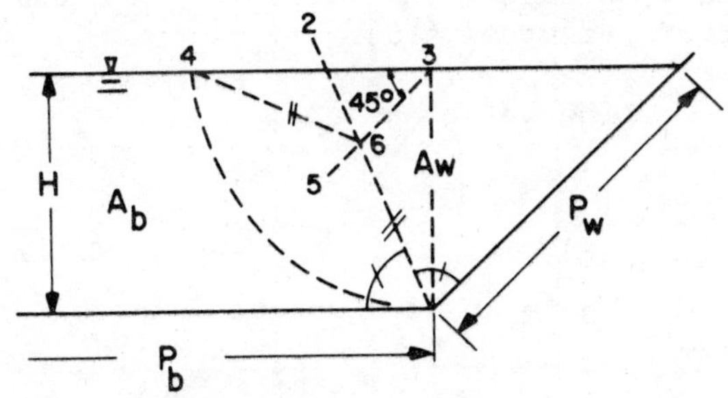

Trapezoidal Channel

Example: Right Side of the Standardized Channel

Scale 1:200

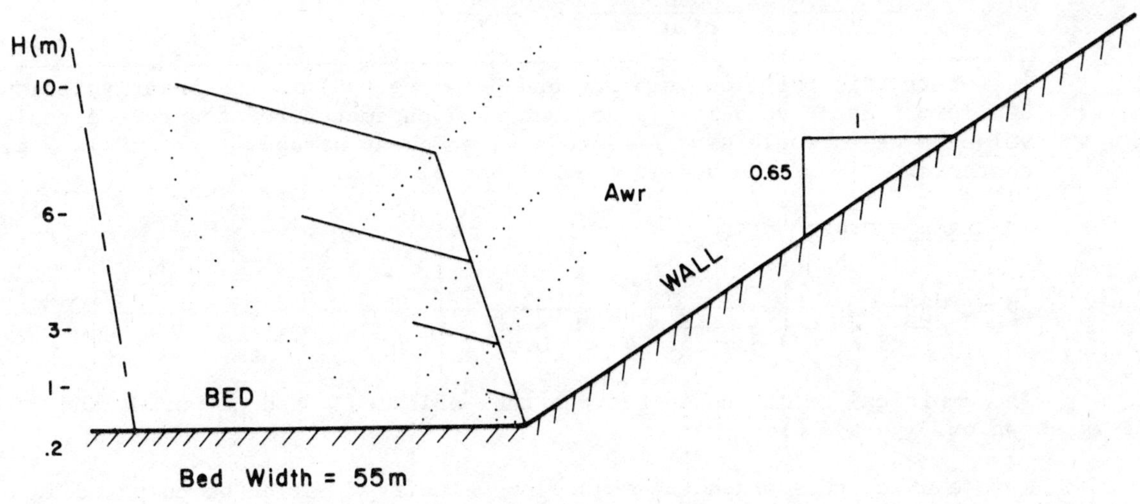

CALCULATION OF r_b

H	A	$A_w = 2A_{wr}$	A_b	$r_b = \frac{A_b}{\text{bedwidth}}$
m	m^2	m^2	m^2	m
0.20	11.0	0.0	11.1	0.20
1.00	56.5	1.0	54.5	0.99
3.00	178.8	9.0	160.8	2.92
6.00	385.4	36.0	313.4	5.70
10.00	703.8	100.1	503.6	9.16

Fig. A.4.5 The side wall correction

$A_b = A'_b + A''_b$ and

$r_b = r'_b + r''_b$

A_b' and r_b' are responsible (in flow equations such a Manning, Nikuradse etc.) for that part of the total head loss due to grain roughness corresponding with a fixed bed roughness, and A''_b and r''_b are responsible for the head loss due to bed forms (dunes, antidunes, ripples etc.).

Bed forms are reciprocally dependent on hydraulic conditions in mobile-bed streams. To resolve this dilemma, Einstein plotted data from experimental flumes and rivers to relate two variables in a unique relationship he called the "bar resistance curve" (Figure A.4.6).

The variables are:

1) $$\mu = \frac{\gamma s - \gamma}{\gamma} \cdot \frac{d_{35}}{r'_b . S}$$

Where μ is a unique function of variables related to grain roughness (index prime').

In Example A.4.1

$$d_{35} = .00062\text{m} \qquad \text{(Figure A.4.3)}$$

$$\mu = \frac{2.65 - 1.00}{1.00} \times \frac{.00062}{r'_b . 0.004} = \frac{.256}{r'_b}$$

2) $\frac{V}{U''_*}$ (where V is the flow velocity and $U''_* = \sqrt{g . r''_b S}$) is the shear velocity due to bed form. Shear velocity is defined as a parameter related to the real fluid velocity which would give rise to a given shear stress τ_o. r''_b is a geometric characteristic of the bed-form component of flow.

In Example A.4.1

$$r''_b = \frac{U''^2_*}{gS} = \frac{U''^2_*}{9.81 \times 0.004} = \frac{U''^2_*}{0.0392}$$

The empirical relationship between the variables μ and $\frac{V}{U''_*}$ or μ and $\frac{U''_*}{V}$ is shown on Figure A.4.6.

Before using this graph the mean flow velocity V has to be computed in order to calculate U''_*.

If the stage-discharge relationship (H/Q) for the test-reach is known, as in the case of a rated gauging station within the reach, the value of V may be obtained from this and the stage-area relationship (H/A).

To estimate V, use may be made of the Manning-Strickler formula for composite beds with weighted friction factor

$$K = \frac{P^{2/3}}{\left[\frac{P_w}{K_w^{1.5}} + \frac{P_b}{K_b^{1.5}} \right]^{2/3}}$$

Where P is wetted perimeter (see Figure A.4.5).

and $V = K r^{2/3} S^{1/2}$

For Example A.4.1

Given $K_w = 20$ m/s$^{1/3}$, $K_b = 36$ m/s$^{1/3}$, S = 0.004

H (m)	P_b m	P_w m	P m	K m/s$^{1/3}$	A m^2	r m	V m/s	Q m^3/s
0.20	55	0.62	55.62	35.6	11.1	0.20	0.77	8.6
1.00	55	3.08	58.08	34.3	56.5	0.97	2.13	120
3.00	55	9.23	64.23	31.8	178.8	2.78	3.98	712
6.00	55	18.46	73.46	29.4	385.4	5.25	5.62	2170
10.00	55	30.77	85.77	27.4	703.8	8.21	7.05	4960

Several values of r'_b are then assumed. They are chosen some 10 to 20% smaller than the bed hydraulic radius r_b (see Figure A.4.5).

r'_b m	$\mu = \frac{0.256}{r'_b}$	$\frac{U''_*}{V}$ (Fig A.4.6)	V m/s	U''_* m/s	r''_b m
0.15	1.71	0.044	0.77	0.034	0.029
0.90	0.28	0.004	2.13	0.0085	0.002
2.50	0.10	0	3.98	0	0
5.00	0.05	0	5.62	0	0
8.50	0.03	0	7.05	0	0

In the example, the calculated bed form resistance is negligible or zero at each stage implying a regular bed without dunes or ripples. Therefore, in the following calculation r'_b will be taken as equal to r_b.

r''_b would become more important when:

i) d_{35} is coarser.
ii) The bed is deeper and narrower.
iii) The slope is smaller.
iv) The selected roughness parameter K is higher.

These three items produce lower velocity, when significant bed forms may develop.

For example;

given $d_{35} = 0.0012$, S = .001, $= \frac{2.65 - 1.00}{1.00} \times \frac{0.002}{r'_b \times 0.001} = \frac{2.0}{r'_b}$, K = 50

and doubling the hydraulic radius r, results in:

r'_b m	$\mu = \frac{2.0}{r'_b}$	$\frac{U''_*}{V}$ (Fig A.4.6)	V m/s	U''_* m/s	r''_b m	r_b m
0.15	13.33	0.11	0.86	0.095	0.92	1.07
0.90	2.22	0.05	2.46	0.123	1.54	2.44
2.50	0.80	0.02	4.96	0.099	1.00	3.50
5.00	0.40	0.008	7.58	0.061	0.38	5.38
8.50	0.24	0.002	10.22	0.020	0.04	8.54

This shows that bed form roughness produces the greater part of the total head loss at lower velocities.

It should be noted that, for the purposes of Example A.4.1, V has been estimated assuming bed-form roughness to be constant over the range of flow. This is not the case, as has been demonstrated by Simons and Richardson (1966) and from much field data. In practice, variations in bed form roughness have a greater influence on estimates of discharge in wide, shallow alluvial channels than in deep narrow channels.

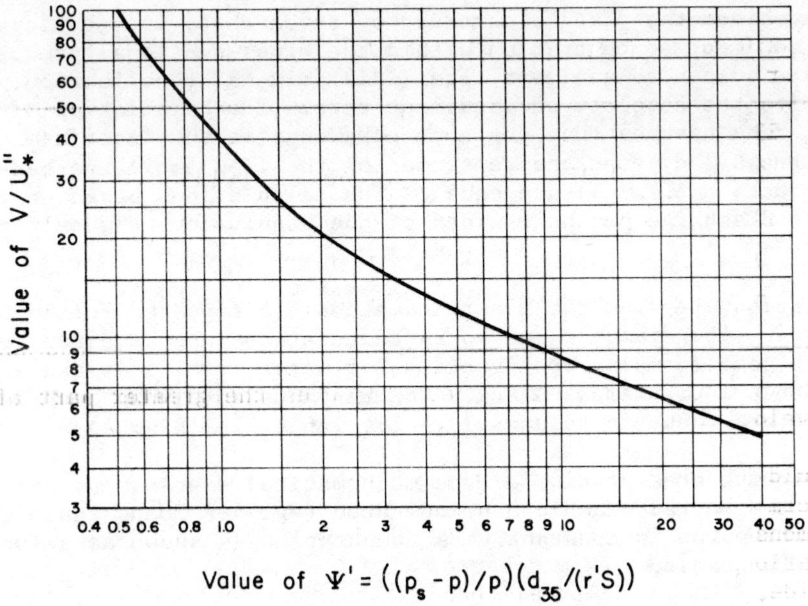

(After Einstein-Barbarossa (Fig. 3) Bar resistance graph)

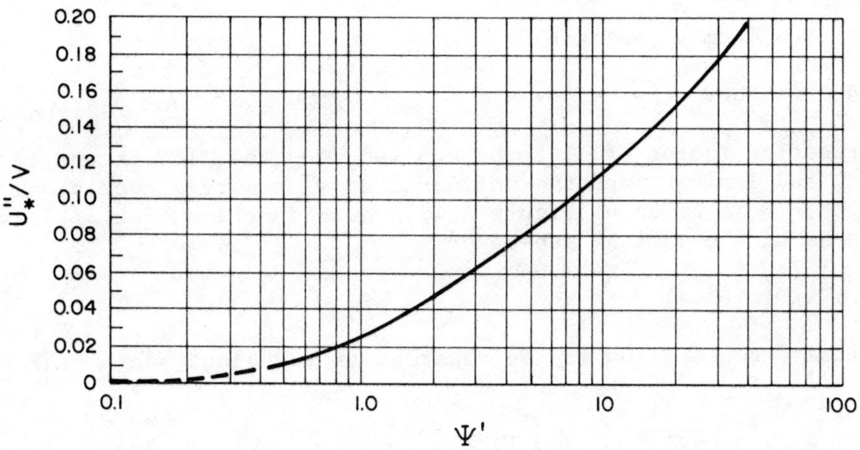

Fig. A.4.6 Bar resistance curve

A.4.1.3 The Einstein Concepts

Einstein combined the results obtained by researchers into the distributions of flow velocity and sediment concentration within the cross-sectional area. From these he derived the relative suspended sediment rate. His original contribution, which forms the next step, deals with the absolute bed-sediment rate. The calculation of the absolute suspended sediment rate assumes that the concentrations at the interface between the bed-layer (of thickness 2 d_i) and the remainder of the flow above the bed-layer up to the free surface (see Section A.4.1.3) are equal. The calculations based on these assumptions are with respect to discharge per unit width of the "equivalent channel" as defined in A.4.1.2.

The calculation steps of the Einstein method are repeated for each combination of water level and grain-size diameter and so 25 calculations are required for the entire procedure. However some formulas are functions of water level only and are grouped in columns 1 to 11, Table 1 of Example A.4.1. Calculations involving both water level and grain-size diameter are shown in columns 1 to 16, Table 2 of Example A.4.1.

In the following presentation of the mathematical expressions, the table and column numbers for the worked example are indicated (eg: 1-7 means Table 1, column 7). Where part of an expression is constant, its computation is shown as a foot-note to the relevant column of the table.

1. Distribution of Flow Velocities

The distribution of velocities in the vertical is given by the following relationships due to Von Karman and Keulegan:

$$V_y = U'_* \cdot \frac{2.30}{\kappa} \log \left(30.2 \frac{y\alpha}{d_{65}}\right)$$

$U'_* = g\, r'_b\, S$ (shear velocity) (1 - 3 = Table 1, column 3)

κ = 0.4 (Von Karman constant)

y = ordinate above the bed

α = correction factor ($\neq 1$) to be applied when the flow is not fully turbulent (transition and laminar) and the influence of viscosity ν conditions the velocity distribution. α is given as Figure A.4.7a as a function of d_{65}/δ', where δ' is the thickness of the laminar sub-layer.

$\delta = 11.6\ \nu/U'_*$ (1-4)

For practical purposes flow may be regarded as turbulent and α taken as unity.

$$\text{then } V_y = 5.75 \sqrt{g\, r'_b\, S} \cdot \log \left[30.2 \frac{y}{d_{65}}\right]$$

For Example A.4.1 the velocity distribution for a water level of 3m, is illustrated on Figure A.4.7b. The calculations are set out below.

Constant term for the example channel:

$$5.75\sqrt{g\, r_b\, S} = 5.75 \sqrt{9.81 \cdot 2.92 \cdot 0.004} = 1.946$$

d_{65} = 0.0022 m, $\alpha = 1$

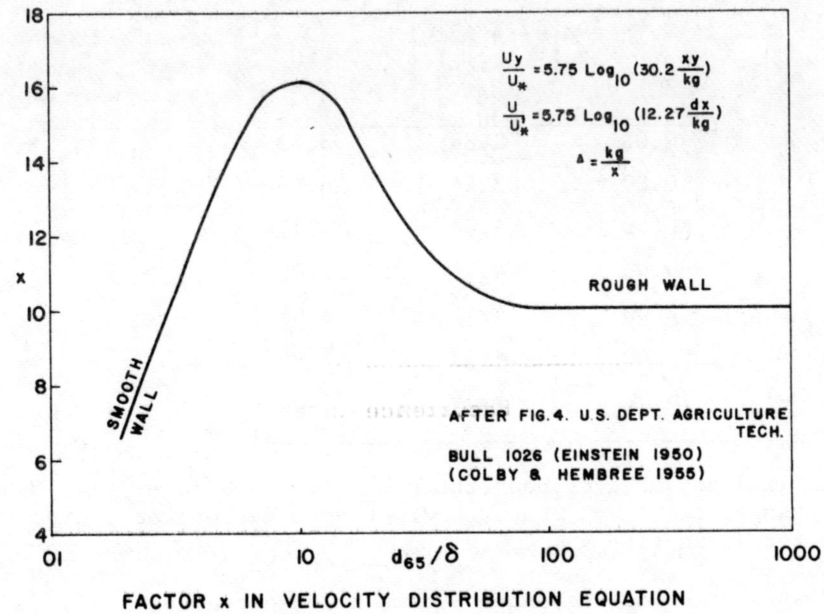

(a) Coefficient a = $f(d_{65}/8)$ distribution

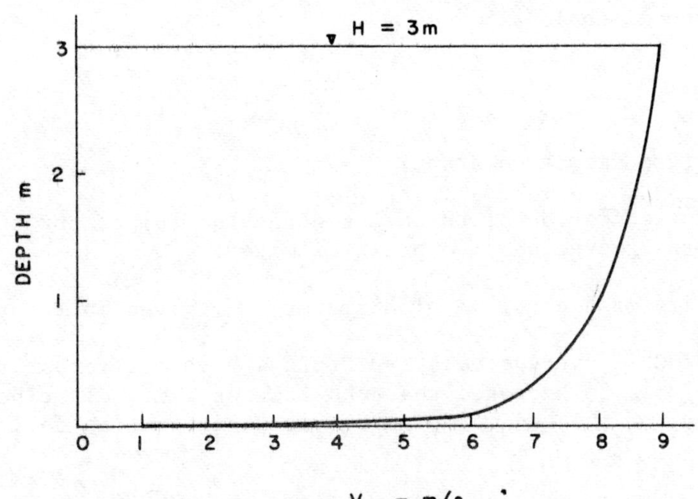

(b) Velocity distribution for H = 3 m

Fig. A.4.7 Einstein method

y m	$\log\left[30.2 \frac{y\alpha}{d_{65}}\right]$	V_y m/s
0.05	2.84	5.52
0.20	3.44	6.69
0.50	3.84	7.47
1.00	4.14	8.05
2.00	4.44	8.64
3.00	4.61	8.98
10.00	5.14	10.00

Note: In the laminar sublayer near the bed $\delta' =$ 0.0314 mm (1-4) the velocity distribution changes to $V_y = 0$ at $y = 0$. (The equation would give $V_y = -\infty$ at $y = 0$). This is discussed in Section A.4.1.3.4.

2. Distribution of Suspended Sediment Concentration

The sediment concentration C_y at any point of the flow cross-section was derived by Vanoni on the basis of turbulence theories.

$$C_y = C_a \left[\frac{2d_{65}\ (H - y)}{y\ (H - 2d_{65})}\right]^z$$

where $z = \frac{w}{\kappa U_*'}$ (non-dimensional)

$U_*' = \sqrt{g\ r_b\ S}$ (1-3)

κ = 0.4 (Von Karman constant)

C_a is the concentration of the bed load at the top of the bed layer of thickness $= 2d_i$ where di is the representative grain diameter.

w is the fall velocity of a particle of diameter d_i, given in Figure A.4.8a.

The graph of C_y/C_a versus relative depth y/H for different values of z is shown in Figure A.4.8b. It is to be noted that the concentration distribution is different for each particle size d_i of the bed material due to variations in fall velocity w with particle diameter.

3. Suspended Sediment Rate

Velocity and concentration distributions are therefore defined in the flow cross-section from the top of the bed layer (y = 0) of thickness $2d_i$ upwards to the water level (y = H). The suspended sediment rate qs_s (s_s = sediment; suspended) is then the integration of $V_y.C_y$ for each grain size d_i over the flow cross-section: the only unknown that then remains is the reference concentration C_a which is then derived from the Einstein bed-load formula (Section A.4.1.3.5).

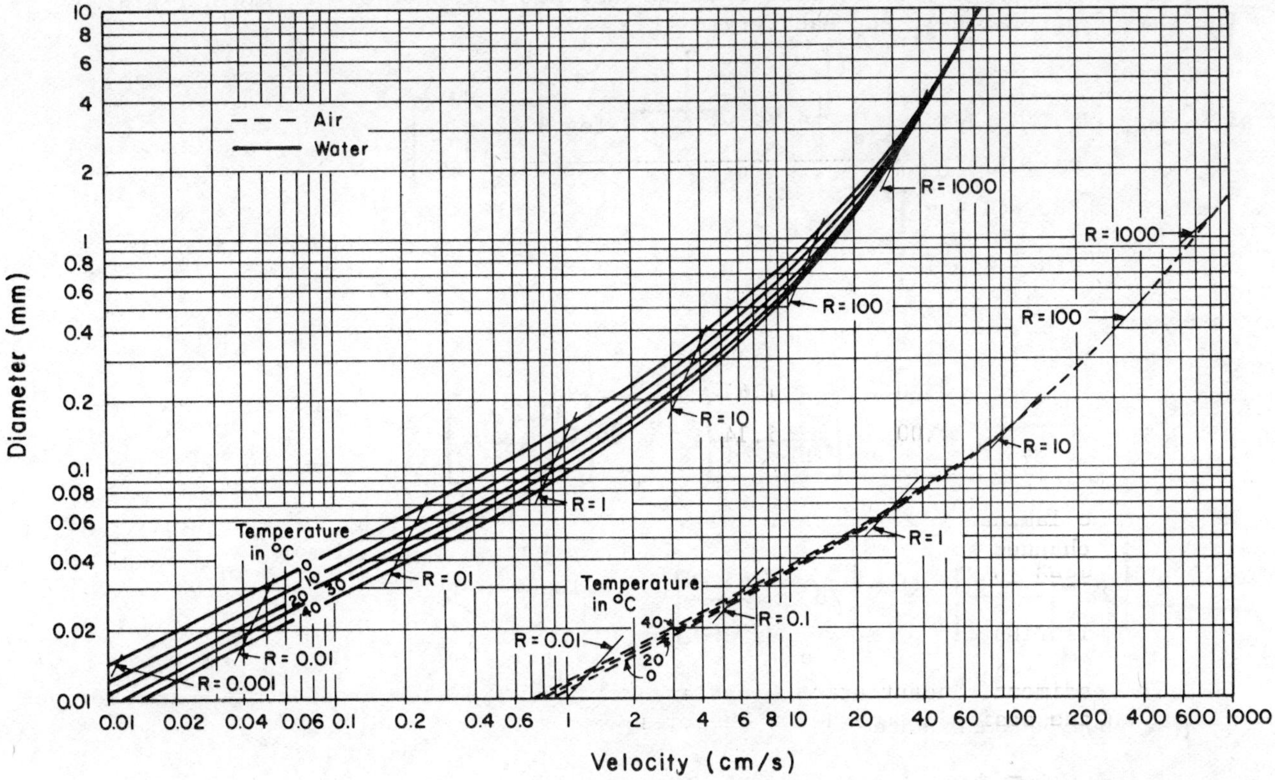

(a)

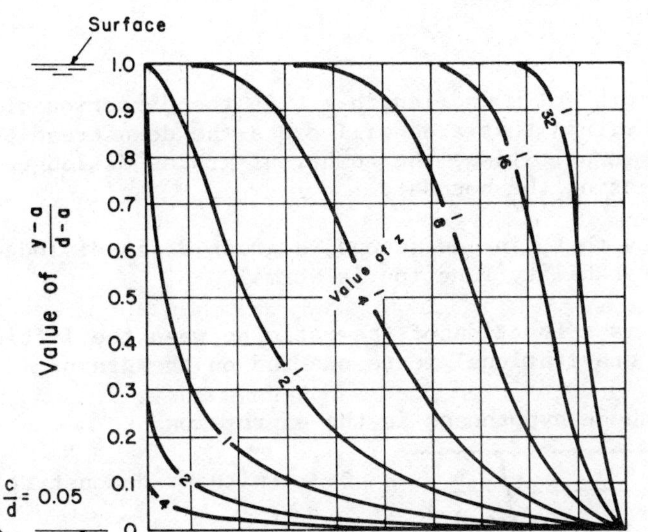

(b)

Fig. A.4.8 (a) Fall velocity of quartz spheres in air and water (Rouse, =937b)
(b) Graph of Rouse suspended load distribution equation for a/d = 0.05 and several values of z

The total suspended sediment rate qs_s defined per unit of bed width (kg/s/m) is the sum of partial qs_{si} for each grain-size d_i.

$$qs_{si} = \int_{2d_i}^{H} V_y \cdot C_y \, dy = \int_{2d_i}^{H} \left[5.75 \sqrt{g.r'_b \, S}.\log 30.2 \frac{y\alpha}{d_{65}} \right]$$

$$\times \; C_a \left[\frac{2d_{65} \; (H - y)}{y(H - 2d_{65})} \right]^z . \; dy$$

which becomes:

$$qs_{si} = 5.75 \sqrt{g \, r'_b \, S} \; . \; H \; . \; C_a \; . \left[\frac{2d_i}{H - 2d_i} \right]^z . \; \log \left[\frac{30.2 \; r'_b}{d_{65}} \right]$$

$$\times \int_{\frac{2d_i}{r'_b}}^{1} \left[\frac{1 - y}{y} \right] dy + \int_{\frac{2d_i}{r'_b}}^{1} \left[\frac{1 - y}{y} \right]^z . \; \log y \; . \; dy$$

4. Bed Sediment Rate

Bed sediment rate qs_b (s_b: bed sediment) applies to the bed layer of thickness $2d_i$. Einstein's analysis is based on physical, dimensional and statistical considerations. His conclusions based on observations are:

- Each grain of diameter d_i rolls or slides an average distance L.
- $L = Cte.d_i \simeq 100 \; d_i$
- On a bed area of width = 1 m and length = L in the direction of flow, every grain which is dislodged within that area will cross the downstream boundary before coming to rest, so in unit time, the number of grains dislodged is equal to the number of grains passing the boundary.
- p is the probability that, in one second, a grain d_i is dislodged. Einstein assumes that the probability function is Normal.
- This probability p is a function of the ratio between the lifting force due to the turbulence and gravitational force exerted on the grain.

The end result of these hypotheses is the expression:

$$\frac{p}{1-p} = C \frac{qs_{bi}}{p_i \gamma_s . g} \sqrt{\frac{\gamma}{\gamma_s - \gamma}} . \sqrt{\frac{1}{g \; d_i^3}} = C \; \Phi \; \text{(without demonstration)}$$

p_i = weight percentage of grain diameter d_i

C is a constant (function of the grain shape)

$$\Phi = \frac{qs_{bi}}{pi \; \gamma_s \; g} . \sqrt{\frac{\gamma}{\gamma_s - \gamma}} . \sqrt{\frac{1}{g \; d_i^3}}$$

and $qs_{bi} = \Phi \cdot p_i \cdot \gamma_s \, g^{3/2} \sqrt{\dfrac{\gamma s - \gamma}{\gamma}} \cdot d_i^{3/2}$

(2-9) (2-8)(2-2) (note 2-8) (2-1)

This first relationship, called Einstein "intensity of bed load transport" is a function of the unknown qs_{bi} and physical and geometrical properties of the grains of diameter d_i.

A second relationship derived by Einstein links the same probability p to the turbulent flow drag forces exerted on the particles and the weight of the particle:

$$p = 1 - \int_{-\beta\Psi - \frac{1}{\sigma_x}}^{+\beta\Psi - \frac{1}{\sigma_x}} e^{-x^2} dx$$

with $$\Psi = \xi \cdot \varepsilon \left[\frac{\text{Log } 10.6}{\text{Log } \dfrac{10.6 \times \alpha}{d_{65}}} \right]^2 \frac{\gamma s - \gamma}{\gamma} \cdot \frac{d_i}{r'_b S.}$$

(2-7)(2-5)(1.8) (1-9)and(1-10) (2-6)

This second expression is called "bed-load equation".

β, C and σ_x are constants that were determined experimentally by Einstein ($\beta = 0.143$; C $= 43.5$; $\sigma = 0.5$)

ξ is a factor that reduces the transport of fine particles because they (2-5) are protected from drag by larger particles, and so from being eroded at the full rate given by the formula. ξ is given in graphical form in Figure A.4.9a as a function of relative grain diameter d_i/X. (2-4) (see definition of X below).

ε has the same function as ξ but is a function of the variable d_{65}/δ
(1-8) (Figure A.4.9b) which defines the importance of the grain diameter within the laminar sublayer.

X reference grain size defined as the smallest particle of the bed which (1-7) is fully affected by the turbulent flow: its definition is then different if the bed is hydraulically rough (turbulent flow) or smooth (transition and laminar flow).

Rough beds: $\dfrac{d_{65}}{\alpha} > 1.80\,\delta$: $X = 0.77 \dfrac{d_{65}}{\alpha}$ (1-7)

Smooth beds: $\dfrac{d_{65}}{\alpha} < 1.80\,\delta$: $X = 1.39$

where δ = thickness of laminar sublayer:

$= 11.6 \dfrac{\nu}{\sqrt{g r_b S}}$ (ν = kinematic viscosity, Figure A.4.4)

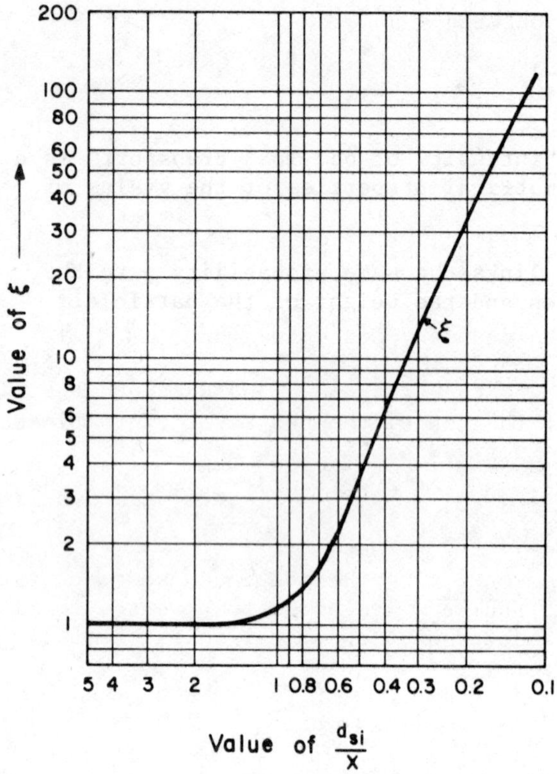

(a) Factor ξ in Einstein's bed load function (Einstein, 1950) in terms of d_{si}/X

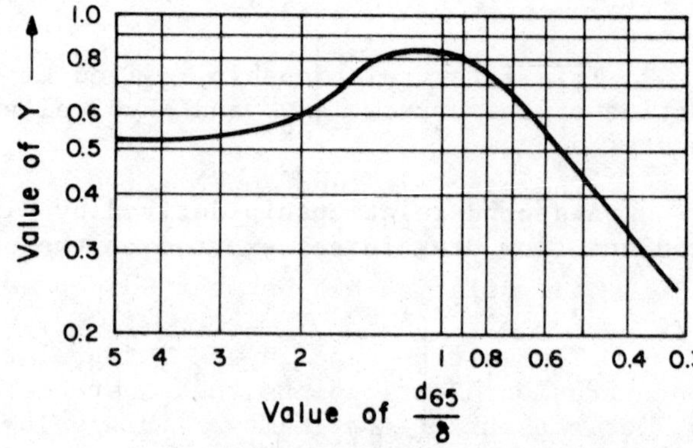

(b) Factor Y in Einstein's bed load function (Einstein, 1950) in terms of d_{65}/δ

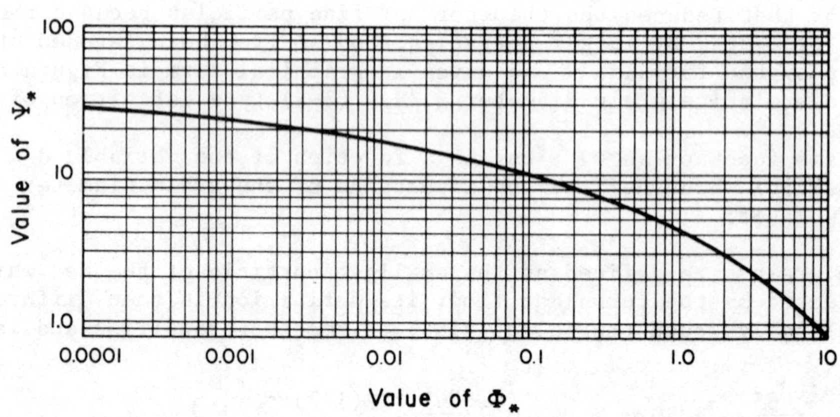

(c) Einstein's $\psi_* - \Phi_*$ bed load function (Einstein, 1950)

Fig. A.4.9 Einstein coefficients

The resulting "bed-load formula" is obtained in equalizing the two expressions of probability p producing an implicit relationship between Φ and Ψ :

$$\frac{C\Phi}{1+\Phi} = 1 - \int_{-\beta\Psi-\frac{1}{\sigma_x}}^{+\beta\Psi-\frac{1}{\sigma_x}}$$

The explicit function $\Phi = f(\Psi)$ is given in graphical form in Figure A.4.9 and is used to derive the partial bed-load rate qs_{bi} (2-9) from Φ (2-8).

5. Relationship Between Suspended and Bed Sediment Concentration

The last concept is the interrelation between the suspended and bed sediment concentration distributions that must be equal at the top of the bed layer of thickness $2d_i$ for reasons of continuity. This value Ca was derived by Einstein as:

$$Ca = \frac{qs_{bi}}{11.6 \cdot 2d_i \sqrt{g\, r'_b\, S}}$$

This value Ca is introduced in the expression giving the suspended sediment rate qs_{si}. Then suspended and bed sediment rates qs_{si} and qs_{bi} are added to give the total sediment rate q_{si} for the representative grain diameter d_i. The final result is the following expression:

$$q_{si} = qs_{bi}\,(2.3 \log \frac{30.2\, \alpha r'_b}{d_{65}} \cdot I_1 + I_2 + 1) \text{ in which}$$

(2-16) (2-19) (1-11) (2-13)(2-14)

integrals I_1 and I_2 are given in graphical form as functions of the exponent z(2-12) and the relative bed layer depth $2d_i/r_b$ (2-10) in Figure A.4.10.

This last expression, incorrectly called the "bed-load function" gives the total sediment rate q_{si} for each representative grain size d_i for each chosen water level.

The total sediment discharge q_s per unit width of the bed is then the sum of the partial sediment rate q_{si}. The international metric system was used through the example calculations and q is then given in units of N/s/m (Newtons per second per metre of bed width).

A.4.1.4 Sediment Rating Curve from Einstein Method

The Einstein method has been verified by its author for both natural rivers and laboratory flumes. However as in any other sediment rate formula the many constants were derived from the particular data sample and the conditions in which measurements were made. It is assumed that the sediment supply is always greater than the sediment transport capacity of the stream, which could not be verified in every case. Nevertheless some confidence may be placed in this method as it is based on physical rather than only statistical principles. Formulas derived from the latter can produce erroneous results when transposed to streams other than those used for their derivation.

The total sediment rate Qs is the product of qs (2-16) and the bed width W (= 55m in the example). The resulting Qs is then given in N/s (Newtons per second) which, when divided by g (= 9.81 m/s^2), produces the result in kg/s (Table 3 Example A.4.1).

The simultaneous discharge Q is the product of velocity V by cross-sectional area A which is the total area $A_w + A_b$ (as defined in Section 15.2.3 and shown on Figure A.4.5). All calculations to determine the sediment rate were made assuming that the area A_b ascribed to the bed is representative of the whole cross section. However, this assumption is not valid for the calculation of the discharge Q because V represents the average of the velocity distribution within the whole cross-section A. Velocity V is calculated by the Nikuradse equation which is consistent with the relationships used in the determination of sediment rate Qs (velocity distribution equation, Section A.4.1.3.1):

$$V = U_*' \;.\; 5.75.\log\left[\frac{12.27\ r_b\ \alpha}{d_{65}}\right]$$

$$V = \underset{(1\text{-}3)}{\sqrt{g\ r_b S}} \;.\; \underset{(3\text{-}1)}{5.75.\log}\left[\frac{12.27\ r_b\ \alpha}{d_{65}}\right] \qquad (3\text{-}2)$$

The calculations of the sediment rating curve based on Einstein's method applied to the example are shown in Table 3, Example A.4.1. Figure 15.2 shows the plotting of the five points which are used to extrapolate the actual measurements of (Q_{si}, Q_i) as described in Section 15.1.

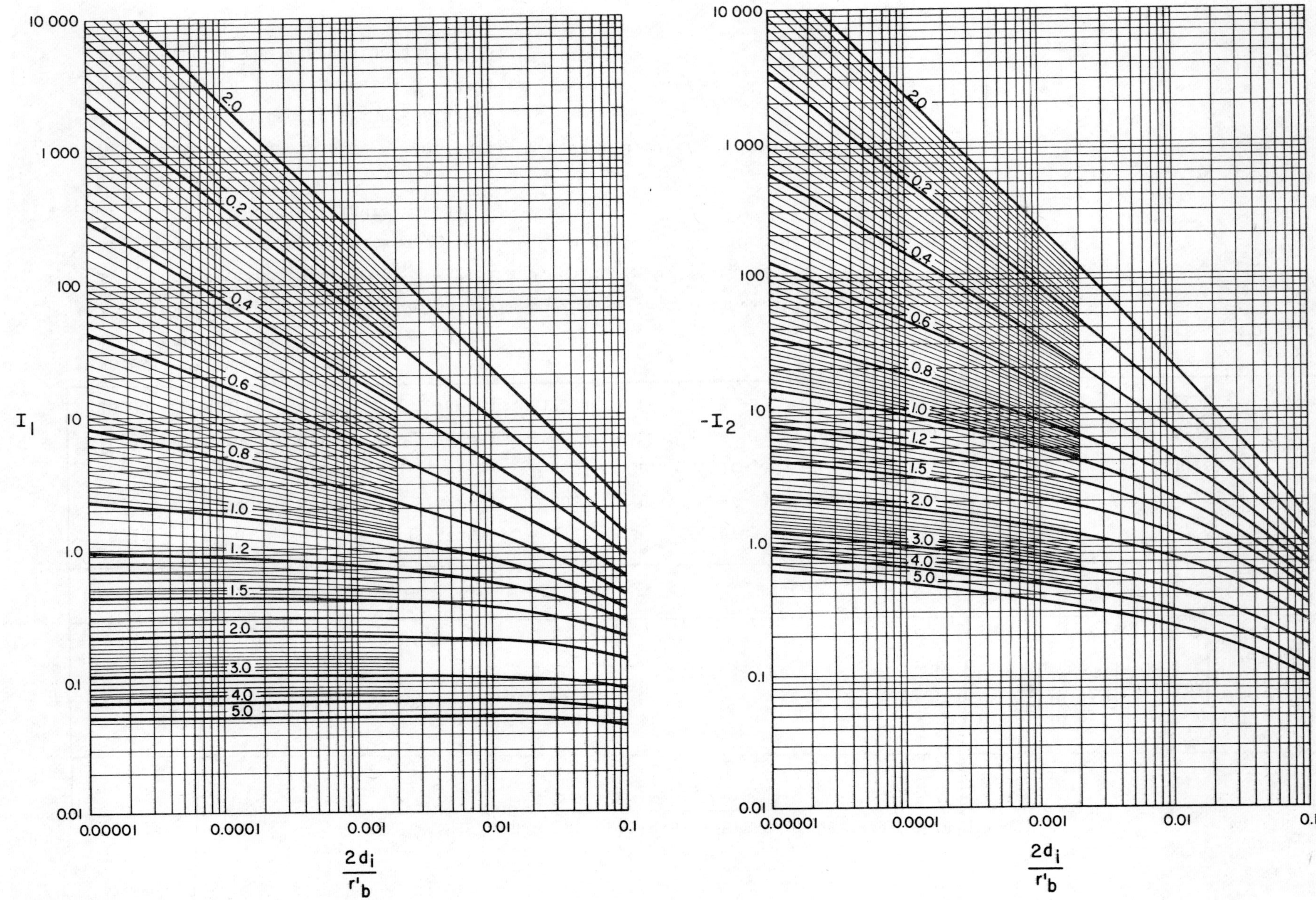

Fig. A.4.10 Integrals I_1 and I_2 of Einstein's method

TABLE 1

Example A.4.1

EINSTEIN METHOD - CALCULATIONS INDEPENDENT OF GRAIN SIZE DIAMETERS

1	2	3	4	5	6	7	8	9	10	11
y	r'_b	U'_*	δ	d_{65}/δ	α	X	ξ	$\frac{10.6X\alpha}{d_{65}}$	$\frac{\log 10.6}{\log\left[\frac{10.6X\alpha}{d_{65}}\right]}$	$2.31\log\left[\frac{30.2\alpha r_b}{d_{65}}\right]$
m	m	m/s	m	-	-	m	-	-	-	-
0.20	0.20	0.086	12.4(-5)	17.7	1	1.69(-3)	0.52	8.14	1.126	7.9
1.00	0.99	0.197	5.42(-5)	40.6	1	1.69(-3)	0.52	8.14	1.126	9.51
3.00	2.92	0.338	3.16(-5)	69.6	1	1.69(-3)	0.52	8.14	1.126	10.59
6.00	5.70	0.473	2.26(-5)	97.3	1	1.69(-3)	0.52	8.14	1.126	11.25
10.00	9.16	0.600	1.78(-5)	123.7	1	1.69(-3)	0.52	8.14	1.126	11.73

Notes:

1. Water Depth
2. Hydraulic radius relative to grain roughness (Section A.4.1.2.2)
3. Shear velocity $U'_* = gr_bS = 9.81 \times 0.004.r_b = 0.198\ r_b$
4. Laminar sublayer thickness $\delta = 11.6\frac{\nu}{U'_*} = 11.6 \times 9.2 \times 10^{-7}/U'_* = 1.067 \times 10^{-5}/U'_*$
5. $d_{65} = 0.0022$ m (Figure A.4.3)
6. $\alpha \sim 1$
7. Reference grain size $X = 0.77d_{65}/\alpha = 0.77 \times 0.0022/1 = 1.69 \times 10^{-3}$
8. Reduction factor (Figure A.4.9)

9.+10. Part of the expression Ψ (2-7)

11. Part of the expression qs_i (2-16)

TABLE 2

EINSTEIN METHOD - CALCULATIONS DEPENDENT ON GRAIN SIZE DIAMETER

Example A.4.1

1	2	3	4	5	6	7	8	9	10	11	12	13	14	15	16
d_i	P_i	r'_b	d_i/X	ξ	$\frac{\gamma s-\gamma}{\gamma} \frac{di}{r'_b \cdot s}$	Ψ	Φ	qs_{bi}	$\frac{2d_i}{r'_b}$	w_{fi}	z_i	I_1	I_2	$2.3\log\frac{30.2\ r'_b}{d_{65}} \cdot I_1+I_2+1$	qs_i and $qs = \Sigma qs_i$
m(10^{-4})	-	m	-	-	-	-	-	$\frac{N}{s.m.}$	-	$\frac{m}{s}$	-	-	-	-	$\frac{N}{s.m.}$
2.5	0.19	0.20	0.148	66.	0.52	22.63	0.000931	$7.31.10^{-5}$	2.50(-3)	0.0344	0.966	1.3	-3.9	7.38	0.001
6.5	0.36	0.20	0.385	7.6	1.34	6.71	0.295	0.184	6.50(-3)	0.0757	2.13	0.2	-0.8	1.78	0.328
20	0.16	0.20	1.18	1.1	4.13	3.00	1.68	2.52	2.00(-2)	0.144	4.04	0.068	-0.24	1.30	3.276
50	0.14	0.20	2.96	1.0	10.31	6.80	0.283	1.47	5.00(-2)	0.231	6.49	0.04	-0.11	1.21	1.779
150	0.15	0.20	8.88	1.0	30.94	20.40	0.00213	0.0614	1.50(-1)	0.402	11.29	0.015	-0.035	1.08	0.066
															5.45
2.5	0.19	0.99	0.148	66.	0.10	4.35	0.873	0.0686	5.05(-4)	0.0344	0.437	22.	-41.	169.2	11.6
6.5	0.36	0.99	0.385	7.6	0.27	1.35	5.15	3.21	1.35(-3)	0.0757	0.961	1.4	-4.8	9.51	30.5
20.	0:16	0.99	1.18	1.1	0.83	0.60	12.92	19.3	4.04(-3)	0.144	1.83	0.24	-1.08	2.20	42.5
50	0.14	0.99	2.96	1.0	2.08	1.37	5.06	26.2	1.01(-2)	0.231	2.93	0.11	-0.44	1.61	42.2
150.	0.15	0.99	8.88	1.0	6.25	4.12	0.989	28.5	3.03(-2)	0.402	5.10	0.05	-0.16	1.32	37.6
															164.4
2.5	0.19	2.92	0.148	66.	0.035	1.52	4.47	0.351	1.71(-4)	0.0344	0.254	180.	-230.	1677.	589.
6.5	0.36	2.92	0.385	7.6	0.092	0.46	16.9	10.6	4.45(-4)	0.0757	0.560	11.5	-27.	95.8	1016.
20.	0.16	2.92	1.18	1.1	0.283	0.21	37.3	55.8	1.37(-3)	0.144	1.065	0.98	-3.9	7.48	417.
50.	0.14	2.92	2.96	1.0	0.706	0.47	16.5	85.4	3.42(-3)	0.231	1.71	0.28	-1.25	2.72	232.
150.	0.15	2.92	8.88	1.0	2.12	1.40	4.93	142.1	1.03(-2)	0.402	2.97	0.11	-0.44	1.72	244.
															2498
2.5	0.19	5.70	0.148	66.	0.018	0.78	9.89	0.777	8.77(-5)	0.0344	0.182	480.	-600.	4803.	3732
6.5	0.36	5.70	0.385	7.6	0.047	0.24	32.6	20.3	2.28(-4)	0.0757	0.400	44.	-83.	413.	8384.
20.	0.16	5.70	1.18	1.1	0.145	0.11	71.7	107.3	7.02(-4)	0.144	0.761	3.35	-10.3	28.4	3047.
50.	0.14	5.70	2.96	1.0	0.362	0.24	32.6	168.8	1.75(-3)	0.231	1.22	0.66	-2.75	5.68	959.
150.	0.15	5.70	8.88	1.0	1.086	0.72	10.7	308.4	5.26(-3)	0.402	2.13	0.18	-0.82	2.21	682.
															16804.
2.5	0.19	9.16	0.148	66.	0.011	0.48	16.2	1.27	5.46(-5)	0.0344	0.143	1000.	-1200.	10531.	13374.
6.5	0.36	9.16	0.385	7.6	0.029	0.15	113.	70.5	1.42(-4)	0.0757	0.315	125.	-185.	1282.	90381.
20.	0.16	9.16	1.18	1.1	0.090	0.07	244.	365.	4.37(-4)	0.144	0.600	9.	-23.5	83.1	30332.
50.	0.14	9.16	2.96	1.0	0.225	0.15	113.	585.	1.09(-3)	0.231	0.963	1.35	-5.4	11.4	6669.
150.	0.15	9.16	8.88	1.0	0.675	0.45	37.3	1075.	3.28(-3)	0.402	1.68	0.3	-1.35	3.17	3408.
															144164.

Notes

1. Grain size diameter (Figure A.4.3)
2. Weight percentage (Figure A.4.3)
3. Bed hydraulic radius (from Figure A.4.5)
4. + 5. Figure A.4.9(a)
6. Part of the expression Ψ
7. Ψ = (2-5) (1-8) (1-10) (2-6)
8. Figure A.4.9(c)
9. $qs_{bi} = \cdot P_i \cdot d_i^{3/2} \cdot s \cdot g^{3/2}$ $= (2\text{-}8).(2\text{-}2).(2\text{-}1)^{3/2} .2650 \times 9.81 \times 1.65$
10. + 11. Fall velocity, Figure A.4.8(a) $\left[1{,}01(-2) = 1.01 \times 10^{-2}\right]$
12. $z = \frac{w}{0.4U'_*}$ with $U'_* = (1\text{-}3)$
13. + 14 Figure A.4.10
15. $2.3\log\left[\frac{30.2\ r'_b}{d_{65}}\right]$ from (1-11)
16. Partial total sediment rate of particles of diameter d_i with the summation qs
17. Values $\Psi < 0.8$ have been extrapolated.

Example A.4.1

TABLE 3

EINSTEIN METHOD - DEFINITION OF SEDIMENT RATING CURVE

	1	2	3	4	5	6	7
d	$\log\left[\frac{12.27 r'_b}{d_{65}}\right]$	V	A	Q	qs	Qs	C
m	-	m/s	m^2	m^3/s	N/m/s	kg/s	kg/m^3
0.2	3.047	1.56	11.1	17.3	5.45	31	1.79
1.0	3.742	4.24	56.5	240.	164.4	922	3.84
3.0	4.212	8.19	179.	1466.	2498.	14000	9.55
6.0	4.502	12.25	385.	4721.	16804.	94212	20.0
10.0	4.708	16.24	704.	11430.	144164.	808260	70.7

Notes:

1. Part of the expression of V
2. $V = U'_* \cdot 5.75 \cdot \log\left[\frac{12.27 r'_b}{d_{65}}\right] = (1\text{-}3) \times 5.75 \times (3\text{-}1)$
3. Cross-sectional area (Figure A.4.5)
4. Q = V.A. = (3-2)(3-3)
5. qs from (2-16)
6. Sediment discharge Qs (kg /s) = qs x W/9.81 m/s^2 = qs x 55 m/9.81
7. Concentration C = Qs/Q = (3-6)/(3-4)

Example A.4.2 Einstein-Brown Formula

This equation is homogeneous so the metric system (S.I. units) can directly be applied.

The basic equations are:

(1) $$\Phi = \frac{40}{\psi 3} \quad \text{for} \quad \frac{1}{\psi} > 0.09$$

(2) $$\Phi = \frac{q_s}{\gamma_s F_1 \sqrt{g\left(\frac{\gamma_s}{\gamma} - 1\right) d_s^3}}$$

(3) $$F_1 = \sqrt{\frac{2}{3} + \frac{36\nu^2}{g d_s^3 \left(\frac{\gamma_s}{\gamma} - 1\right)}} - \sqrt{\frac{36\nu^2}{g d_s^3 \left(\frac{\gamma_s}{\gamma} - 1\right)}}$$

(4) $$\frac{1}{\psi} = \frac{\tau_o}{(\gamma_s - \gamma)\, d_s} = \tau_*$$

(5) $$\tau_o = \gamma r_b S$$

(6) From (2): $$q_s = \Phi \cdot g\, \gamma_s \cdot F_1 \sqrt{g\left(\frac{\gamma_s}{\gamma} - 1\right) d_s^3}$$

(7) From (4) and (5): $$\frac{1}{\psi} = \frac{\gamma\, r_b S}{(\gamma_s - \gamma)\, d_s} = \frac{\gamma\, r_b S}{(\gamma_s - \gamma)\, d_s} = \frac{\gamma}{\gamma_s - \gamma} \cdot \frac{S r_b}{d_s}$$

(8) From (3) $$F_1 = \sqrt{\frac{2}{3} + \underbrace{\frac{36 \times (9.2)^2 \times 10^{-14}}{9.81 \times (0.00338)^3 \times 1.65}}_{= \alpha}} - \sqrt{\alpha}$$

$$\alpha = 0.00004875$$

$$F_1 = \sqrt{0.666\,.. + 0.00004875} - \sqrt{0.00004875} = 0.8095$$

(9) From (6)

$$q_s = \Phi \times 9.81 \times 2650 \times 0.8095 \times 9.81 \times 1.65 \times 0.00338^3 \text{ N/m/s}$$

$$q_s = \Phi \times 16.64 \text{ N/m/s}$$

(10) From (7)

$$\frac{1}{\psi} = \frac{\gamma}{(\gamma_s - \gamma)} \cdot \frac{S}{d_s} \cdot r_b = \frac{1000 \times 0.004}{(2650 - 1000) \times 0.00338} \cdot r_b$$

$$\frac{1}{\Psi} = 0.7178 \cdot r_b$$

Tabulation

r_b	$\frac{1}{\Psi} = 0.7178 r_b$	$40\left(\frac{1}{\Psi}\right)^3 = \Phi$	$q_s = \Phi \times 16.64$	$Q_s = \frac{55m \times q_s}{9.81}$	Q	Concentration C
m	-	-	N/m/s	kg/s	m^3/s	kg/m^3
0.20	0.144	0.118	1.969	11.0	17.3	0.64
0.99	0.711	14.354	238.9	1335	240	5.6
2.92	2.096	368.3	6128	34357	1085	32
5.70	4.091	2739.6	45588	255590	4721	54
9.16	6.575	11369.9	189195	1060726	11430	93

Example A.4.3 - Engelund-Hanson Formula

Basic equations:

(1) $$fe\,\Phi_e = 0.1\tau_*^{2.5}$$

(2) $$fe = \frac{2\,\tau o}{\gamma\,V^2}$$

(3) $$\Phi\,e = \frac{qs}{\gamma s\sqrt{\left(\frac{\gamma s}{\gamma} - 1\right) g d_s^3}}$$

(4) $$\tau = \frac{\tau o}{(\gamma_s - \gamma)\,d_s}$$

(5) From (1) to (4)

$$q_s = 0.05\,\gamma_s\,V^2\sqrt{\frac{d_s}{g\left(\frac{\gamma s}{\gamma} - 1\right)}} \times \left[\frac{r_b\,S}{(\gamma s - \gamma)\,d_s}\right]^{3/2}$$

$$\tau o = r_b S$$

$$qs = 0.05\,\gamma_s V^2\sqrt{\frac{d_s}{g\left(\frac{\gamma s}{\gamma} - 1\right)}}\left[\frac{r_b S}{(\gamma_s - \gamma)d_s}\right]^{3/2} \text{ N/m/s}$$

Numerically

$$q_s = 0.05 \times 2650 \times 9.81 \times V^2.\sqrt{\frac{0.00085}{9.81 \times 1.65}} \times \left[\frac{r_b \times 0.004}{1.65 \times 0.00085}\right]^{3/2} \text{ N/m/s}$$

$$q_s = 1299.8 \times V^2 \times 0.007247 \times (r_b)^{3/2} \times 4.8165 \text{ N/m/s}$$

$$q_s = 45.37 \times V^2 \times r_b^{3/2} \text{ N/m/s}$$

r_b m	V m/s	q_s N/m/s	$Q_s = \frac{55m \times q_s}{9.81}$ kg/s	Q m^3/s	C kg/m^3
0.20	1.56	9.88	55	17.3	3.18
0.99	4.24	803	4502	240	18.8
2.92	6.07	8341	46764	1085	43
5.70	12.25	92652	519456	4721	110
9.16	16.24	331729	1859847	11430	163

Example A.4.4 Inglis-Lacey Formula

Basic equation:

$$q_s = 0.562 \frac{\nu^{1/3} g^{1/3} V^5 . \gamma . g}{w . g^2 . y} = \frac{\nu^{1/3} . V^5 . \gamma \times 0.562}{w . y . g^{2/3}} \quad N/m/s$$

Numerically $w = 9 \times 10^{-2}$ m/s, $\nu = 9.2 \times 10^{-7}$ m^2/s

$$q_s = \frac{0.562 \times (9.2 \times 10^{-7})^{1/3} \times V^5 \times 1000}{9 \times 10^{-2} \times y \times 9.81^{2/3}}$$

$$q_s = \frac{0.562 \times 0.00973 \times V^5 \times 1000}{0.09 \times y \times 4.583}$$

$$q_s = \frac{V^5}{y} \times 13.25 \quad N/m/s$$

y m	V(1) m/s	q_s N/m/s	Q_s kg/s	Q m^3/s	C kg/m^3
0.20	1.56	612	3431	17.3	198
1.00	4.24	18157	102000	240	425
3.00	6.07	36395	204000	1085	188
6.00	12.25	109179	3415000	4721	723
10.00	16.24	1496739	8392000	11430	734

(1) V is as calculated as shown in Table 3 of Example A.4.1 for the Einstein Bed-Load Function.

Example A.4.5 - Meyer-Peter Formula

Basic Equations:

$$q_s^{2/3} = \left[250\ q^{2/3}\ .\ S\right] - 42.5\ d_s$$

q_s is given in kg/s/m

$$q = \frac{Q}{W}\ m^3/s/m$$

$$q_s^{2/3} = \left[250\ \frac{Q^{2/3}}{W^{2/3}}\ .\ S\right] - 42.5\ d_s$$

$$q_s = \left[250\ q^{2/3}\ S - 42.5\ d_s\right]^{3/2}$$

Numerically

$$q_s^{2/3} = \left[\frac{Q^{2/3} \times 250 \times 0.004}{55^{2/3}}\right] - \left[42.5 \times 0.00338\right]$$

$$q_s = \left[q^{2/3} - .05445\right]^{3/2}$$

Q m^3/s	$q^{2/3}$	q_s kg/m/s	$Q_s = 55m \times q_s$ kg/s	C kg/m^3
17.3	0.463	0.261	14.4	0.83
240	2.670	4.23	233	0.97
1085	7.301	19.51	1073	0.99
4721	19.46	85.48	4701	1.00
11430	35.09	207.4	11407	1.00

EXAMPLE A.4.6

COMPUTATION OF TOTAL BED SEDIMENT LOAD - COLBY METHOD

(1)	(2)	(3)	(4)	(5)	(6)	(7)	(8)	(9)	(10)	(11)	(12)	(13)	(14)
Q	A	V	d	W	Qsu	Qsu	Cr	Qsm	Cm	Cm / Cr	Correction Factor C	Qsu·C	Q_{sT}
m^3/s	m^2	m/s	m	m	t/m/day	kg/s	ppm	kg/s	ppm			kg/s	kg/s
10	11	.91	0.7	25	30	19.4	700	45	2500	3.6	2.2	43	90
100	50	2.0	1.7	47	460	298	3500	935	5500	1.57	1.5	447	1400
500	170	2.9	3.7	x	1300	843	6500	7990	9200	1.42	1.4	1180	9200
1000	250	4.0	5.0	x	2000	1296	9000	20140	12000	1.33	1.3	1685	22000

Note: t = tons

Col.
(1) From discharge rating curve.
(2) From area rating curve.
(3) Col.(1)/Col.(2)
(4) Depth from water to surface sand bed.
(5) Water surface width.
(6) Figure A.4.11(a) Q_{su} value x 3.281 (conversion from tons/ft to tons/m)
(7) (Col.(6) x 2240 x 55)/(86400 x 2.2) (conversion from tons/m/day to kg/s - assuming 55m sand bed width)
(8) From Figure A.4.11(b).
(9) From measured suspended sediment rating curve - Figure 15.1
(10) Col.(9) x 1000/Col(1)
(11) Col.(10)/Col.(8)
(12) From Figure A.4.11(c)
(13) Col.(6) x Col.(12)
(14) Col.(9) + Col.(13).

NB. Figures in Col.(14) have been rounded off.

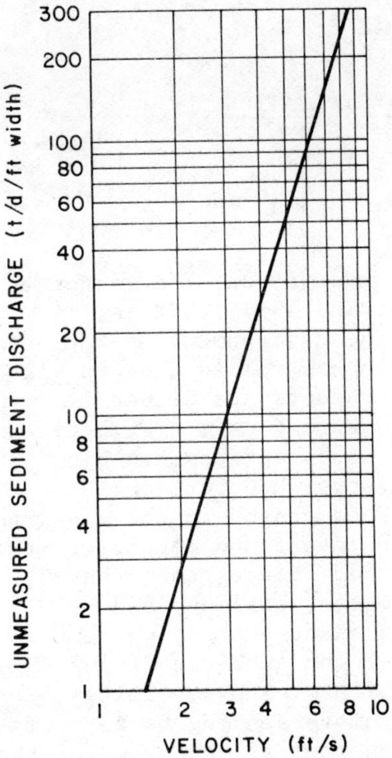

(a) Colby's '1957) graph of unmeasured sediment discharge against mean velocity (Fig. 2110 in ASCE Manual)

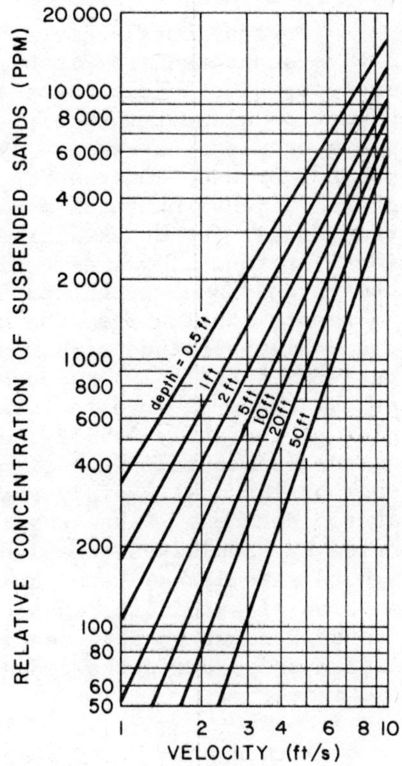

(b) Colby's (1957) chart of relative concentration against mean velocity (Fig. 2111 in ASCE Manual)

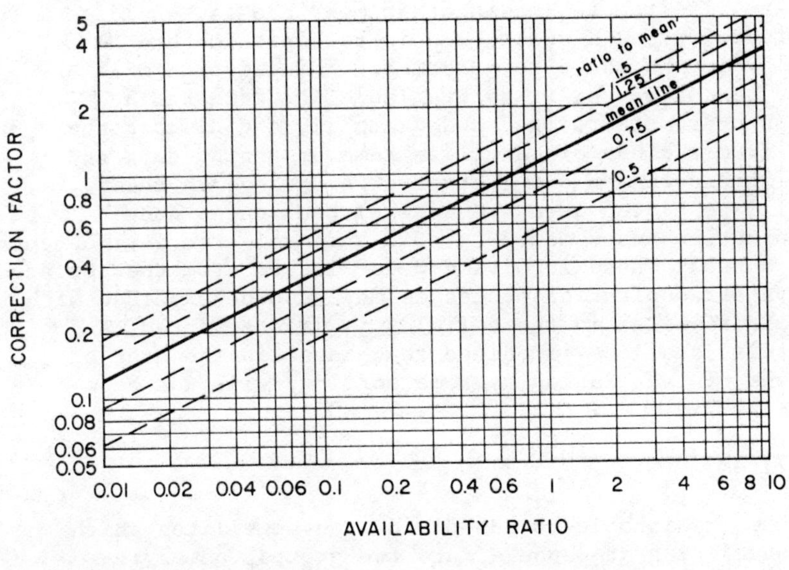

(c) After Colby's (1957) chart of availability ratio against correction factor (Fig. 2112 in ASCE Manual)

Fig. A.4.11

ANNEX 5

THE USE OF SATELLITE REMOTE SENSING IN THE PRECIPITATION HYDROMETEOROLOGY OF ARID ZONES

A.5.1 Introduction

Environmental remote sensing has been defined as "The observation of a target by a device separated from it by some distance". (Barrett and Curtis, 1976). It is, therefore, clearly different from conventional observation of the environment of Man, or any part or element thereof, for, traditionally such observation has relied upon in situ sensors or sensor systems - systems in which the observing instruments are either touching, or even immersed in, the object of interest. Remote sensing as a distinctive approach in environmental monitoring is, in one way, as old as man's interest in his surroundings, for we use our eyes, ears, and skin as remote sensors; however, as a scientific tool it was effectively born with the invention of photography in the mid-19th century, and as a recognised discipline it's rise has been both recent and meteoric: the term remote sensing seems to have been coined as recently as 1960. Since ground-based, low and high altitude airborne remote sensing platforms (e.g. masts, balloons and aircraft respectively) had been used widely prior to 1960 it is clear that it was the birth of space exploration, first by high-altitude rockets (e.g. the American Viking series of the early 1950's), and then by spacecraft (manned orbiters) and satellites (unmanned spacecraft in more permanent orbits) which prompted remote sensing to take off as one of the most rapidly-expanding fields of scientific/ technological endeavour of the last two decades. Since the first experimental satellite-borne remote sensing packages carried by satellites were launched into orbit by the USA in 1959 (on the Explorer and Vanguard satellites) very rapid progress has been made towards the development and operation of satellite remote sensing systems. Today their variety and flexibility rival those of present in situ sensing systems - although it is much more profitable to consider these different types of systems as complements to each other rather than competitors.

Although much progress has been made with the development of remote sensing systems for environmental monitoring from ground-based platforms (e.g. meteorological radar), aircraft (e.g. infra-red line scan), and rockets (recoverable aero-film multispectral cameras), it is in satellite remote sensing that most promise would seem to reside for applications in hydrometeorology as a whole, and in precipitaton hydrometeorology in particular, especially in the short-term future. This section will summarise the satellite systems and sensor capabilities available for research and operations relating to precipitation hydrometeorology, and exemplify and discuss the types of work which may now profit from the use of satellite remote sensing data and techniques. It will conclude with some notes on our further expectations in these directions. Emphasis throughout will be placed upon the special challenges and difficulties associated with precipitation assessments in arid and semi-arid environments. Other aspects of hydrometeorology, notably those involving water losses from the surface of the Earth via evaporation and evapotranspiration processes may also be examined with benefit from satellite altitudes, but research in these fields is less advanced, and operational applications of the results are less immediate than those in the area of water gains. Consequently, questions of evaporation hydrometeorology will not be addressed further in this account.

A.5.2 Satellite Remote Sensing System

In recent years it has become fashionable to sub-divide the satellites which observe the Earth surface and its enveloping atmosphere into two groups, namely:-

a. Environmental satellites, and

b. Earth resource satellites.

These are differentiated primarily on the basis of the spatial resolution of the data they supply, though, largely because of data-flow constraints, the normal concomitant of higher spatial resolution is lower frequency of coverage of any given area. Therefore, for example, the 80m resolution of the multispectral scanners (MSS) on Landsat 1, 2 and 3 has been accompanied by an 18-day imaging frequency, whilst the lower (1km) resolution of the high-resolution picture transmission (HRPT) system of the US National Oceanic and Atmospheric Administration (NOAA) satellites has been coupled with the much higher imaging frequency of 12 hours.

For precipitation studies it is clear that the Earth resource satellites, as exemplified by the Landsat satellites, have little to offer, though they have considerable potential and/or realised capabilities elsewhere in hydrology where highly repetititive data are not required, e.g. in the identification of coastal fresh-water submarine springs, and in land surface hydrogeological mapping. Our attention will be focussed more profitably on environmental satellites. These can be further subdivided into:

a. Polar orbiting satellites, and

b. Geostationary (equatorial orbiting) satellites.

In the first case above, the satellites occupy relatively low-level orbits (usually between 500-1500km above the surface of Earth), crossing the equator at high angles so that each orbit takes such a satellite close to the North and South Poles. These satellites are usually sun-synchronous, i.e. their orbits are so organised, that, as the Earth rotates on its polar axis within the orbital ellipses, each new orbit results in the presentation of a new strip of the global target to the satellite in such a way that the relationship between Earth surface and sun-angle is kept relatively constant. Characteristic orbital periods are c.100 min., 14-15 orbits being required for each satellite to investigate the entire globe under daylight conditions every 24 hours. Of course, since such satellites spent one-half of each orbit over the night-time side of the globe, each unit area on the surface is thus viewed twice every 24 hours, once in day-time and once at night. Crossing times at the equator differ from system to system, but NOAA satellites have crossed at about 0900 and 2100 hours local time.

Other noteworthy families of American polar-orbiting environmental satellites have included Tiros and Essa, their Russian counterparts including Molniya, Meteor, and a few of the huge and varied Cosmos series. Present American families include the civilian Tiros-N (replacements for NOAA operational satellites), Defense Meteorological Satellite Program (DMSP) Block V-D satellites, and the on-going research and development Nimbus family.

In the second case above, the satellites are placed into orbit at approximately 35,400km in the plane of the equator, and advance in the same direction as the rotation of the Earth. This type of orbit is geosynchronous, and geostationary in that it ensures that any satellite occupying it appears to be fixed or stationary above a point on the Earth's surface. Such a satellite is able to record the same geographic field of view throughout the diurnal cycle, commonly at intervals of 30 min.

The earliest geostationary satellites were the experimental Applications Technology Satellites (ATS) of the early 1970s. They were followed by the Synchronous Meteorological Satellites (SMS). Currently, five geostationary satellites are providing a complete circumglobal coverage between latitudes of effectively about 50°N and S of the equator. These include two American Global Operational Environmental Satellites (GOES -1 and 2), one covering the eastern Americas and the western Atlantic, and the other the western Americas and eastern Pacific; a European Space Agency (ESA) satellite, Meteosat, located over West Africa; an American satellite over the Indian Ocean controlled by ESA, and named GOES-IO; and a Japanese satellite, Himawari, located over Indonesia.

In satellite studies of precipitation hydrometeorology analyses of rain-bearing cloud systems and the monitoring of associated precipitation have figured most prominently. For these, both polar-orbiting and geostationary satellites have been shown to have their uses, especially through their imaging of the Earth in visible (VIS) and infrared (IR) regions of the electromagnetic spectrum. Resulting data have the following characteristics:-

a. VIS data. These depend physically on the albedo or percentage reflectivity of the target. Consequently clouds appear bright in VIS images, there being a general, but by no means invariable, relationship between the brightness and thickness of cloud. At present VIS images are obtainable during daytime only.

b. IR data. These depend physically on the radiation temperatures of the target. Cold surfaces radiate little, so their radiation temperatures are low, whilst warmer surfaces radiate more, and have higher radiation temperatures. IR images are processed to facilitate fairly direct comparisons with VIS images: in the IR images high cloud tops are bright, middle cloud tops are less bright, and low level tops are dim.

It is on such factual and physical bases that a proper understanding and appreciation of satellite contributions to precipitation hydrometeorology may be built.

A.5.3 Applications of Satellite Data in Precipitation Hydrometerology

A.5.3.1 Improved Rainfall Mapping: Methods and Procedures

Soon after the first specialised weather satellite was placed into Earth orbit on 1st April 1960 the first attempts were made to assess rainfall from the evidence of weather satellite data. Recently there has been an increase in interest in the possibility that satellites might be employed with benefit in rainfall monitoring, prompted by the growing realisation that, in many areas, conventional (raingauge) data are tending to become less, rather than more, plentiful and dependable. (See Barrett, 1977a). Very broadly there are two different ways in which rainfall might be assessed from satellite altitudes.

A.5.3.2 Physically Direct Methods

Here the radiation signals intercepted and evaluated by a satellite sensor system emanate from the falling rain itself. Rainfall can be "seen" by satellite systems sensing in certain microwave (radar) regions of the electromagnetic spectrum, as has been demonstrated by the passive Electrically Scanning Microwave Radiometers (ESMR) on recent Nimbus satellites, recording naturallyemitted radiation with frequencies of 19.35 GHz and 37.0 GHz. (See e.g. Allison et al, 1974, Rao et al, 1976). Unfortunately the resulting assessments of rainfall rates have been more readily established over sea than land on account of the stronger background signals emanating from the latter, although some progress is being made in the extraction of rainfall rates over land (Fowler et al, 1979). Active microwave systems (radars, not radiometers) might function better over both land and sea, particularly on account of the higher resolution data they might return (perhaps c.1km cf. the 25-50km of the ESMR systems). However, no such system has been flown yet on an orbiting spacecraft. Indeed, there are no firm plans even for an operational ESMR-type system. Consequently, for potential operational uses of satellite data in precipitation hydrometeorology through the forseeable future we must look to other ways of assessing rainfall and rain-cloud systems.

A.5.3.3 Physically Indirect Methods

Clouds are very prominent features of satellite images in the VIS and IR regions of the electromagnetic spectrum. A wide range of studies by many workers in different research centres, notably in North America, Western Europe, the U.S.S.R. and Australia have aimed at the identification of appropriate relationships between VIS and IR cloud image characteristics and observed rainfall, expressed either in terms of instantaneous rain rates or rainfall totals accumulated through various unit periods of time. Comprehensive reviews of these studies have been prepared by Martin and Scherer (1973), and Barrett and Martin (in press). In every case some cloud characteristic or characteristics are considered to be related to rainfall, as indicated by raingauges and/or radar observations. It is such techniques which, although less satisfactory than physically direct methods may ultimately prove to be, offer the best prospects for immediate use.

In the U.S.A. much research has gone into the study of relations between rainfall rates and satellite image brightness, at first in the VIS (e.g. Sikdar, 1972; Grosh et al., 1973) and more recently and extensively in the IR (e.g. Griffith et al., 1978): the IR approach has the strong advantage that IR observations can be made by satellites by night as well as by day. Given the continuous coverage of the Americas by geostationary SMS and GOES satellites since 1974, and the assurance that such coverage will be continued in the future, most American studies have relied on geostationary, rather than polar orbiting satellite data.

Unfortunately, the likelihood of continuing geostationary satellite coverage elsewhere is much less strong, and it would seem wise to base operations in these other regions - which include the principal arid and semi-arid areas of the world - on data from polar orbiters. Indeed, this argument has received very recent recognition by the National Oceanic and Atmospheric Administration (NOAA) of the U.S. Department of Commerce, who are embarking on a new programme (code-named 'Agristars') to develop a global rainfall-assessment method for primary use in crop production forecasting, based not on geostationary, but polar-orbiting satellite data.

The problems and potentials of satellite rainfall monitoring in the dominantly arid continent of Australia were recognised as early as 1969 (see Barrett, 1970, 1971). Since then considerable polar orbiting method development has taken place in the University of Bristol, largely at the behest of different divisions and units of FAO and associated bodies. This work has been based on data from polar orbiters, and would therefore, seem to be the most appropriate to outline in detail at this juncture, especially since the chief trials of "the Bristol Method" have been carried out in the arid zone.

The aims of the work may be summarised as follows:-

a. To design a satellite-improved rainfall mapping method through the simultaneous exploitation of the more desirable attributes of both satellite and conventional observations, namely the relative accuracy of raingauge observations of rainfall (which represent total accumulations of rainfall through time, but only at the point locations occupied by the raingauges), and the more complete areal indications of rain afforded by satellite images (which provide better pictures of the spatial distributions of individual rainfall events).

b. To develop a satellite-improved rainfall mapping method which could be implemented more or less immediately in most parts of the world. This aim alone has necessitated dependence on polar orbiting satellite data, for the reasons mentioned earlier. However, an additional point in favour of the polar orbiters is that the linear resolutions of their data have been better than those from geostationary satellites by factors ranging generally from about 2.5-7.5.

c. To develop a method for improved rainfall monitoring which could be implemented at modest cost. Since questions of cost are highly significant for developing countries and/or agencies of limited jurisdiction, the Bristol Method has been designed to employ human analysts/interpreters rather than expensive computing and automatic image analysis equipment. We believe that, at present, "eyeball" methods of rainfall assessment involving cloud type identification in satellite images can be implemented more accurately and efficiently for the whole range of situations likely to be encountered in operational contexts than "objective" methods tested by others for a more limited range of situations.

d. To test the Bristol Method in selected areas, and, through verification of the results, to assess its potential utility for operational use. The areas of study of direct interest in the present context include parts of the Sultanate of Oman (Barrett, 1977b), and North-West Africa (comprising Algeria, Libya, Morocco and Tunisia; see Hielkema and Howard, 1976; Barrett, 1977c and 1977d; and Barrett and Lounis, in press). These studies will be referred to further in the account which follows.

e. To implement the Bristol Method operationally in an appropriate form (for it is flexible in its use, especially in so far as spatial resolution is concerned), i.e. to take the concept of satellite-improved rainfall monitoring from the status of a research technique to that of a fully-fledged operational procedure. At one time it was anticipated that this transition would be effected first for water-resource evaluation purposes in the north-east of Oman; it now seems much more likely that the transition will first take place in north-west Africa, under the auspices of the regional Desert Locust Control Commission, as a means of improving the efficiency and effectiveness of desert locust survey and control.

Viewed at its simplest, the Bristol Method assumes that rainfall (R) is a function of several aspects of satellite imaged cloud, such that:-

$$R = f\left[C_a, C_t\ (R_p, R_i)\ , A\right]$$

where C_a is cloud area (in tenths) in grid squares of selected size, C_t is cloud type (upon which the probability (R_p) and intensity (R_i) of associated rain depend), and A is altitude above a chosen datum. In evaluating R_p and R_i suitable ranks of clouds are drawn up and accorded values on dimensionless scales depending on meteorological expectations, experience, and local evidence. The products of (C_a, C_t, A) are <u>cloud indices</u>, which operationally must be translated into rainfall estimates through some pre-established form of relationships. For this reason this type of method has become known as a <u>cloud indexing method</u>. Similar methods, indeed all branches from the 1970 Bristol root, have been developed by Follansbee (1973) and Follansbee and Oliver (1975) for a variety of climatic zones, and by Park (1976) in a proposal for an FAO global crop production programme.. A cloud-indexing method based on geostationary satellite data involving a substantial degree of automation is being used by Earthsat Corporation in support of wheat production estimates in the USA and elsewhere (see Merritt (ed), 1976).

The basic philosophy of the Bristol Method is that <u>satellite image data can be used to improve rainfall maps between the locations of the available raingauge ("ground-truth") stations, against whose rainfall observation/satellite cloud index relationships the clouds in the conventional data-remote areas are assessed.</u> Thus, the satellite data are employed in the "extension" of the available conventional data, an approach which has been advocated in association with other meteorological variables also (see Giddings, 1976, Barrett (ed), 1979).

The structure of the Bristol Method in its present form is schematised in Figure A.5.1. Key steps in the Method may be summarised as follows:-

Preparatory Stages

a. The study region is subdivided into regular grid squares ("cells") whose size should be related to the purpose of the study in hand, the scale of the study region, the resolution of the available satellite imagery, available funding etc. These cells include some which contain dependable rainfall gauges ("gacells"), and others which do not: for the latter ("satcells"), rainfall must be estimated from the evidence of the satellite imagery. Each satcell is considered to be related to a given gacell or groups of gacells for calibration purposes, based on climatic affinities.

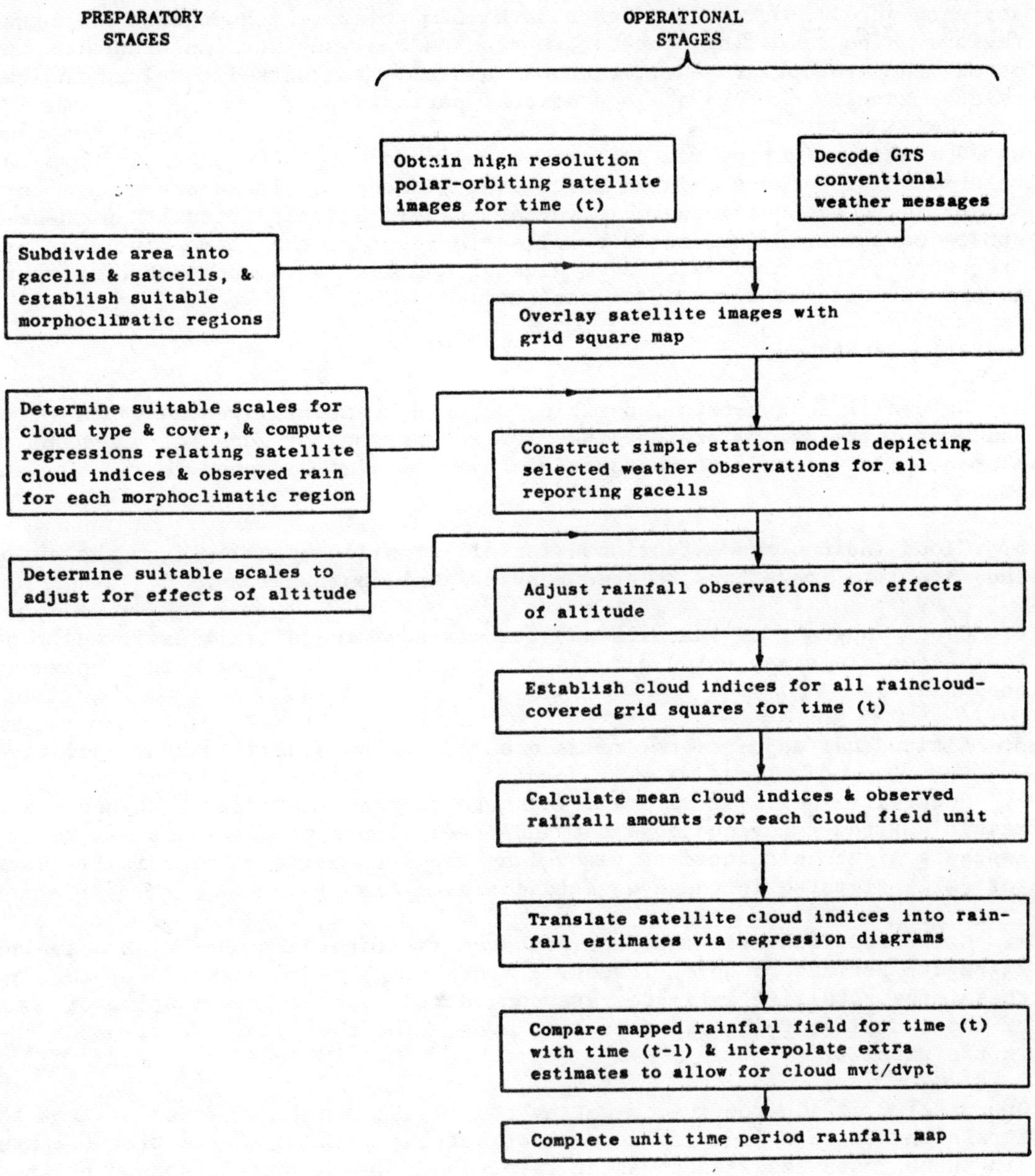

Fig. A.5.1 Flow diagram for the Bristol method in its present form

b. Weather and climate records are examined so that appropriate cloud ranks and scales may be established for cloud indexing.

c. Regression analyses are undertaken for a selected historical period to establish the form of normal statistical relationships between appropriate cloud indices and observed rainfall for a selected unit period (usually 12 or 24 hours) for each gacell, or group of gacells within relatively homogeneous (morphoclimatic) regions. The resulting best-fit regression analyses furnish standards which permit the appropriate estimation of rainfall from satellite cloud indices for related gacells through the operational period(s).

d. Suitable weighting factors are established for altitude to improve the relationships between related satcells and gacells. These are constants; their effect is to account for simple variations of rainfall intensity with height above sea-level, as evidenced by climatic records: the common increase of cloud frequency and/or cloud area over high ground will be accommodated by routine assessments of cloud cover for each cell.

Operational Stages

a. Rainfall, and wherever possible, other conventional weather data (usually daily or twice-daily) are obtained for the network of gacells, and selected elements plotted on a standard transparent overlay designed to fit the satellite imagery.

b. Cloud indices are established for all significant cells from the evidence of the satellite images and related conventional observations.

c. Cloud indices for satcells are translated into rainfall estimates through the regression diagrams, using gacells or groups of gacells as guides wherever possible.

d. Altitudinal adjustments are made before each rainfall map is completed.

e. Successive pairs of rainfall maps are compared in order that under-estimates resulting from raincloud movement and/or development between satellite imaging instants might be reduced or removed by the interpolation of suitable estimates for cells affected in these ways (See Figure A.5.2).

f. Completed isohyetal maps are prepared for each unit period, and subsequent aggregate periods (5 days, 1 week, 1 month, etc) as required. In preparing these, the satellite estimates are considered to be representative of the centres of the satcells; raingauge data are invoked for the exact locations of the gauges in the gacells.

One particular sub-routine is worthy of special mention. This involves the use of the regression diagrams (see (c) above) for satcell rainfall estimation (Figure A.5.3). In those instances when apparent rain cloud over a satcell does not extend over an appropriate gacell, the rainfall estimate for that satcell is established on the basis of the computed regression relationship for the area in question. However, where cloud does extend over a suitable gacell it is possible, and may be desirable, to "float" the satcell estimate up or down to account for the fact that, on different occasions, clouds with similar appearances in the satellite images, or otherwise yielding the same cloud indices, may give quite different falls of rain. In the present method a meteorological assumption is made, namely that the rainfall associated with a given mass of cloud is uniformly more or less than the norm evidenced by the regression relationship between cloud indices and rainfall totals. Hence, the regression diagrams can be conceived as stacks of curves parallel to the computed regression. Given a particular gacell relationship between an established cloud index (a) and reported rain (c) which is, say, considerably above the regression, (as in Figure A.5.3) it may be considered that the

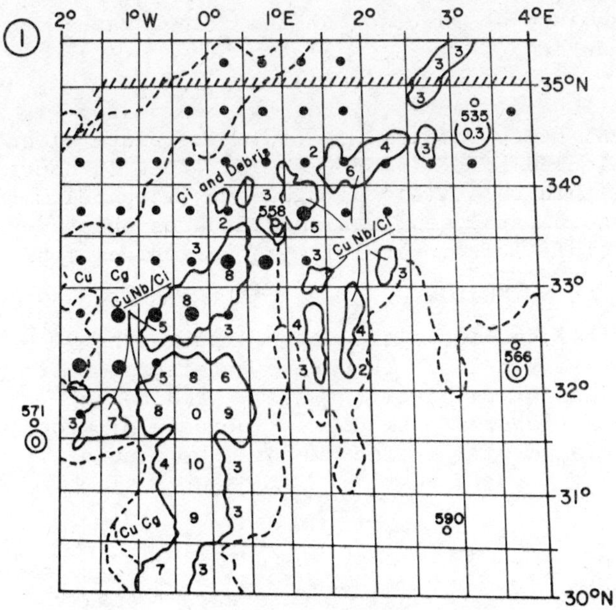

///////	:	Boundary of recession area
————	:	Major cloud boundaries
– – – –	:	Non-rain cloud boundaries
° 571	:	Station numbers
8	:	Cloud amount per $1/2°$ grid square
Altitude	:	• 1000 - 1500 m ● 1500 - 2000 m
(0.3)	:	12 hr rainfall reports

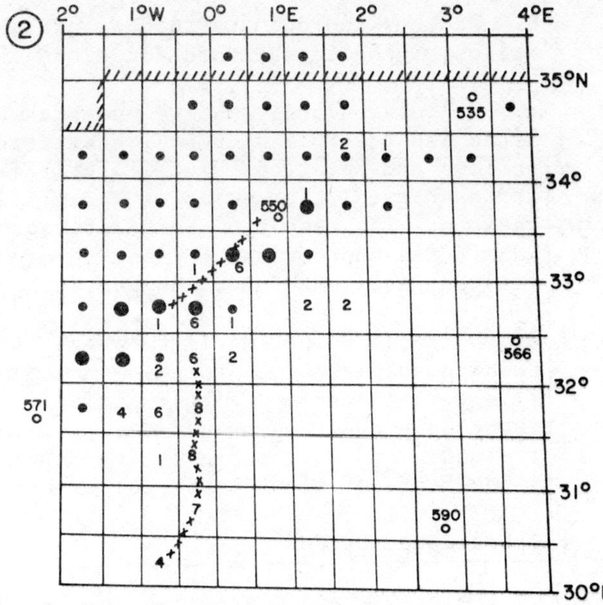

//////	:	Boundary of recession area
° 571	:	Station numbers
6	:	Rainfall for 12 hr period estimated from clouds at 0800Z, using regression method
xxxxxxx	:	Axes of chief rain cloud systems
Altitude	:	• 1000 - 1500 m ● 1500 - 2000 m

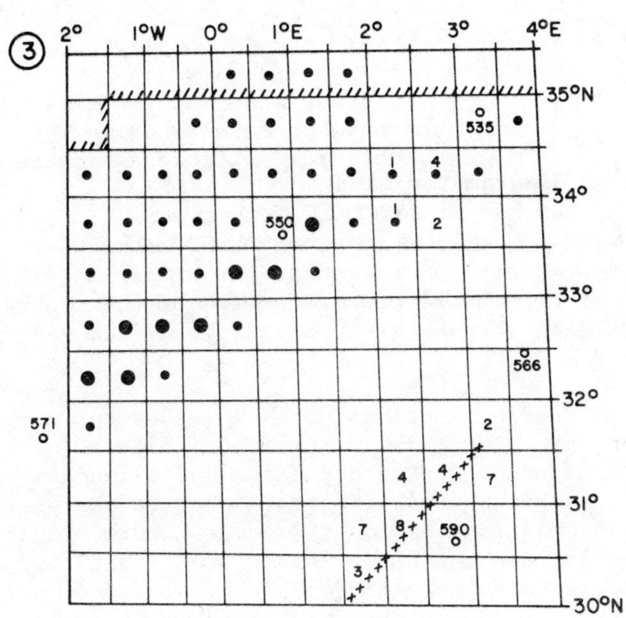

//////	:	Boundary of recession area
° 571	:	Station numbers
8	:	Estimated rainfall for 12 hr period (mm)
xxxxxxx	:	Axis of rain cloud system
Altitude	:	• 1000 - 1500m ● 1500 - 2000

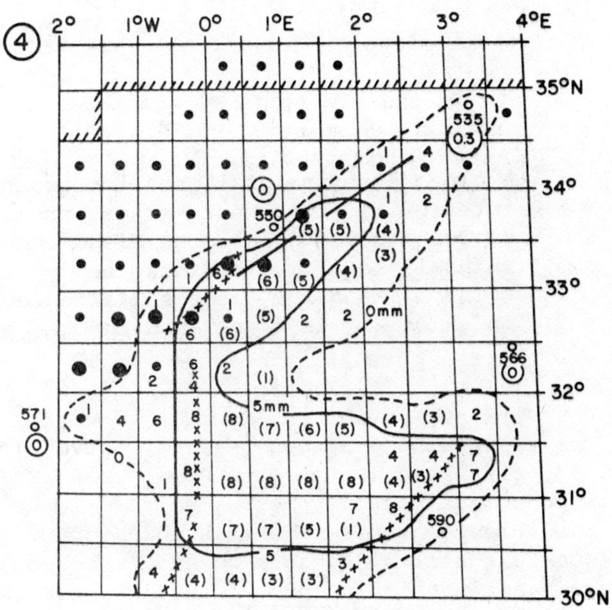

xxxxxxx	:	Axes of rain cloud system
———— / – – – –	:	Final isohyets including account of system movement
(2)	:	Interpolated rainfall estimates to account for system movement
2	:	Rainfall for 12 hr periods estimated through regression method, disregarding system movement

Fig. A.5.2

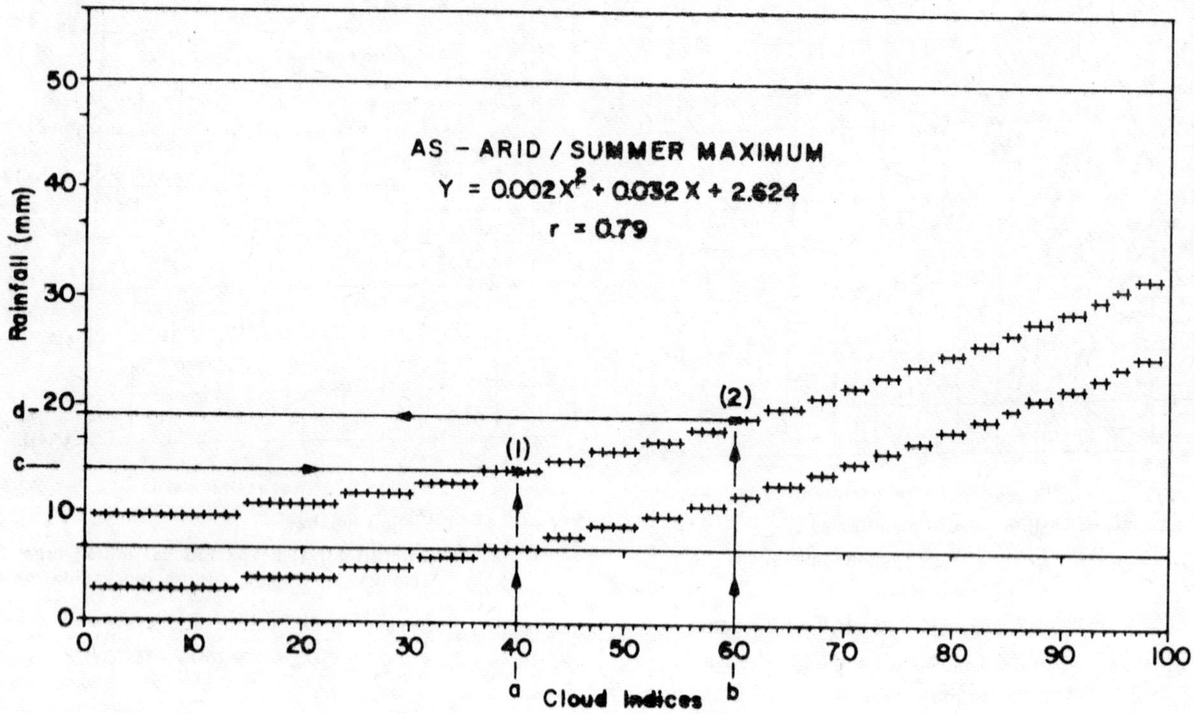

Fig. A.5.3

A cloud index/rainfall total regression diagram for part of the arid interior of north-west Africa, showing its use in rainfall estimation. If for a given gacell the cloud index is (a) and the observed unit period rainfall total is (c) then the overlay regression is moved up through point (1) and for a satcell affected by the same rain cloud system an index of (b) is translated into an estimate of (d).

FIGURE A.5.3

Stages in the completion of a unit period rainfall map for an event in central Algeria to allow for rain system movement/development between satellite imaging instants:

(a) an analysis of the raincloud revealed by a NOAA satellite image at t_1;

(b) a rainfall estimation map based on (a);

(c) a rainfall estimation map for t_2, in this case 12 hours after t_1;

(d) a completed rainfall map for (t_1 + t_2).

equivalent relationship for a satcell affected by the same raincloud system will be located on the curve parallel to the regression through the established gacell point. Thus, once the operative curve has been identified, associated satcell cloud (b) can be easily interpreted in terms of an appropriate rainfall estimate (d) through that curve rather than the (climatic mean) regression line.

In practice, this floating procedure is facilitated by the use of two regression diagrams for each morphoclimatic region, one of which is used as a transparent overlay above the other. The transparent regression is moved up or down above the other until the point representing the required cloud index/ observed rainfall relationship for one or more gacells ((1) in Figure A.5.3) is intersected by the overlain regression. Using a setsquare it is then a simple task to identify the intersection of associated satcell cloud indices and the overlain regression ((2) in Figure A.5.3), and to translate each in turn into a suitable rainfall estimate.

Given the nature of these procedures, the Bristol Method may be considered most appropriately to be an extension of classic operational meteorology, and can be taught most quickly and effectively to trained meteorologists. Since the required inputs (weather observations, satellite images) are most likely to be available in a meteorological office practical expediency suggests that the best site for the operational execution of the Method would be in or near such an establishment.

A.5.4 Improved Rainfall Mapping: Results from the Arid Zone.

For comparative purposes, results may be cited here from experience gained with the Bristol Method in two regions of widely different size, where very different objectives applied.

A.5.4.1 Rainfall Evaluation on North-East Oman

Operational tests of the Bristol Method were carried out here to assess the utility of the method for evaluating rainfall in a comparatively small area of rugged terrain, and to provide information for subsequent raingauge network design related to intended extensions of agriculture and coastal industry, both of which would necessitate an increase in the exploitation of ground water resources. (Barrett, 1976, 1979b). In this comparatively small area, 130 grid squares were deployed, each measuring 0.2° latitude by 0.2° longitude. The unit time period was 24 hours, dictated by the available raingauge observations. About one-quarter of the cells contained gauges.

Analyses of the resulting monthly maps suggested that this method coped well with both spatial and temporal variations of rainfall. The following, contrasting, results will serve as examples:-

a. July 1974 (Figures A.5.4A and A.5.4B) was a month of widespread shower activity, especially in the central mountainous area. More importantly, it was a month in which the available raingauge distribution was poorly suited to the distribution of active convective outbreaks over the north-west/south-east trending mountain range. This is reflected in the satellite-assisted map (Figure A.5.4B), which reveals much more extensive rainfall than the raingauge map (Figure A.5.4A).

b. June 1974 (Figures A.5.4C and A.5.4D) in marked contrast, was a month of quite localised convectional rain showers. Here the satellite-improved map (Figure A.5.4D) is different only in minor respects.

c. 1974 (Figures A.5.4E and A.5.4F). The most obvious contrast between this pair of maps is the greater detail of the satellite-assisted product (Figure A.5.4F). However, close inspection reveals that, although the satellite map identifies five centres of relatively heavy rainfall along the Jebel, compared with only three in the raingauge map, elsewhere the satellite map indicates lower rainfall than otherwise anticipated.

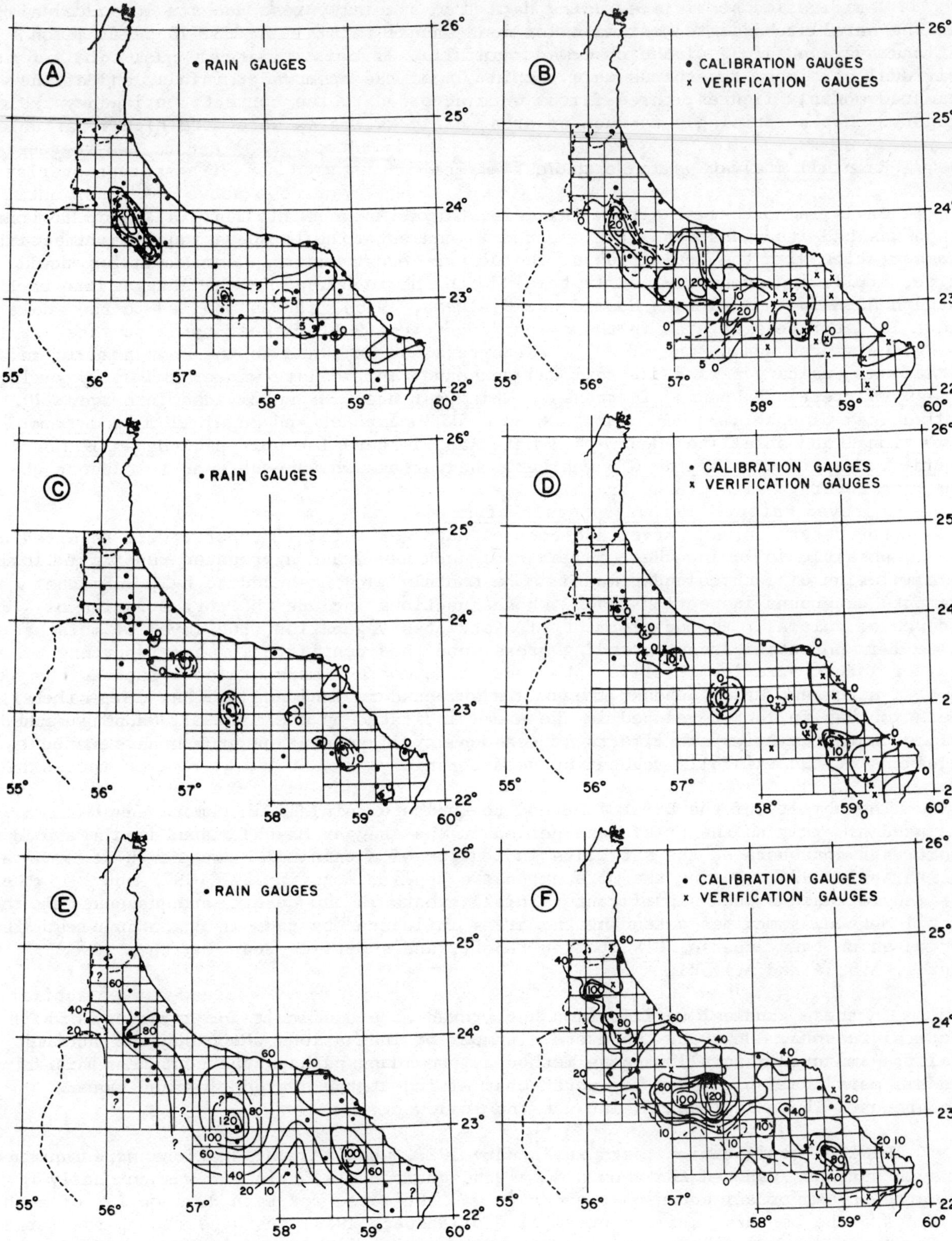

Fig. A.5.4 Monthly (A-D) and annual (E, F) rainfall maps for north eastern Oman from gauge data (A, C, E) and gauge + satellite data (B, D, F). For further details and discussion see text.

Verification statistics, using data from specially reserved stations, revealed that the satellite-assisted estimate for mean annual rainfall at the 15 verfication stations averaged 97.9% of the observed rainfall. Of the estimated annual totals for individual stations, two-thirds were within 20% of the observed rainfall, and of the estimated monthly figures, three-fifths were accorded to the correct contingency table category. Not surprisingly in such an area, daily estimates were more highly variable.

A.5.4.2 Rainfall Evaluation in North-West Africa

It is in north-west Africa (Algeria, Libya, Morocco and Tunisia) that the Bristol Method has had its widest application, and is nearest to full operational implementation, in association with the activities of the Desert Locust Control Commission for that region. Following successful early trials with the technique in the Ahaggar Massif of southern Algeria in 1976 (see Howard and Hielkema, 1976), it was applied to the whole desert locust recession area in those four countries (90% of their total surface area) during the winter and spring of 1976-77 (Barrett, 1977d), and to the same area in an operational training period for national personnel in the late winter and early spring of 1978-79 (Barrett and Lounis, in press). This vast area was subdivided into over 2700 grid squares of unit size 0.5° latitude by 0.5° longitude, and rainfall mapped from conventional and satellite (Noaa - 5 & Meteosat) imagery for unit periods of 12 hours (the minimum permitted by the GTS weather observations which were used to support the satellite coverage with ground truth).

For desert locust survey and control purposes rainfall reports were required as soon as possible following the completion of each week, and aggregated reports following the completion of each calendar month. The monthly maps are used to help plan the movements of ground inspection teams, whose functions include the corroboration of evidence of rainfall, the development of associated vegetation, the level of moisture in the soil in potential locust breeding areas, and the identification and reporting of signs of locust activity of different kinds. The monthly maps are also used as the bases on which high-resolution Landsat images are selected to indicate whether post-rainfall vegetation has in fact developed in the areas indicated by the rainfall mapping method: surface characteristics and effects of seasonality of vegetation growth have strong influences on the rainfall/ vegetation relationship.

The ability of the Bristol Method to indicate rainfall in remote desert areas was evidenced as early as the brief test period in the Ahaggar Massif, when special field report data corroborated the satellite indication of a rainfall event north of Tamanrasset on 29th April 1976. During the more extensive applications in 1977, 1978 and 1979 there were many occasions when significant rainfall events in the desert were assessed by the Bristol Method, sometimes extending the areas influenced by them in comparison with the maps based on (very sparse) GTS station report, and sometimes reducing them. (See Figures. A.5.5A and A.5.5B).

In these studies, verification has proved more difficult for a variety of practical reasons. However, there are a number of indications which suggest that the satellite-improved rainfall mapping method is providing perhaps the best near real-time rainfall maps available today for north-west Africa at the stated time and space resolutions. These indications include the following:-

a. Limited verification tests in January 1977, when several stations were reserved for such purposes, showed that, on a multi-categoric basis, rainfall was correctly estimated on 94% of all occasions, and rain was estimated for rain days on 76% of such occasions. The omnibus "skill score" (cf. Follansbee, 1976) was 0.80 though the large number of rain/zero estimate occasions in this arid region encourages such a flattering result.

b. Contemporaneous field observations have confirmed some satellite-based conclusions that rain has fallen; (e.g. in the Ahaggar Massif north of Tamanrasset on 29th April 1976 (Hielkema and Howard, 1976).

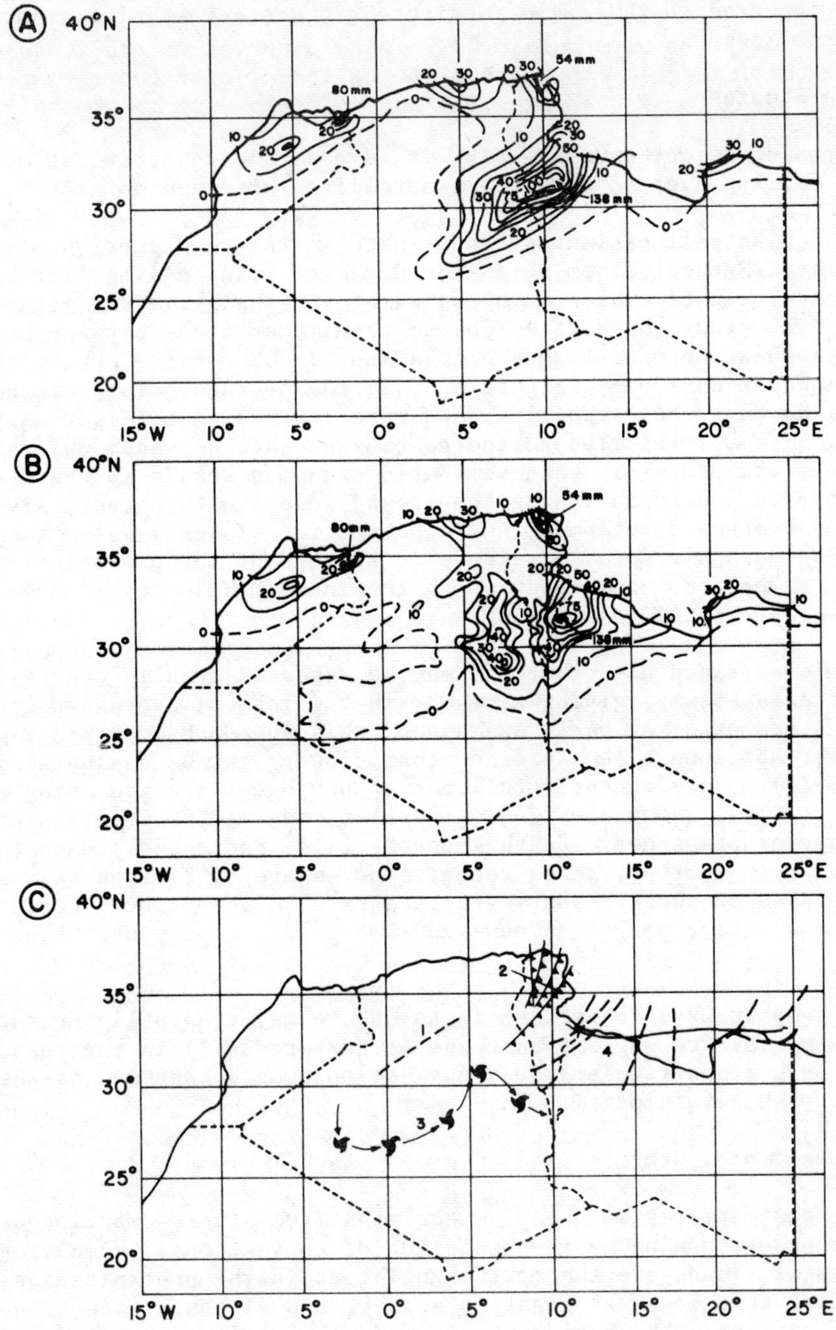

Fig. A.5.5 Simplified maps of rainfall in north-west Africa for 29 March-4 April 1977; (a) evidenced by gauge data only; (b) from gauge + satellite data; and (c) a depiction of the synoptic systems revealed by the satellite to have caused the rain. The original maps were gridded and assessed for over 2700 squares.

Conversely, there has been no evidence that significant but unpredicted growths of vegetation have occurred in the region monitored in north-west Africa during the detailed mapping exercises.

A.5.4.3. Depth/Area Studies

There are two different ways in which it is possible to arrive at estimates of the volume of rain falling over an area where satellites are used as aids.

a. Gauge data and satcell estimates can be used in the construction of isohyetal maps (as in the Bristol Method), from which the depth of rain falling over selected areas (e.g. large catchments) can be evaluated by planimetry. A gravimetric method was used in a follow-up to the Oman study (Barrett 1976b) to planimeter areas between neighbouring isohyets in the maps shown here as Figure A.5.6A and A.5.6B for the entire study area. On the basis of raingauge data only (Figure A.5.4A) the average point rainfall was assessed as 50.6mm; however when satellite data were invoked also (Figure A.5.4A) this figure fell to only 38.4mm. This led to the conclusion that conventional rainfall mapping and dependent procedures in that semi-arid area may result in the over-estimation of water feeding streams and replenishing the ground water, with potentially serious implications for the planned development and exploitation of local water resources. Similar results might be anticipated elsewhere in regions of abrupt relief if its effects on rainfall are not taken into consideration in the interpolation of isohyets between relatively broadly-spaced rainfall stations.

b. Satellite-viewed cloud distributions may be assessed areally over selected catchments in an integrated way, giving a result in the form of average point rainfall, or some depth/area statement. An early example of this approach employed fixed weights or indices for different cloud types in order that cloud might be evaluated in terms of rain (Follansbee, 1973); more recently calibration has been effected using river gauging data (e.g. Amorocho, 1975), raingauges and/or weather radar (e.g. Griffith et al., 1978), and weather radar and/or atmospheric depth sounding (i.e. radiosonde) data (Wylie, personal communication). However, these recent studies are of limited relevance here, for they have been based on geostationary stellite data, have not been concerned with arid or semi-arid land areas, and have focussed almost exclusively on convectional clouds.

Although the approaches summarised in (a) above may eventually be superceded by those in (b), it seems fair to suggest that the latter are still in the research and development stage, and, in particular could not yet cope as adequately as the former with frontal and other trough situations.

A.5.4.4. Severe Convectional Storms

Some of the most interesting results emerging from studies mentioned in (b) above have related to our understanding of the behaviour of convectional (cumuliform/cumulonimbus) clouds. These types of clouds are the most significant in the precipitation hydrometeorology of arid and semi-arid zones at least in summer, and although attention on these so far has focussed primarily on the humid tropics and sub-tropics some findings, and procedures based on them, may be outlined now. Perhaps the most extensive studies have been those of the Florida Area Cumulus Experiment (FACE). The most pertinent results have been reported by Woodley et al. (1972) and Griffith et. al (1976), and may be summarised as follows (see Scofield and Oliver, 1977):-

a. Bright clouds in VIS imagery produce more rainfall than darker clouds.

b. Bright clouds in the VIS, and clouds with cold tops in the IR imagery which are expanding in areal coverage (in early and mature stages of development), produce more rainfall than those which are not expanding.

c. Decaying clouds produce little or no rainfall.

d. Clouds with cold tops in the IR produce more rainfall than those with warmer tops.

e. Clouds with cold tops that are becoming warmer produce little or no rainfall.

f. Merging of cumulonimbus clouds increases the rainfall rate of the merging clouds.

g. Most of the significant rainfall occurs in the upwind (at anvil level) portion of a convective system. The highest and coldest clouds form where the thunderstorms are most vigorous, and the rain heaviest. These cold clouds get thinner downwind and look warmer in IR imagery as the anvil material blows away from its origin over the updraft.

These facts have been incorporated into a scheme for estimating convective rainfall from satellite imagery, devised by members of the Applications Group of the National Environmental Satellite Service (NESS) of NOAA. This provides a rather complex "decision tree" to guide the analyst of geostationary satellite VIS and IR imagery to estimate precipitation associated particularly with severe storms and hurricanes. As such, it is of potential value to hydrologists especially in flash flood situations. Given a supply of geostationary satellite data it might serve as a useful adjunct, or sub-routine, to approaches like the Bristol Method to improve their ability to cope with the more extreme convective rainfall events: these are, perhaps, the events least well accommodated by more general methods founded on climatology and statistics.

Although the decision tree approach has been developed primarily for the assessment of severe storms over the more humid regions of the U.S.A. to provide information in situations when gauge data cannot be obtained because of power failures etc. Scofield and Oliver propose that it could be fine-tuned for local terrain differences and climate. For example, they say that in mountaineous terrain in the drier parts of the United States, it is known that thunder storms drop much more rain on the upper parts of the mountains than in the valleys. Fine-tuning of the estimations in such an area would take into account the normal distribution of rainfall with height (for convective rains) if known, or just the height if the climatology is not known.

A.5.4.5 Rain System Distribution Behaviour

If there is a need for satellite assistance in the study of rain-clouds and precipitation over a nation as well-equipped with rain-gauges and weather radar as the coterminus United States, there must be an overwhelming need in many other less well-documented regions in the arid zone. Background work for both the Oman and North-West African projects discussed earlier quickly revealed considerable gaps in our knowledge and understanding of the synoptic and sub-synoptic scale weather and cloud systems responsible for rainfall events of various intensities and magnitudes, amongst which the severe convectional storms figured prominently.

It is in this sense in particular that satellites can be of value in hydro-meteorology far beyond the narrow confines of mere rainfall evaluation. Clearly a better comprehension of the atmospheric patterns and processes which result in rain in the arid zone should be beneficial to both meteorological and hydrological forecasting, and would be essential for the design of rainfall augmentation schemes, e.g. those based on cloud seeding. Some interesting indications relevant to these points have emerged in the course of the rainfall-mapping exercises in Africa and the Arabian peninsula:-

a. It is known that important rainfall events in the Sahara are associated with the tails of fronts from Mediterranean depressions, with mid-troposopheric troughs bringing energy and moisture from the West African region, and with narrow lower troposopheric troughs where horizontal shearing effects are strong, e.g. between neighbouring cells of high pressure.

However, the satellite imagery suggests that mobile, sub-synoptic scale masses of partly-organised convective clouds not mentioned in the literature occur sometimes even in the truly arid regions (See Figure A.5.5C). The significance of these cloud masses in terms of rain has not yet been explored. Satellite data have also revealed that convective cloud occurs from time to time especially over higher ground in the southern Sahara. Both types of cloud outbreaks - the mobile and the relatively stationary but diurnally very variable - could best be elucidated further through specific studies based on half-hourly or hourly data from Meteosat.

b. Satellite imagery clearly reveals the growth and extent of convective clouds over the abrupt mountain ranges in north-east Oman, and could be used to identify those days and localities which might be best-suited to cloud seeding operations such as those suggested for that region by Simpson (1976).

Satellites would appear to offer the best - and in many conventional data-sparse areas, perhaps the only - means whereby the nature and distribution of rain-bearing systems might become better known, with concomitant advantages for hydrometry, e.g. raingauge network design.

A.5.4.6 Conclusions

After more than two decades of experiment and operations with environmental satellites there can be no doubt that satellite remote sensing systems are here to stay. Although their best-developed applications are in synoptic meteorology and climatology, there is equally no doubt that there are several other broad areas of environmental science in which satellites can play a useful, perhaps even vital part. Precipitation hydrometeorology is one of these.

Although most work in this field has been carried out in regions of abundant rainfall, both theory and some practical experience indicates that the use of satellite data for such purposes should not be limited to humid and sub-humid zones. Indeed, the particular sparseness of conventional meteorological observation and rainfall measuring networks in the arid and semi-arid zones suggest that satellites might be even more useful complements to the in situ observatories here than elsewhere. In some ways the problems of interpreting the satellite data accurately are greatest where rainfall is often from heavy but scattered showers. However, the same is true of conventional data, for rainfall observations from relatively isolated raingauges are notoriously unrepresentative of true areal accumulations of rain. Consequently, it seems fair to suggest that the potential for satellite assistance in the dry zones of the world is at least as great as it is elsewhere. Since the chief problems seem likely to involve calibration and verification, and the key limitations will involve accuracy and detail, satellite methods should not be seen as a universal panacea for existing deficiencies in rainfall monitoring in the arid zone. The satellite approach is most necessary where large areas are poorly monitored on the ground; but it is most successful in absolute terms where it is used to support good networks of raingauges in relatively small (but not very small) areas - whence the existing data are the more nearly satisfactory. For some applications short period rainfall assessments are desired but it is unrealistic to expect that (even geostationary) satellite imagery can be interpreted readily in terms of rainfall totals for periods of less than 12 or 24 hours in the absence of enough continuous-recording raingauges to calibrate the clouds. And, of course, if a suitable supply of satellite imagery cannot be arranged this remote sensing approach must be ruled out.

All such considerations combine to indicate that the potential use of satellite remote sensing in support of any particular exercise in precipitation hydrometeorology should be examined critically before a decision to proceed is made. But once such a decision has been taken, useful results should accrue in the meteorology of rainfall systems and in rainfall mapping, with benefits for many operations in pure and applied hydrology including irrigation and drainage design and management, river stage and hydrograph prediction, ground water recharge estimation, the evaluation of hydrology cycles at different scales, and the conservation of water resources.